Dick

How to Improve Rubber Compounds

John S. Dick

How to Improve Rubber Compounds

1800 Experimental Ideas for Problem Solving

2nd Edition

HANSER

Hanser Publishers, Munich Hanser Publications, Cincinnati

The Author:
John S. Dick,
2915 Stanley Rd., Fairlawn, OH 44333, USA
John.Dick@dynisco.com
www.rubberchemist.com

Distributed in North and South America by:
Hanser Publications
6915 Valley Avenue, Cincinnati, Ohio 45244-3029, USA
Fax: (513) 527-8801
Phone: (513) 527-8977
www.hanserpublications.com

Distributed in all other countries by
Carl Hanser Verlag
Postfach 86 04 20, 81631 München, Germany
Fax: +49 (89) 98 48 09
www.hanser-fachbuch.de

Library of Congress Cataloging-in-Publication Data

Dick, John S.
How to improve rubber compounds : 1800 experimental ideas for problem solving / John S. Dick. -- 2nd edition.
pages cm
ISBN 978-1-56990-533-3 (hardcover) -- ISBN 978-1-56990-534-0 (e-book) 1. Rubber. 2. Rubber, Artificial. I. Title.
TA455.R8D53 2014
678'.23--dc23
2013037811

Bibliografische Information Der Deutschen Bibliothek
Die Deutsche Bibliothek verzeichnet diese Publikation in der Deutschen Nationalbibliografie; detaillierte bibliografische Daten sind im Internet über <http://dnb.d-nb.de> abrufbar.

ISBN 978-1-56990-533-3
E-Book ISBN 978-1-56990-534-0

Production Management: Steffen Jörg
Coverconcept: Marc Müller-Bremer, www.rebranding.de, München
Coverdesign: Stephan Rönigk
Typesetted by Manuela Treindl, Fürth
Printed and bound by Kösel, Krugzell
Printed in Germany

Disclaimer

The information contained in this book is experimental in nature and is directed towards individuals with advanced scientific training and rubber compounding experience. The author, review panel members, editors, and publisher make no representation or warranties, either expressed or implied, with respect to the accuracy, currency, completeness, or suitability of the information contained in this book and will not be liable for any damages, costs or liability arising out of any use of material contained in this book for any purpose including, without limitation, technical inaccuracies and typographical errors.

Preface

Rubber compounding is an art as well as a science. This book is intended to be a companion for the earlier book titled *Rubber Technology: Compounding and Testing for Performance.* As indicated in the preface of the earlier edition of this book, the art in rubber compounding is to achieve the best trade-offs in properties and to be able to meet product performance requirements at acceptable production costs.

The objective of this book is to provide the experienced and well-trained rubber compounder with new experimental ideas, which he or she may be able to use to improve certain specific rubber compound properties. Of course, when compounders are considering implementing some of these general experimental ideas for their specific compounds, not only should they thoroughly check out the effects on the target compound property, but also how this change will be affecting other compound properties as well. Generally, any specific change in a rubber compound will certainly affect a wide range of other properties for better or for worse. There always appear to be a wide range of "unintended consequences," which have to be thoroughly checked out through laboratory testing, factory trials, controlled field evaluations, and so forth. All companies should have a formal approval process and sign-off procedures before any compounding change is allowed in production. Many times, the effective improvement of a rubber compound requires not just one change but an intelligent selection of multiple changes in order to achieve the best overall compromise (best trade-off) in compound properties to more effectively meet the product's performance requirements.

The new edition now contains over 1800 experimental ideas to improve specific rubber compound properties, which are organized in five chapters covering improving cured physical properties, degradation resistance, measurable processability properties, qualitative processing attributes, and tire performance properties. In addition, this book includes an appendix on rubber blends commonly used for specific product applications and an appendix on commonly used cure systems. The majority of these experimental ideas were found from a thorough review of the general rubber literature. Another set of experimental ideas was obtained from the companion book mentioned above. Also, a very important source of additional experimental ideas from the first edition was the contributions of the eighteen-member Review Panel for this book (listed in Section 1.6). The additional ideas published in the 2nd editon are the result of an extensive review of over one thousand recent jounal articles and conference presentations from rubber societies in the United States, Europe, China, and India.

The experimental ideas given in this book may not work in all specific situations. Changes to improve one target compound property will certainly affect other properties as well, for better or for worse, and this book does not purport to show how these other properties are affected. Also, this book does not purport to address safety and health issues. The information contained in this book is experimental in nature and is meant for individuals with advanced scientific training and rubber compounding experience.

September 2013, John S. Dick

www.rubberchemist.com

Contents

1 Introduction

Rubber compounding is a highly complex endeavor that involves many interactions and includes many ways to achieve the target properties and economic goals while maintaining an acceptable trade-off for these characteristics. The purpose of this book is to provide the reader with various experimental ideas that may guide him or her to develop better compounds and solve technical problems.

1.1 Complexity of Compound Ingredient Selection

Rubber formulations generally are not simple and usually involve the careful selection of several material systems, each of which can have a major impact on the commercial rubber product. The following are examples of some of these systems:

- Base rubber or rubber blend
- Filler/oil system
- Cure system
- Antidegradant system
- Tackifying system (if applicable)
- Adhesion system (if applicable)
- Flame retardant system (if applicable)
- Blow system (for cellular rubber applications)
- Special process-enhancing chemical additives

The selection of a base rubber (elastomer) has the greatest effect on the properties of a compound. There are well over 30 major rubber types. However, literally hundreds of different grades may be available for any one major category of

RT: *Rubber Technology, Compounding and Testing for Performance*. 2nd ed. Edited by J. S. Dick, Hanser Publications, 2009 ■ GEN: General References from the Literature ■ RP: Review Panel (Advisors for this book) ■ IMPORTANT: The information contained herein is experimental in nature and is meant for individuals with advanced scientific training and rubber compounding experience. All safety precautions should be taken when replicating or applying this information. The author makes no representation or warranty, either expressed or implied.

rubber, each with a different effect. For example, there are at least 150 different grades of SBR, 50 of BR, 150 of EPDM, 280 of NBR, 100 of silicone rubber, and 75 of fluoroelastomers. Each of these different grades imparts different properties to the compound.

To add to this complexity, many commercial rubber formulations are based on blends of two or more elastomers. For example many rubber compounds are based on blends of SBR and BR, natural rubber (NR) and BR, or various blends of SBR, NBR HIIR, CR, EPDM, and so forth. Blending different rubber polymers is an effective way to achieve certain compound properties. However, many times these blended polymers, although compatible (covulcanizable), are not very miscible, or soluble, in each other. Therefore, continuous and discontinuous phases or domains are established from mixing. These different phases have different affinities for various fillers such as carbon black during mixing. These differences in distribution of carbon black can have a unique effect on the properties of the compound.

The selection of an effective filler/oil system for the formulation can also have a very strong influence on compound properties. Many different fillers and reinforcing agents can be used in rubber compounding to enhance physical properties. A wide variety of petroleum-based processing oils and hundreds of synthetic ester plasticizers can also be used. Currently more than 42 standard commercial grades of carbon black are available to the rubber compounder. Some rubber formulations will use more than one grade of carbon black to balance the compound performance properties, adding further to the complexity.

Literally thousands of combinations of organic accelerators and sulfur are used to achieve specific vulcanizate properties. In addition to the common sulfur cures, other cure systems such as peroxide or resin cures can be applied. For effective resistance to in-service degradation, hundreds of different antioxidant and/or antiozonant combinations are used commercially. Different commercial grades are available even for very simple and "straightforward" compounding ingredients such as sulfur or zinc oxide. For example, different sulfur grades with different mesh sizes and chemical treatments are available to ensure better dispersion in certain formulations. Zinc oxide grades are used with different average particle sizes and treatments to enhance certain physical properties.

1.2 Industrial Complexity

As noted in the preceding, there are virtually an infinite number of ways to select raw materials and write recipes to meet customers' specifications. It is very costly for corporations to develop rubber compounds to meet these customer requirements. Therefore, most rubber fabricators tend to be quite secretive regarding these formulations to maintain what they consider their competitive advantages. Hence the rubber industry possesses literally hundreds of thousands of quite different recipes to meet customer needs.

However, this industrial complexity does not end with just the sheer number of different recipes used in commerce. Even if two companies were using identical recipes and identical raw materials, they could still be producing products with different properties. For example, it is quite possible for the components of the same recipe to be mixed on different types of mixers or to use a different mixing sequence that results in a different state of mix. Identical compounds that are cured under different conditions of time, temperature, or pressure can also acquire different vulcanizate properties.

1.3 Complexity of Ingredient Property Relationships

Another difficulty in rubber compounding is the problem of ingredient property relationships. One form of interaction is experienced by the compounder when he or she tries to formulate to a specific property such as durometer hardness, for example. Increasing or changing the curative package to achieve a higher cured crosslink density (X_1) will result in a higher durometer hardness value (Y_1). On the other hand, increasing the loading of carbon black (X_2) will also increase this durometer hardness (Y_1) for a different reason. Significantly reducing oil loading (X_3) will again increase hardness for still another reason. This is very common in rubber compounding. In addition, these different independent variables usually show an interaction because their effects are not necessarily additive or linear. Changes in the cure system, the filler-oil system, the base rubber, the rubber blending, the method of mixing, and so forth (... X_n) will usually affect a specific property such as durometer hardness (Y_1). However, any one or a combination of these deliberate compound changes (independent variables) will also affect in very different ways a very wide variety of *other* compound properties such as ultimate tensile strength (Y_2), ultimate elongation

(Y_3), percent rebound (Y_4), aging resistance (Y_5), tear resistance (Y_6), flex fatigue resistance (Y_7), oil resistance (Y_8), ... (Y_n). Another way of describing this type of interaction from a specific compound change is to refer to it as a *multiple response.* There is an old compounder's "saying" that "you can never change just one thing" (RP: R. J. Del Vecchio). These different compound changes that increased hardness also changed elongation, tensile strength, oil resistance, compression set, flex fatigue, and so forth in very different ways for better or worse regarding product performance. The same is true for processability properties that result as well. Processability properties are almost always affected for better or worse by a compound change(s)–another unintended consequence. To add to this complexity is the fact that many of these multiple responses are nonlinear and usually give both better and poorer compound performance properties simultaneously. This type of interaction resulting in multiple responses is extremely common in rubber compounding. Therefore when using this book to develop ideas on how to change or optimize a specific rubber compound property such as durometer hardness, you are virtually guaranteed to affect just about every other compound property that can be measured. Many of these effects may result in poorer rubber compound performance.

Thus for this principle "...of having multiple responses for a single compound change," the compounder must develop an understanding of how these responses can be different for each selected compound change. For example, some ingredient changes are independent of other ingredient changes (no interaction), whereas others show a great deal of interaction. For example, a change in the concentration of two carbon blacks is usually just the sum of the effects of the individual carbon blacks by themselves. But changes in concentration of a combination of sulfur and accelerator can result in many different responses to hardness, and other properties, than the individual response to separate concentration variations of the sulfur only or accelerator only. This is called *chemical interaction* or synergy (RP: R. J. Del Vecchio). This book provides ideas on how to optimize specific compound properties "one property at a time." As indicated earlier, however, when implementing any change in recipe or process to improve a single compound property, it will *always* change other compound properties *for better or for worse.* If the compounder has many other compound or processing properties to consider as well, the complexity of predicting them all from a single recipe change or process change can be very difficult (RP: R. J. Del Vecchio). A skillful compounder can usually work out the necessary acceptable compromise among properties to meet all the compounding property targets. This ability comes from experience and intuition and usually from a series of trial-and-error experiments or a "one change at a time" approach until the correct balance of compound properties is achieved. This is the so-called "black art" of compounding (RP: R. J. Del Vecchio). This book is intended to aid the compounder in this approach.

A more modern approach to compounding is to use the statistical method of *design of experiments* (DOE) to achieve this optimal set of properties. To carry out a DOE is much simpler than it was a few years ago because more effective user-friendly software is now available. In a DOE, multiple recipe and process changes can be studied simultaneously to analyze multiple responses and generate a mathematical solution. These DOEs will require more initial effort (typically 8 to 16 compound variations to be formulated, mixed, and tested for all relevant properties in the same period of time). However, this DOE technique is more effective at finding the best recipe and/or process combinations to meet all the compound property requirements, which may be better than changing one variable at a time (RP: R. J. Del Vecchio).

Whichever approach is used ("one variable at a time" or DOE), this book should help the compounder in trying different ideas. From reviewing the ideas given in this book, the compounder may develop a better "feel" and understanding of how to possibly change the recipe or process to meet all the property targets.

Even though this book provides many diverse experimental approaches for changing or improving a specific individual compound property, it *does not* necessarily predict the negative effects for the other compound properties that might result from making this change(s). However, by reviewing the other sections of this book that discuss those other properties of concern; the compounder may develop an appreciation of how a specific recipe change or process adjustment might affect these other targeted compound properties. Thus the various experimental ideas given in this book may suggest a combination of experimental variables that the compounder might try out in order to find the best balance in desired compound properties (RP: R. J. Del Vecchio).

Because trying to improve one single rubber compound property will *always* result in changing other rubber compound properties, for better or for worse, the rubber compounder should review the many ways to change or improve a targeted compound property to find the formulation or processing modification that gives the best trade-off against the other required compound characteristics.

In this book, we present experimental ideas on various possible ways to improve or optimize a specific compound property. However, whenever you implement a recipe or process change to modify one property, you are virtually *guaranteed* to change many other compound properties, many times in the wrong direction. It is a truism that almost all compound or process changes implemented to modify one compound property for the "good" will also change many other properties in both favorable and detrimental ways. In other words, trying to change one property may result in unintended consequences for other properties. This is where the *art* (or black art) of rubber compounding is applied. This "art" reflects a compounder's skill in finding the best compromise among properties to meet all the required compound specifications, which may require the compounder to make more than one change to achieve the best overall balance. By reviewing many of the suggestions in this book, the compounder may develop a better intuition for how best to achieve this compromise or trade-off with compound properties. In short, a good compounder should acquire a combination of knowledge and skill from the sciences of chemistry, physics, statistics, and both mechanical and process engineering. The application of these sciences is the true art of compounding (RP: R. Dailey).

As noted earlier, this book provides a large number of diverse experimental ideas for enhancing a specific compound property. It does *not* describe the possible ill effects on other compound properties. However, by reviewing the experimental ideas for these other properties, the reader may develop a limited appreciation for some (but certainly not all) of these consequences. In fact, the compounder may try a combination of experimental ideas to achieve his or her rubber compounding goals.

1.4 Experimental Ideas

The experimental ideas provided in this book are *general* suggestions that may or may not work in your *specific* compound. Because of the diversity and scope of rubber compounds used in commerce today, these general statements may not always be applicable to every compound, including yours. Certain precautions should always be taken.

- For all ideas and suggestions provided here, consider any possible health, safety, or environmental hazards that might be created at your facility, including all relevant U.S. Occupational Safety and Health Administration (OSHA) and Environmental Protection Agency (EPA) issues.
- Check out the effectiveness of these experimental ideas in your compound(s) by implementing these ideas in a laboratory-scale mix. Measure the effects of this compound change(s) on your specific formulation and target property as well as all other relevant compound properties that determine the effectiveness of the compound in your customer's end-use application. It is important to check these experimental ideas thoroughly for your specific compound because no two compounds are identical. The only way to know if the idea

may work for your compound is to test it initially in a laboratory and compare the results against a "control" compound. Be sure to measure effects on the subject property as well as effects on all the other relevant properties vs. your control compound. Your compound change may have improved flex fatigue vs. your control, for example; however, it may also be detrimental to such properties as aging resistance and so forth compared to the control.

- Remember that laboratory mixing is not equivalent to factory mixing. Likewise, laboratory cures are not equal to factory cures. In other words, a compound change that may show improvements in the laboratory may not work in the factory. If the laboratory trials are successful, then try limited and controlled factory trials and product field trials to validate this experimental idea or proposed change. This will help to determine if there are any "hidden" problems that could not be detected at the laboratory scale. Further, these evaluations should be followed through the entire product service life to make sure there are no problems that might hurt the product's performance throughout its use or create any product safety issues.

1.5 Sources of Ideas

There are three basic sources of the experimental ideas given in this book:

1. The first primary source of experimental ideas is from the second edition of the earlier "companion" book published in 2009, *Rubber Technology: Compounding and Testing for Performance*, edited by John S. Dick, and published by Hanser Gardner Publications, Cincinnati. These sources are designated by the prefix "RT:" throughout this book.
2. The second source of experimental ideas is from a review of the general literature. At the end of each idea is cited a specific reference(s). These sources are designated by the prefix "GEN:" throughout this book.
3. The third source of experimental ideas is from the direct contribution of members of the "Review Panel" cited in Section 1.6 of this book. These sources are designated by the prefix "RP:" throughout this book.

The reader should thoroughly review these cited references as well as other references not cited in this book before beginning any experiments. These "other references" should be both of a general and specific nature. Thus looking up other references that are not cited in this book but could be relevant to the problems is greatly encouraged.

The following "experimental ideas" for the most part are generalizations. However, rubber compounding is very complex and such generalizations almost *never* work across vastly different formulations. Therefore all these ideas should be thoroughly researched and checked out as outlined in Section 1.4.

1.6 Review Panel

The following are members of the Review Panel who have reviewed the text of this book and made suggestions for additions, deletions, or changes. In addition, they have individually provided further experimental ideas for the enrichment of this book.

Name	Affiliation	E-mail Address
Roger Dailey	Goodyear Chemical Co.	roger_dailey@goodyear.com
R. J. Del Vecchio	Technical Consulting Services	techconsultserv@juno.com
Ronald W. Fuest	Uniroyal Chemical (Retired)	ron@fuest.net
James R. Halladay	Lord Corp.	james_halladay@lord.com
Fred Ignatz-Hoover	Flexsys America L.P.	ignatz-hoover@flexsys.com
Mark Lawrence	Indspec	mlawrence@indspec-chem.com
John M. Long	DSM	john.long@dsm.com
Oscar Noel	Luzenac America	oscar.noel@america.luzenac.com
Leonard L. Outzs	DuPont Dow Elastomers	Leonard.L.outzs@dupont-dow.com
Thomas D. Powell	Degussa Corp.	Thomas.powell@degussa.com
Charles P. Rader	Advanced Elastomer Systems (Ret.)	charles_rader@msn.com
Ronald Schaefer	Dynamic Rubber Technology	rschaeferdrt@msn.com
Kelvin K. Shen	Borax/Luzenac, Inc.	kelvin.shen@borax.com
John Sommer	Elastech	johngsommer@att.net
James Stevenson	Honeywell	james.stevenson@honeywell.com
Byron H. To	Flexsys America	byron.h.to@flexsys.com
Walter Waddell	ExxonMobil	walter.h.waddell@exxon-mobil.com
Meng-Jiao Wang	Cabot Corp.	meng-jiao_wang@cabot-corp.com

2 Improving the Physical Properties of Cured Rubber Compounds

2.1 Increasing Tensile Strength

In the rubber industry, ultimate tensile strength is considered a fundamental material property. This test parameter measures the ultimate strength of the cured vulcanizate. Even though a typical rubber product probably never will be stretched anywhere close to its ultimate tensile strength, many rubber product users still consider it an important indicator of the overall quality of the compound. Therefore, tensile strength is a very common specification property that a compounder typically will need to find a way to meet even though its relevance to the specific product end use is usually poorly defined.

A compounder might consider trying the following ideas on an ***experimental*** *basis if an additional improvement in tensile strength is needed. Also, all relevant literature sources, including the ones cited below, should be researched and read.* ***Caution:*** *these general experimental ideas may not work in all specific situations. Changes to increase tensile strength will certainly affect other properties as well, for better or for worse, and this book does not purport to show how these other properties are affected. Also, this book does not purport to address safety and health issues.*

Generally

To achieve the highest tensile strength, generally one should start with strain crystallizing base elastomers such as natural rubber (NR), polychloroprene (CR), isoprene rubber (IR), hydrogenated nitrile rubber (HNBR), or polyurethane (PU). RP: J. R. Halladay.

Natural Rubber

Compounds based on NR generally impart better tensile strength than similar compounds based on CR. RT: Chapter 6, "Elastomer Selection," R. School, p. 133.

Of the various grades of NR, ribbed smoked sheet no. 1 gives the highest tensile value. RT: Chapter 7, "General Purpose Elastomers and Blends," G. Day, p. 156.

Ribbed smoked sheet no. 3 reportedly may actually impart higher ultimate tensile strength than ribbed smoked sheet no. 1, at least in blackloaded compounds. RP: J. M. Long.

For NR-based compounds, avoid using a chemical plasticizer (peptizer) such as dibenzamido-diphenyl disulfide or pentachlorothiophenol (PCTP) because it may reduce the ultimate tensile strength of the compound. RT: Chapter 14, "Ester Plasticizers and Processing Additives," C. Stone, p. 366.

Polychloroprene

Polychloroprene (CR) is a strain-crystallizing polymer that imparts high tensile strength even without dependency on high loadings of reinforcing fillers. In fact, sometimes the ultimate tensile strength can actually be increased by reducing the filler loading. Selecting a CR with a higher molecular weight may also increase the ultimate tensile strength. RP: L. L. Outzs.

Consider xanthate-modified types of CR for higher levels of tensile strength and elongation. GEN: Nobuhiko Fujii (Denka Corp.), "Recent Technical Improvements of CR and ER in Industrial Applications," Presented at the Southern Rubber Group, March, 2006.

SBR

Use emulsion SBR polymerized at +5°C (41°F) (or lower) instead of +50°C (122°F) for a significant improvement in tensile strength. RT: Chapter 7, "General Purpose Elastomers and Blends," G. Day, p. 149. RP: J. M. Long.

Emulsion SBR usually provides higher tensile strength than solution SBR. RT: Chapter 7, "General Purpose Elastomers and Blends," G. Day, p. 156.

Substituting an oil-extended high molecular weight SBR for an unextended SBR while making compound adjustments to keep the same oil level may improve the ultimate tensile strength. RP: J. M. Long.

NBR

Use nitrile rubber (NBR) grades with higher bound acrylonitrile (ACN) content for higher compound tensile strength. RT: Chapter 8, "Specialty Elastomers," M. Gozdiff, p. 194. Increasing ACN content to an optimal level of about 41 percent should optimize the tensile strength for some NBR compounds. GEN: Andy Anderson, "Keeping It Real with NBR and HNBR Polymers," Zeon Chemicals, Presented at the Energy Rubber Group, September 12–15, 2011.

Consider NBR grades with narrower molecular weight distribution for higher compound tensile strength. RT: Chapter 8, "Specialty Elastomers," M. Gozdiff, p. 197.

Molecular Weight Effect

Using raw NBR elastomers with higher Mooney viscosity and molecular weight may increase the compound's ultimate tensile strength through an optimum before it decreases. GEN: D. Ghang, "Investigation of the Structure-Property Relationships of Improved Low Compression Set Nitrile Rubbers," *Rubber Chemistry and Technology*, March-April, 1981, Vol. 54, p. 170.

Carboxylated Elastomers

Consider using carboxylated elastomers in place of noncarboxylated versions, that is, carboxylated NBR (XNBR) in place of NBR, or XHNBR in place of HNBR. RP: J. R. Halladay.

Use XNBR with the appropriate zinc oxide in place of conventional NBR to achieve a higher compound tensile strength. RT: Chapter 8, "Specialty Elastomers," M. Gozdiff, p. 199.

The hydrogenation of carboxylated nitrile (XNBR) to HXNBR reportedly will improve both room-temperature ultimate tensile strength as well as high-temperature tensile strength. GEN: John E. Dato (Lanxess), "Hydrogenated Nitrile Rubber for Use in Oilfield Applications," Paper No. 57 presented at the Fall Meeting of the Rubber Division, ACS, Cincinnati, OH, October 10, 2006.

EPDM

Use semicrystalline EPDM grades (with higher ethylene content) to impart higher ultimate tensile strength. RT: Chapter 8, "Specialty Elastomers," R. Vara, J. Laird, p. 191.

Increasing the molecular weight of EPDM can result in an improved ultimate tensile strength. GEN: Arnis Paeglis, "Use of High Molecular Weight EPDM in Blends to Enhance Properties," Paper No. 22 presented at the Spring Meeting of the Rubber Division, ACS, May 17-19, 2004, Grand Rapids, MI.

EPDM Catalysts

Single-site constrained geometry (CGC) metallocene catalyst technology for commercial polymerization has made it possible to efficiently produce commercial grades of EPDM with very high ethylene content and higher crystallinity. These special grades with higher ethylene content impart higher ultimate tensile strength to the vulcanizate. In addition, by substituting a higher concentration of ENB for propylene at constant ethylene content, still higher ultimate tensile strength values can be achieved using this unique technology. GEN: D. Parikh, M. Hughes, M. Laughner, L. Meiske, R. Vara, "Next Generation of Ethylene Elastomers," Presented at ACS Rubber Div. Meeting, Fall, 2000.

Consider using Buckminsterfullerene C_{60} in an EPDM with exposure to ultraviolet light to develop tensile strength significantly higher than that generated by free-radical crosslinking from dicumyl peroxide. This forms a "nanocomposite" from the EPDM/C_{60} combination. GEN: G. Hamed, "Reinforcement of Rubber," *Rubber Chemistry and Technology*, July–August, 2000, Vol. 73, p. 524.

Reactive EPDM

Substituting a 2% maleic-anhydride modified EPDM in place of unmodified EPDM in a blend with NR reportedly will improve the ultimate tensile strength of the NR/EPDM compound. GEN: A. Coran, "Blends of Dissimilar Rubbers: Cure-Rate Incompatibility," *Rubber Chemistry and Technology*, May–June, 1988, Vol. 61, p. 281.

Gas-Phase EPDM

Try compounding with an ultra-low Mooney viscosity gas-phase polymerized EPDM with very high ethylene content and high filler loading. (Higher ethylene content reportedly increases the ultimate tensile strength values to the vulcanizate.) GEN: A. Paeglis, "Very Low Mooney Granular Gas-Phase EPDM," Presented at ACS Rubber Div. Meeting, Fall, 2000, Paper No. 12.

Gel

Raw synthetic rubber such as SBR usually contains stabilizers. However, mixing SBR compounds at high temperatures such as 325°F (163°C) or above can cause the formation of both redispersible gel (or "loose" gel that can be milled out) and so-called "tight" gel (which cannot be milled out and remains insoluble in the appropriate solvents). Both of these types of gel can reduce the ultimate tensile strength of the rubber compound. Therefore it might be prudent to restrict how high the dump temperature should be. GEN: R. Mazzeo, "Preventing Polymer Degradation During Mixing," *Rubber World*, February 1995, p. 22.

Cure

Optimize the ultimate crosslink density from curing. RT: Chapter 3, "Vulcanizate Physical Properties, Performance Characteristics, and Testing," J. S. Dick, p. 47.

Avoid undercuring the compound. RT: Chapter 3, "Vulcanizate Physical Properties, Performance Characteristics, and Testing," J. S. Dick, p. 49.

Avoid overcuring the compound. RT: Chapter 3, "Vulcanizate Physical Properties, Performance Characteristics, and Testing," J. S. Dick, p. 49.

Avoid porosity from curing caused by low pressure, using volatile ingredients, and so forth. RT: Chapter 3, "Vulcanizate Physical Properties, Performance Characteristics, and Testing," J. S. Dick, p. 49.

Reportedly natural rubber based compounds cured with CBS instead of MBS or TBBS may have higher ultimate tensile strength and even higher ultimate elongation. GEN: M. Brendan Rodgers, Donald Tracey, Walter Waddell, "Production, Classification and Properties of Natural Rubber," Paper No. 37 presented at the Spring Meeting of the Rubber Division, ACS, San Antonio, TX, May 16, 2005.

For some rubber compounds based on specialty elastomers such as HXNBR or some fluoroelastomers, applying a post-cure may improve the ultimate tensile strength. GEN: Richard Pazur, L. P. Ferrari, E. C. Campomizzi (Lanxess), "HXNBR Compound Property Improvements Through the Use of Post Cure," Paper No. 70 presented at the Spring Meeting of the Rubber Division, ACS, 2005; "New DAI-EL Fluoroelastomers for Extreme Environments," Daikin America, Inc., Presented at a meeting of the Energy Rubber Group, September 15–18, 2008, San Antonio, TX.

Step-Down Cure

For autoclave-cured products, avoid porosity and low tensile strength by slowly reducing pressure toward the end of the cure cycle. This is called a "step-down cure." RP: L. L. Outzs.

Cure Time and Temperature

Consider curing at a lower temperature for a longer time to avoid extra network sulfur and achieve better sulfur crosslink density so as to achieve higher ultimate tensile strength. GEN: Flexsys Technical Bulletin, "High Temperature Curing Compounding," 1996–1997.

Processing

Avoid contaminating the compounding ingredients, raw elastomers, or the mixing process with dirt. RT: Chapter 3, "Vulcanizate Physical Properties, Performance Characteristics, and Testing," J. S. Dick, p. 49.

Polyurethane

For very high ultimate tensile strength properties, base the compound on either polyester or polyether urethane polymers. RT: Chapter 6, "Elastomer Selection," R. School, p. 126.

The ultimate tensile strength for a two-part cast polyurethane system usually can be increased by adjusting the curative ratio. The curative ratio refers to the relative amounts of prepolymer and curative. The amount of curative [such as methylene-bis-orthochloroaniline (MBCA)] needed to match exactly the isocyanate groups on the prepolymer is referred to as a "100% theory" or "100% stoichiometry." For example, if the curative is reduced by 5%, it is referred to as "95% theory" or "95% stoichiometry." On the other hand, if the curative is increased by 5% from theory, it is referred to as "105% theory" or "105% stoichiometry." Generally, going to a lower stoichiometry, such as 95%, may improve ultimate tensile strength. RT: Chapter 9, "Polyurethane Elastomers," R. W. Fuest, p. 251.

Usually, but not always, selecting an ester-type polyurethane provides an advantage for tensile strength in demanding applications. RT: Chapter 9, "Polyurethane Elastomers," R. W. Fuest, p. 257.

Consider blending polyurethane with a conventional rubber-based compound to improve tensile strength. GEN: T. Jazlonowski, "Blends of Polyurethane Rubber with Conventional Rubber," Paper No. 46 presented at the Spring Meeting of the Rubber Division, ACS, April, 1999, Chicago, IL.

Silicone

To ensure good tensile strength, avoid basing a compound on either silicone or fluorosilicone rubber. RT: Chapter 6, "Elastomer Selection," R. School, p. 136.

AEM

To significantly increase the ultimate tensile strength for terpolymers of ethylene-acrylic elastomer (AEM) (under the trade name of Vamac, from DuPont), which uses a traditional diamine cure system based on hexamethylenediamine carbamate (HMDC) and DPG, consider the addition of dicumyl peroxide and a 1,2-polybutadiene (Ricon, 152) to reportedly increase the ultimate tensile strength. GEN: H. Barager, K. Kammerer, E. McBride, "Increased Cure Rates of Vamac, Dipolymers and Terpolymers Using Peroxides," Presented at ACS Rubber Div. Meeting, Fall, 2000, Paper No. 115.

Silicone–EPDM

Consider using compounds based on the new silicone–EPDM in place of the traditional silicone to achieve better ultimate tensile properties. GEN: M. Chase, "Roll Coverings Past, Present, and Future," Presented at Rubber Roller Group Meeting, New Orleans, May 15–17, 1996, p. 7.

Reactive BR

Using zinc oxide to crosslink a gum carboxylic butadiene rubber that possesses one carboxyl group for approximately every 100 carbons on the backbone can produce a cured rubber that has higher tensile strength than those produced using normal sulfur cures. This becomes an "ionomeric elastomer" in which ionic crosslink sites form nanometer-sized domains. GEN: G. Hamed, "Reinforcement of Rubber," *Rubber Chemistry and Technology*, July–August, 2000, Vol. 73, p. 524.

Mixing

Increase the percent dispersion of reinforcing fillers such as carbon black through better mixing techniques. RT: Chapter 3, "Vulcanizate Physical Properties, Performance Characteristics, and Testing," J. S. Dick, p. 49. GEN: S. Monthey, T. Reed, "Performance Differences Between Carbon Blacks and CB Blends for Critical IR Applications," *Rubber World*, April, 1999, p. 42.

Avoid any contamination or undispersed material of relatively large particle size. RT: Chapter 3, "Vulcanizate Physical Properties, Performance Characteristics, and Testing," J. S. Dick, p. 49.

Prudent use of processing aids with a filler system can be used to achieve better dispersion and thus higher ultimate tensile strength. GEN: Hermann-Josef Weidenhaupt, Kishor Katkar, "Novel Zinc-Free Processing Aid for Silica Compounds," Paper No. 34 given at the IRE 2011 Rubber Conference, Chennai, India.

Phase Mixing

For an SBR/BR blend, a lower tensile strength was reported when there was a higher concentration of carbon black in the SBR phase through phase mixing techniques. GEN: W. Hess, C. Herd, P. Vegvari, "Characterization of Immiscible Elastomer Blends," *Rubber Chemistry and Technology*, July–August, 1993, Vol. 66, p. 329.

It has been reported that through phase mixing techniques of an NR/BR blend, one can achieve an increase in the amount of carbon black dispersed in the BR phase, which may increase the ultimate tensile strength. Other reports have found different effects. Also, Hess reported a large drop in tensile strength if all the carbon black is concentrated in the NR phase. GEN: E. McDonel, K. Baranwal, J. Andries, *Polymer Blends*, Vol. 2, Chapter 19, "Elastomer Blends in Tires," Academic Press, 1978, p. 282; W. Hess, C. Herd, P. Vegvari, "Characterization of Immiscible Elastomer Blends," *Rubber Chemistry and Technology*, 1993, Vol. 66, p. 329.

Compounds based on a blend of base elastomers with different compatibilities may have an improvement in ultimate tensile strength from an effective selection of compatibizers (such as diblock copolymers). GEN: "Effects of Copolymers as Compatibilizer on Blends," Paper presented at the Spring Meeting of the Rubber Division, ACS, April, 1999, Chicago, IL.

Rubber compounds are commonly based on a blend of different elastomers. These different elastomers do not possess the same solubility parameters and establish continuous and discontinuous phases at the microscopic level from mixing. Carbon black affinity for different elastomers is reported as follows:

BR > SBR > CR > NBR > NR > EPDM > IIR

Through phase mixing techniques, the concentation and distribution of carbon black can be controlled between domains in order to maximize tensile strength. GEN: Eric S. Castner (Goodyear Tire and Rubber Co.), "Where's the Filler?: Morphology Control for Improved Dynamic and Mechanical Properties," Paper No. 13 presented at the Fall Meeting of the Rubber Division, ACS, October 5–6, 2004.

Blending

Compounds based on polyoctene and NR give high tensile strength values, although generally not as high as those based on polyurethane. RT: Chapter 6, "Elastomer Selection," R. School, p. 126.

NR/IR

Although high molecular weight is reported to have a positive effect on tensile strength and elongation, strain-induced crystallization as found with high *cis*-IR and NR shows a greater positive effect. RT: Chapter 7, "General Purpose Elastomers and Blends," G. Day, p. 156.

NR/EPDM Blends

Consider covulcanizing an NR/EPDM blend with a sulfur/peroxide cure. The NR will impart better tensile strength to the compound. GEN: S. Tobing, "Co-vulcanization in NR/EPDM Blends," *Rubber World*, February, 1988, p. 33.

NBR/PVC

Adding a given quantity of millable polyurethane incrementally to an NBR/PVC-based compound reportedly will improve the compound's tensile strength. GEN: T. Jablonowski, "Blends of PU with Conventional Rubbers," *Rubber World*, October, 2000, p. 41.

The use of SBR 4503 (an SBR that is a divinylbenzene-crosslinked, hot-polymerized emulsion polymer at 30% bound styrene) with an NBR/PVC blend was reported to improve the ultimate tensile strength. GEN: J. Zhao, G. Ghebremeskel, J. Peasley, "SBR/PVC Blends with NBR as a Compatibilizer," *Rubber World*, December, 1998, p. 37.

TPV

The performance of thermoplastic vulcanizate (TPV) for tensile strength is derived directly from the high degree of crosslinking in the elastomer phase. Select a TPV with a higher crosslink density in the rubbery domain from dynamic vulcanization to improve tensile strength. RT: Chapter 10, "Thermoplastic Elastomers," C. P. Rader, p. 274. RP: C. P. Rader.

In addition, it is important to achieve an extremely fine particle size ($< 1\ \mu m$ in diameter) for these crosslinked elastomer particles to impart good tensile strength to the TPV. These crosslinked particles must be evenly dispersed throughout the matrix of the TPV. Select a TPV with a very fine particle size for the rubber phase to improve tensile strength. RT: Chapter 10, "Thermoplastic Elastomers," C. P. Rader, p. 274. RP: C. P. Rader.

Thermoplastic elastomers often exhibit anisotropy, especially in the case of higher shear rate processes such as injection molding, which can affect the ultimate tensile strength depending on the direction in which the rubber sample is taken in relation to the flow direction. RT: Chapter 10, "Thermoplastic Elastomers," C. P. Rader, p. 274.

Ground Rubber

If ground rubber is used to extend a rubber compound, select one with a higher mesh size (finer particle size) to avoid a drop-off in tensile strength. The finer that the average particle size of the ground rubber is, the lower will be the decrease in the ultimate tensile strength. RT: Chapter 11, "Recycled Rubber," K. Baranwall, W. Klingensmith, p. 291.

Avoid extending tire compounds with ground rubber tire (GRT) extenders because these additives will decrease the compound's tensile strength. However, the larger particle size GRT degrades tensile strength even more. GEN: A. Naskar, S. De, A. Bhowmick, P. Pramanik, R. Mukhopadhyay, "Characterization of Ground Rubber Tire and Its Effect on Natural Rubber Compounds," *Rubber Chemistry and Technology*, November-December, Vol. 73, p. 902.

Fillers

Generally, increasing the surface area of a filler, using the smaller particle sized fillers, such as a carbon black or silica, will increase the tensile strength. GEN:

R. Mastromatteo, E. Morrisey, M. Mastromatteo, H. Day, "Matching Material Properties to Application Requirements," *Rubber World*, February, 1983, p. 26. RP: J. R. Halladay.

Consider using a small loading of well-dispersed carbon nanotubes to improve ultimate tensile strength. GEN: Gary D. D'Abate (Pinnacle Elastomeric Technology), "Compounding Materials," Presented at a meeting of the Energy Rubber Group, September, 2011, Galveston, TX.

Proper selection of an organic surfactant with the proper organically modified nanoclay (based on montmorillonite or MMT) was used to improve the tensile strength of an EPDM compound. GEN: P. J. Yoon, Carl McAfee, "Effects of Nanoclay Structure on the Mechanical Properties of EPDM," Paper No. 16 presented at the Fall Meeting of the Rubber Division, ACS, Oct. 10–12, 2006, Cincinnati, OH.

Careful use of a natural rubber based nanocomposite using nanoclays such as modified montmorillonite was reported to improve ultimate tensile strength. GEN: A. K. Bhowmick, M. Bhattacharya, and S. Mitra, "Exfoliation of Nanolayer Assemblies for Improved Natural Rubber Properties: Methods and Theory," Paper No. 103 presented at the Fall Meeting of the Rubber Division, ACS, October 13–15, 2009, Pittsburgh, PA.

Carbon Black

Increasing carbon black loading to an optimum will usually increase ultimate tensile strength, as long as it can be dispersed. This optimum loading occurs at a lower level for finer particle sized carbon black. RT: Chapter 12, "Compounding with Carbon Black and Oil," S. Laube, S. Monthey, M-J. Wang, p. 308. RP: J. M. Long, M-J. Wang.

Increasing the surface area of the carbon black used in a compound will usually increase ultimate tensile strength. RT: Chapter 12, "Compounding with Carbon Black and Oil," S. Laube, S. Monthey, M-J. Wang, pp. 308, 317.

Improving the dispersion of carbon black through longer mix cycles will usually improve ultimate tensile strength of the compound. RT: Chapter 12, "Compounding with Carbon Black and Oil," S. Laube, S. Monthey, M-J. Wang, p. 308.

Adding ball-milled dispersions of reinforcing carbon black to natural rubber latex before coagulation and then compounding will yield higher tensile strength than simply adding free carbon black directly into the Banbury mixer. GEN: R. Alex, K. Sasidharan, T. Kurlan, A. Kumarchandra, "Carbon Black/Silica Masterbatch from Fresh Natural Rubber Latex," Paper No. 27 given at IRC 11, Mumbai, India.

Substitute a very small amount of carbon black with halloysite nanotubes (HNTs), which is adequately dispersed, in order to improve a rubber compound's ultimate tensile strength, often with an improvement in ultimate elongation at break as well. GEN: Minna Poikelispaa, Amit Das, Wilima Dierkes, Jyki Vuorinen (Tampere University of Technology, Finland), Paper No. 53 presented at the Fall Meeting of the Rubber Division, ACS, October 10–13, 2011, Cleveland, OH.

Silica

Consider using precipitated silica with higher surface area to increase significantly the ultimate tensile strength of a rubber compound. To improve tensile properties even more, consider using precipitated silica treated with a silane coupling agent. RT: Chapter 13, "Precipitated Silica and Non-black Fillers," W. Waddell, L. Evans, p. 333.

Nonreinforcing Filler Effects

To achieve the highest ultimate tensile strength, one should avoid using nonreinforcing or "extending fillers" such as clay, calcium carbonate, whiting, talc, ground sand, and so forth. RP: J. R. Halladay.

Clay

To improve the ultimate tensile strength of a clay-loaded rubber compound, consider using a clay that is treated with a silane coupling agent. RT: Chapter 13, "Precipitated Silica and Non-black Fillers," W. Waddell, L. Evans, p. 333.

Oil Effects

Minimize the use of plasticizers to achieve high ultimate tensile strength. RP: J. R. Halladay.

Sulfur

When curing compounds based on NBR polymers, sometimes it is difficult to achieve good dispersion of conventional "rubber maker's sulfur." Therefore magnesium carbonate treated sulfur is used as the curative for nitrile-based compounds because it disperses more effectively in this polar elastomer. If good dispersion of curatives is not achieved, then tensile strength properties will be hurt. RT: Chapter 16, "Cures for Specialty Elastomers," B. H. To, p. 398.

Sulfur-Based Curatives

Make sure that accelerators [such as MBTS with a high melting point about 167 to 179°C (332.6 to 354.2°F), which is usually above the dump temperature with most internal mixers] are fine enough in particle size (below 100 μm) so as not to adversely affect tensile strength. If the particle size of these rubber chemicals with high melt points is not fine enough, the ultimate tensile values will be lower (as much as 50%), which could affect the service life of the rubber product. The reason for this is that large particles cause nonhomogeneous distribution of crosslinks in the rubber matrix. It is reportedly also possible for accelerators to have too fine a particle size, which can cause a decrease in ultimate tensile as well (as much as 10% or more). GEN: W. Helt, "Accelerator Dispersion," ASTM D11 Meeting, July 28, 1990.

Consider the use of an experimental accelerator, bis-(diisopropyl)thiophosphoryl disulfide (DIPDIS), in synergistic combinations with conventional thiazole-type accelerators to produce NR vulcanizates with higher ultimate tensile strength. GEN: S. Mandal, R. Datta, D. Basu, "Studies of Cure Synergism, Part 1: Effect of bis(diisopropyl) thiophosphoryl Disulfide and Thiazole-Based Accelerators in the Vulcanization of Natural Rubber," *Rubber Chemistry and Technology*, September-October, 1989, Vol. 62, p. 569.

Polysulfide Crosslinks

Use a cure system that will give a higher preponderance of polysulfide crosslinks (a conventional cure) to achieve higher tensile strength (compared to an EV cure with a preponderance of mono- and disulfide crosslinks). RP: J. R. Halladay.

Polychloroprene Cures

Consider using a traditional ethylene thiourea (ETU) cure with all the appropriate safety and health precautions for a significant improvement in ultimate tensile strength when curing a polychloroprene-based compound. RT: Chapter 16, "Cures for Specialty Elastomers," B. H. To, pp. 400–401, 406.

Peroxide Cures

Using a peroxide masterbatch instead of the traditional "peroxide extended on an inert filler" can greatly improve the ultimate dispersion of the peroxide in the compound during mixing, which results in more uniform physical properties. RT: Chapter 17, "Peroxide Cure Systems," L. Palys, p. 417.

For peroxide cures, use a coagent as well to add unsaturation to the system, which results in higher crosslink density than the peroxide would have generated if used without the coagent. The reason for this is that it is more efficient and easier to crosslink with a free radical to unsaturation than by hydrogen abstraction to a saturated polymer chain. Because the use of a coagent can lead to different types of crosslink networks, improved tensile strength can result. RT: Chapter 17, "Peroxide Cure Systems," L. Palys, pp. 431–432.

Zinc diacrylate metallic coagent (Saret® 633) is sometimes used in peroxide cures to improve the vulcanizate's tensile strength. GEN: R. Costin, W. Nagel, "Coagents for Rubber-to-Metal Adhesion," *Rubber & Plastics News*, March 11, 1996, p. 14.

Using a coagent with a peroxide cure may increase tensile strength. GEN: P. Dluzneski, "Peroxide Vulcanization of Elastomers," *Rubber Chemistry and Technology*, July–August, 2001, Vol. 74, p. 451.

Antioxidants

In some cases, adding antioxidants early in a mix cycle may help prevent the deleterious effects of the mixing mastication on the polymer structure and molecular weight. GEN: H. Kim, G. Hamed, "On the Reason that Passenger Tire Sidewalls Are Based on Blends of Natural Rubber and *cis*-Polybutadiene," ACS Rubber Div. Meeting, Fall, 1999, Paper No. 184.

Resins

For elastomer bases such as SBR, NBR, or CR with low loadings of nonreinforcing fillers such as clay, whiting, lithopone, and so forth, adding 15 to 25 phr of a hydrocarbon resin such as a coal tar resin may improve the ultimate tensile strength. GEN: F. O'Connor, J. Slinger, "Processing Aids: The All-Inclusive Category," *Rubber World*, October, 1982, pp. 21–23.

Orientation

The ultimate tensile properties of vulcanizates, and in particular thermoplastic elastomers, can indeed be affected by the degree of orientation and anisotropy that might be present. GEN: B. Gedeon, "Anisotropy in Thermoplastic Elastomers," *Rubber World*, October, 1998, p. 34.

Double Networking

It is well known that plastics can be drawn or stretched during processing to create orientation, directional (anisotropic) properties, and improved directional physical properties on cooling below the melt temperature (T_m) or glass transition temperature (T_g) of the plastic. On the other hand, rubber compounds are quite different. Normally any orientation during the processing of a rubber compound will decay somewhat after processing. However, "double networking" during the curing process is a unique method of introducing permanent chain orientation into a rubber compound. This is achieved first by lightly crosslinking or partially curing the rubber compound in a normal manner, which establishes the first network. This lightly cured rubber compound is stretched to an extension ratio (or set) of α_0 and crosslinked further. After this second cure, the vulcanizate is released. From the release, this second formed crosslinked network "resists the retraction of the first network." This results in a vulcanizate with a residual extension ratio (or set) of α_r.

A "dumbbell" specimen cut parallel to the stretch direction may have higher tensile strength than one produced by the conventional single-network cure. GEN: G. Hamed, M. Huang, "Tensile and Tear Behavior of Anisotropic Double Networks of a Black-Filled Natural Rubber Vulcanizate," *Rubber Chemistry and Technology*, November–December, 1998, Vol. 71, p. 846.

Fiber Orientation

Overall, some of the best fibers to select for improving tensile strength are unregenerated wood cellulose fiber with an average length-to-diameter ratio of about 100 : 1 to 200 : 1 and ideally treated with a resorcinol formaldehyde vinyl pyridene latex for adhesion. GEN: L. Goettler, K. Shen, "Short Fiber Reinforced Elastomers," *Rubber Chemistry and Technology*, 1983, Vol. 56, p. 575.

The use of aramid fibers in rubber compounds has been reported to improve the strength of a rubber compound. GEN: Debbie Banta, Weatherford Co., "Can Nanotechnology Provide Innovative, Affordable Elastomer Solutions to Oil and Gas Service Industry Problems?," Presented at the Energy Rubber Group, January 19, 2012, Houston, TX.

Ionomers

By adding an "ionolyzer" such as zinc stearate to an "ionic elastomer," which is a metal-sulfonated EPDM compound, one can significantly improve the tensile strength to a degree approaching the strength of polyurethanes. GEN: W. MacKnight, R. Lundberg, "Elastomeric Ionomers," *Rubber Chemistry and Technology*, July–August, 1984, Vol. 57, p. 652.

Ionic Crosslinks

"Ionically crossslinked" cured compounds reportedly have good tensile strength because the crosslink sites can "slip," thus migrating without tearing. GEN: T. Kempermann, "Sulfur-Free Vulcanization Systems for Diene Rubber," *Rubber Chemistry and Technology*, July–August, 1988, Vol. 61, p. 422.

Stress Crystallization

Compounding from stress crystallizing polymers such as NR and polychloroprene will improve compound tensile strength. GEN: F. Eirich, *Science and Technology of Rubber*, Chapter 9, "The Rubber Compound and Its Composition," M. Studebaker, J. Beatty, Academic Press, 1978, p. 367.

2.2 Increasing High-Temperature Tensile Strength

Although a given rubber compound may meet a room-temperature ultimate tensile strength requirement, it might not meet a particular high-temperature tensile strength requirement that a customer might specify. This new requirement might be needed because of a customer's concern regarding high-temperature product performance. Meeting these special requirements can be a challenge.

*The following **experimental** ideas might help a compounder meet these requirements. Also, all relevant literature sources, including the ones cited below, should be researched and read. **Caution:** these general experimental ideas may not work in all specific situations. Changes to increase high-temperature tensile strength will certainly affect other properties as well, for better or for worse, and this book does not purport to show how these other properties are affected. Also, this book does not purport to address safety and health issues.*

Silicone

Under extremely high temperatures, silicone rubber may often impart better high-temperature tensile strength than most organic elastomers. RP: J. R. Halladay.

SBR

High-temperature stress-strain properties for an SBR compound reportedly can be improved by adding NR as a 50:50 blend. RT: Chapter 7, "General Purpose Elastomers and Blends," G. Day, p. 157.

EPDM

Through new proprietary Ziegler-Natta catalyst technologies, unique high-temperature crystallinity through ethylene sequencing in EPDM reportedly gives better "high-temperature" tensile strength. With this ethylene sequencing, some crystallinity undergoes multiple crystallinity transitions up to 75°C (167°F). GEN: S. Brignac, H. Young, "EPDM with Better Low-Temperature Performance," *Rubber & Plastics News*, August 11, 1997, p. 14.

CR

For polychloroprene-based compounds, use a W-type neoprene with a precipitated silica up to 40 parts with perhaps 2 phr of polyethylene glycol (PEG). RP: L. L. Outzs.

RT: *Rubber Technology, Compounding and Testing for Performance*. 2nd ed. Edited by J. S. Dick, Hanser Publications, 2009 ■ GEN: General References from the Literature ■ RP: Review Panel (Advisors for this book) ■ IMPORTANT: The information contained herein is experimental in nature and is meant for individuals with advanced scientific training and rubber compounding experience. All safety precautions should be taken when replicating or applying this information. The author makes no representation or warranty, either expressed or implied.

Silica Effects

In some situations, between 10 and 20 phr of precipitated silica may increase the hot tear and hot tensile strength of a rubber compound. RP: J. R. Halladay.

Nanoclays and Carbon Nanotubes

The use of nanoclays or carbon nanotubes in rubber compounds has been evaluated and considered to improve the performance of rubber compounds used in oil drilling operations to have less deterioration under the higher temperatures and higher pressures that are commonly encountered in "downhole" applications. GEN: Debbie Banta, Weatherford Co., "Can Nanotechnology Provide Innovative, Affordable Elastomer Solutions to Oil and Gas Service Industry Problems?," Presented at the Energy Rubber Group, January 19, 2012, Houston, TX.

2.3 Increasing Ultimate Tensile Elongation

Very often a customer will specify requirements on just how far a cured rubber compound can be stretched without breaking. This is another basic material property that is determined from stress-strain testing of a cured "dumbbell" specimen in accordance with standard test conditions denoted by ASTM or ISO procedures.

The following experimental ideas can be considered for increasing the ultimate elongation of the compound. Also, all relevant literature sources, including the ones cited below, should be researched and read. ***Caution:*** *these general experimental ideas may not work in all specific situations. Changes to increase ultimate tensile elongation will certainly affect other properties as well, for better or for worse, and this book does not purport to show how these other properties are affected. Also, this book does not purport to address safety and health issues.*

SBR

Use emulsion SBR polymerized at −10°C (14°F) [instead of +50°C (122°F)] for a significant increase in tensile ultimate elongation. RT: Chapter 7, "General Purpose Elastomers and Blends," G. Day, p. 149.

NR

Of the various grades of NR, peptized CV60 gives the highest elongation. RT: Chapter 7, "General Purpose Elastomers and Blends," G. Day, p. 156.

Polyurethanes

Ultimate elongation for a two-part cast polyurethane system usually can be increased by adjusting the curative ratio. The curative ratio refers to the relative amounts of prepolymer and curative. The amount of curative [such as methylene-bis-orthochloroaniline (MBCA)] needed to match exactly the isocyanate groups on the prepolymer is referred to as a "100% theory" or "100% stoichiometry." Thus if curative is reduced by 5%, it is referred to as "95% theory" or "95% stoichiometry." On the other hand, if the curative is increased by 5% from theory, it is referred to as "105% theory" or "105% stoichiometry." Generally going to a higher stoichiometry, such as 105%, may improve ultimate elongation. RT: Chapter 9, "Polyurethane Elastomers," R. W. Fuest, p. 251.

NBR

Selecting an NBR with a percent bound acrylonitrile content between 33 and 41% may impart the highest percent ultimate elongation values as well as the highest ultimate tensile strength to the rubber compound. GEN: Andy Anderson (Zeon), "Keeping It Real with NBR and HNBR Polymers," Presented at a meeting of the Energy Rubber Group, September 12–15, 2011, Galveston, TX.

Polychloroprene and Fillers

Use an inorganic filler with a larger particle size in place of a finer particle size carbon black in CR formulations to increase ultimate elongation. In addition, substitute thermal blacks in place of reinforcing or semireinforcing carbon blacks to increase ultimate tensile elongation. RP: L. L. Outzs.

Consider xanthate-modified types of CR for higher levels of tensile strength and elongation. GEN: Nobuhiko Fujii (Denka Corp.), "Recent Technical Improvements of CR and ER in Industrial Applications," presented at Southern Rubber Group, March, 2006, Asheville, NC.

TPE and TPV

Sometimes thermoplastic elastomers and thermoplastic vulcanizates exhibit anisotropy, especially with higher shear rate processes such as injection molding, which can affect the ultimate elongation and ultimate tensile strength depending on the direction in which the rubber sample is taken in relation to the flow direction. RT: Chapter 10, "Thermoplastic Elastomers," C. P. Rader, p. 274.

Carbon Black

Decreasing the surface area of the carbon black used in a compound will usually increase the ultimate elongation of the compound. RT: Chapter 12, "Compounding with Carbon Black and Oil," S. Laube, S. Monthey, M-J. Wang, p. 308.

Decreasing the structure of the carbon black used in a compound will usually increase the ultimate elongation of the compound. RT: Chapter 12, "Compounding with Carbon Black and Oil," S. Laube, S. Monthey, M-J. Wang, p. 308.

Decreasing the loading of the carbon black used in a compound will usually increase the ultimate elongation of the compound through an optimum. RT: Chapter 12, "Compounding with Carbon Black and Oil," S. Laube, S. Monthey, M-J. Wang, p. 308.

In general, decreasing the loading level of a reinforcing filler in the compound will usually increase the ultimate elongation and decrease the ultimate tensile strength. GEN: Gary D. D'Abate (Pinnacle Elastomeric Technology), "Compounding Materials," Presented at a meeting of the Energy Rubber Group, Sept. 13, 2011, Galveston, TX.

Talc

An equal substitution by volume of fine particle size talc for part of the carbon black reportedly results in an increase in ultimate elongation with possibly only a small effect on ultimate tensile strength and possibly an increase in low strain modulus. GEN: *Luzenac America's Handbook*, "Mistron Vapor in Rubber Compounds," 1980. RP: O. Noel.

Curing

All other things being equal in the composition of the rubber compound, a decrease in the cured crosslink density of a given compound may generally cause the percent ultimate elongation to increase, generally with a corresponding decrease in cured compound static modulus as well. GEN: William Boye (Sartomer Co.), "The Utility of Coagents in the Radical Cure of Elastomers," Presented at a meeting of the Energy Rubber Group, 2008, Houston, TX.

Sulfur Cure

An advantage of sulfur vs. peroxide cures is that it is easier to produce higher ultimate elongations in sulfur-cured compounds. RT: Chapter 17, "Peroxide Cure Systems," L. Palys, p. 434.

Generally high-sulfur cures will impart higher ultimate elongation values than what is imparted by low-sulfur cures. RP: J. R. Halladay.

Gel

Raw synthetic rubber such as SBR usually contains stabilizers. However, mixing SBR compounds at high temperatures such as 325°F (163°C) or above can cause the formation of both redispersible gel (or "loose" gel that can be milled out) and so-called "tight" gel (that cannot be milled out and remains insoluble in the appropriate solvents). Both of these types of gel can reduce the cured ultimate elongation of the rubber compound. Therefore, it might be prudent to restrict how high the dump temperature should be. GEN: R. Mazzeo, "Preventing Polymer Degradation During Mixing," *Rubber World*, February 1995, p. 22.

Fibers

If one is trying to avoid low ultimate elongation values, avoid using a loading of short aramid, cellulose, jute, or other fibers in the compound. GEN: K. Watson, A. Frances, "Elastomer Reinforcement with Short Kevlar Aramid Fiber for Wear Applications," *Rubber World*, August, 1988, p. 20.

20 Microns Nano Minerals Limited (India) claims that using their product "Vaporlink" (a fibrous lath-like crystal stucture nanomicron filler) with other

fillers will significantly improve ultimate elongation. GEN: Presentation by 20 Microns Nano Minerals Limited at IRE 2011, Chennai, India.

Carbon Nanotubes

The careful use of carbon nanotubes in rubber compounds has been reported to increase ultimate elongation without the loss of ultimate tensile strength of the rubber compound. GEN: Debbie Banta, Weatherford Co., "Can Nanotechnology Provide Innovative, Affordable Elastomer Solutions to Oil and Gas Service Industry Problems?," Presented at the Energy Rubber Group, January 19, 2012, Houston, TX.

Substitute a very small amount of carbon black with halloysite nanotubes (HNTs) that are adequately dispersed in order to improve a rubber compound's ultimate tensile strength, commonly with an improvement in ultimate elongation at break as well. GEN: Minna Poikelispaa, Amit Das, Wilima Dierkes, Jyki Vuorinen (Tampere University of Technology, Finland), Paper No. 53 presented at the Fall Meeting of the Rubber Division, ACS, October 10–13, 2011, Cleveland, OH.

Nanoclays

Proper selection of an organic surfactant with the proper organically modified nanoclay (based on montmorillonite or MMT) was used to improve the ultimate tensile elongation as well as the tensile strength of an EPDM compound. GEN: P. J. Yoon, Carl McAfee, "Effects of Nanoclay Structure on the Mechanical Propertiers of EPDM," Paper No. 16 presented at the Fall Meeting of the Rubber Div. ACS, October, 2006.

Organosilane with Silica-Loaded Compounds

Use of an organosilane with a precipitated hydrated silica may enhance the ultimate elongation and tensile strength of the rubber compound. GEN: Cynthia M. Flanigan, Laura Beyer, David Klekamp, David Rohweder (Ford Motor Co., Dearborn, MI), "Comparative Study of Silica, Carbon Black, and Novel Fillers in Tread Compounds," Paper No. 34 presented at the Fall Meeting of the Rubber Division, ACS, October 11–13, 2011, Cleveland, OH.

Mixing

Increasing carbon black dispersion will usually increase ultimate elongation. GEN: W. Hess, "Characterization of Dispersions," *Rubber Chemistry and Technology*, July–August, 1991, Vol. 64, p. 386.

Phase Mixing

A special four-step phase mixing sequence was used for a carbon black/silica loaded tire tread compound based on BR, SSBR, and BIMS, which significantly increased both the cured ultimate elongation and the ultimate tensile strength. GEN: W. Waddell, R. Poulter (Exxon), "Phase Mixing of Brominated Isobutylene-co-*para*-methylstyrene with Precipitated Silica to Enhance the Properties of a Tire Tread Compound," Paper No. 32 presented at the Spring Meeting of the Rubber Division, ACS, April 13, 1999, Chicago, IL.

Molecular Weight Effects

Using raw NBR elastomers with a lower Mooney viscosity and molecular weight may increase the compound's ultimate elongation. GEN: D. Ghang, "Investigation of the Structure-Property Relationships of Improved Low Compression Set Nitrile Rubbers," *Rubber Chemistry and Technology*, March-April, 1981, Vol. 54, p. 170. The same is reportedly true for emulsion SBR, solution SBR, BR, and IR. RP: R. Dailey.

State of Cure

Generally a lower state of cure results in a higher ultimate elongation. RP: J. R. Halladay.

2.4 Increasing Hardness and Modulus

Basically, "modulus" is considered a measure of the stiffness of a material. Stiffness and hardness are usually closely correlated. However, what is called "modulus" in the rubber industry is actually stress at a specified elongation as determined from a stress-strain test using a tensile tester. This stress is usually measured at a strain of 100, 300, or 400%. On the other hand, the strain applied when measuring rubber hardness with a durometer is much lower than 100%. Because of this great difference in the magnitude of deformation applied in the hardness test vs. the so-called "modulus" measurement with the tensile tester and the well known nonlinearity of rubber, these two parameters many times do not correlate exactly.

The compounder might consider some of the following ***experimental*** *ideas when trying to increase the compound's cured hardness or "modulus." Also, all relevant literature sources, including the ones cited below, should be researched and read.* ***Caution:*** *these general experimental ideas may not work in all specific situations. Changes to increase hardness and/or modulus will certainly affect other properties as well, for better or for worse, and this book does not purport to show how these other properties are affected. Also, this book does not purport to address safety and health issues.*

Mixing

When mixing butyl rubber compounds containing carbon black, consider the use of high- temperature mixing (heat treatment) to promote greater pigment-polymer interaction. RT: Chapter 8, "Specialty Elastomers," G. Jones, D. Tracey, A. Tisler, p. 175. This may also work with some other polymer bases. RP: J. M. Long.

It was reported by Boonstra and Medalia that sometimes overmixing a rubber compound (excessive work history in the mixing of a batch) can reduce the cured modulus values. So in order to increase cured modulus, consider not "overmixing" the stock. GEN: B. Boonstra, A. Medalia, "Effect of Carbon Black Dispersion on the Mechanical Properties of Rubber Vulcanizates," *Rubber Chemistry and Technology*, 1963, Vol. 36, No. 1, p. 115.

Molecular Weight Effects

Using raw NBR elastomers with higher Mooney viscosity and molecular weight may increase the compound's tensile modulus. GEN: D. Ghang, "Investigation of the Structure-Property Relationships of Improved Low Compression Set Nitrile Rubbers," *Rubber Chemistry and Technology*, March-April, 1981, Vol. 54, p. 170.

SBR

Small amounts of blocked styrene in what is normally random SBR may improve room-temperature hardness. RT: Chapter 7, "General Purpose Elastomer and Blends," G. Day, p. 148.

Use ESBR polymerized at 5°C (41°F) (or lower) instead of 50°C (122°F) for an increase in 300% modulus. RT: Chapter 7, "General Purpose Elastomers and Blends," G. Day, p. 149.

Increasing bound styrene content in SBR will increase room-temperature compound hardness and modulus. RT: Chapter 7, "General Purpose Elastomers and Blends," G. Day, p. 157.

NR

Epoxidized natural rubber (ENR) can increase compound modulus compared to untreated NR. RT: Chapter 7, "General Purpose Elastomers and Blends," G. Day, p. 156.

Adding ball-milled dispersions of reinforcing carbon black to natural rubber latex before coagulation and then compounding will yield higher cured modulus and hardness than simply adding free carbon black directly into the Banbury mixer. GEN: R. Alex, K. Sasidharan, T. Kurlan, A. Kumarchandra, "Carbon Black/Silica Masterbatch from Fresh Natural Rubber Latex," Paper No. 27 presented at India Rubber Expo, IRC 11, January 19, 2011, Chennai, India.

IR

Changing the IR microstructure predominantly from *cis* to *trans* will increase the room-temperature modulus, as long as the temperature is below the T_m or melt point. RT: Chapter 7, "General Purpose Elastomers and Blends," G. Day, p. 156.

EPDM

Use semicrystalline EPDM grades (with higher ethylene content) to impart higher cured hardness. RT: Chapter 8, "Specialty Elastomers," R. Vara, J. Laird, p. 191.

Use a base elastomer such as EPDM that can accept a very high loading of carbon black to achieve greater hardness. RT: Chapter 12, "Compounding with Carbon Black and Oil," S. Laube, S. Monthey, M-J. Wang, p. 310.

Consider using the new EPDM grades based on over 80% ethylene content using the new constrained-geometry metallocene catalyst technology. GEN: R. Vara, C. Grant, C. Daniel (DuPont Elastomers LLC), "Techniques for Achieving High Hardness EPDM Formulations," Paper No. 78 presented at the Fall Meeting of the Rubber Division, ACS, October 16, 2001, Cleveland, OH.

Gas-Phase EPDM

Try compounding with an ultra-low Mooney viscosity gas-phase polymerized EPDM with very high ethylene content and high filler loading. (A higher ethylene content usually imparts higher hardness values to the vulcanizate.) GEN: A. Paeglis, "Very Low Mooney Granular Gas-Phase EPDM," Presented at ACS Rubber Div. Meeting, Fall, 2000, Paper No. 12.

NBR

Use an emulsion NBR that is produced from a coagulating system based on calcium chloride for a higher tensile modulus. RT: Chapter 8, "Specialty Elastomers," M. Gozdiff, p. 195.

Consider NBR grades with narrower molecular weight distribution for higher compound modulus. RT: Chapter 8, "Specialty Elastomers," M. Gozdiff, p. 197.

Room-temperature hardness might be increased by using an NBR with high bound ACN content of approximately 45 to 50%. GEN: R. Grossman, Q & A, *Elastomerics*, January, 1989.

Using NBR grades with higher ACN content should normally result in higher hardness values. GEN: "A Comparative Evaluation of Hycar Nitrile Polymers," Manual HM-1, Revised, B. F. Goodrich Chemical Co.

When selecting from nitrile elastomer grades with concerns for higher cured durometer values, consider NBR grades that have higher ACN content. GEN: R. Del Vecchio, E. Ferro, "Effects of NBR Polymer Variations on Compound Properties," Presented at ACS Rubber Div. Meeting, Spring, 2001, Paper No. 21.

Heat aging of some compounds based on NBR may increase the hardness of the cured compound 11 points or so. GEN: R. Del Vecchio, E. Ferro, "Effects of NBR Polymer Variations on Compound Properties," Presented at Rubber Div. ACS Meeting, Spring, 2001, Paper No. 21.

CR

The use of faster crystallizing polychloroprene (CR) in place of slower crystallizing CR will usually increase cured hardness. Adding 5 to 15 phr of a high-styrene resin can also increase hardness. RP: L. L. Outzs.

XNBR

Use XNBR with the appropriate zinc oxide in place of conventional NBR to achieve a higher compound modulus. RT: Chapter 8, "Specialty Elastomers," M. Gozdiff, p. 199.

CPE

Increasing the chlorine content in chlorinated polyethylene (CM) usually results in higher compound cured crosslinked density for thiadiazole-cured stocks. However, decreasing chlorine content in CM polymers usually results in higher compound cured crosslinked density for peroxide-cured stocks. RT: Chapter 8, "Specialty Elastomers," L. Weaver, p. 213.

Polyurethane

Two-part cast polyurethane systems can be compounded to provide a high degree of hardness, up to 75 D, while retaining some rubber-like quality, and 85 D with properties similar to those of a hard plastic. In fact, normally most cast polyurethane (PU) systems will have hardnesses well above 50 (Shore A). [Polyurethane formulations that result in cured durometer values below 50 A may achieve this by using a benzoate ester plasticizer for polyester-based PU systems or phthalate ester plasticizers, such as dioctyl phthalate (DOP), for polyether-based PU systems]. RT: Chapter 9, "Polyurethane Elastomers," R. W. Fuest, pp. 252–255.

Castable polyurethane fabricators using MDI prepolymer systems may be able to increase their product's hardness by using aromatic diol chain extenders such as bis-(hydroxyethyl) ether of resorcinol (HER) and hydroquinone (HQEE). The use of these chain extenders will create hard domains that have higher melting points. GEN: R. Durairaj, "Chain Extenders Increase Heat Tolerance," *Rubber & Plastics News*, November 29, 1999.

PNR

Polynorbornene would not normally be selected as the base elastomer when trying to compound for a very high cured hardness. On the other hand, PNR can be used with appropriate oil loadings to achieve very low cured hardness values–down to Shore A values of 15 to 20. RT: Chapter 8, "Specialty Elastomers," C. Cable, p. 224.

Crosslink Density

In general, increasing the crosslink density will increase the hardness and modulus of the vulcanizate. GEN: D. Campbell, A. Chapman, "Relationships Between Vulcanizate Structure and Vulcanizate Performance," Malaysian Rubber Producers Research Association, Brickendonbury, Hertford, UK.

The modulus of a rubber vulcanizate is related to the crosslink density. Crosslink density in turn is crudely related to the square root of the product of the concentration of sulfur and the concentration of accelerator. GEN: W. Helt, "Accelerator Dispersion," ASTM D11 Meeting, Philadelphia, PA, July 28, 1990.

Sulfur Cures

Consider the use of an experimental accelerator, bis-(diisopropyl) thiophosphoryl disulfide (DIPDIS) with synergistic combinations with conventional thiazole-type accelerators to produce NR vulcanizates with higher crosslink density and modulus. GEN: S. Mandal, R. Datta, D. Basu, "Studies of Cure Synergism, Part 1: Effect of Bis(diisopropyl) Thiophosphoryl Disulfide and Thiazole-Based Accelerators in the Vulcanization of Natural Rubber," *Rubber Chemistry and Technology*, September-October, 1989, Vol. 62, p. 569.

To maximize cured hardness and modulus for a semi-EV or EV cure, consider curing for a longer time at a lower temperature rather than at a higher temperature for a shorter time because the longer cure may result in a higher ultimate crosslink density and greater hardness. GEN: M. Lemieux, P. Killcoar, "Low Modulus, High Damping, High Fatigue Life Elastomer Compounds for Vibration Isolation," *Rubber Chemistry and Technology*, September-October, 1984, Vol. 57, p. 792.

If zinc oxide is used at levels below 3 phr in a sulfur-cured compound, cured modulus values tend to be lower. GEN: W. Hall, H. Jones, "The Effect of Zinc Oxide and Other Curatives on the Physical Properties of a Bus and Truck Tread Compound," Presented at ACS Rubber Div. Meeting, Fall, 1970.

Increasing the concentration of stearic acid in a sulfenamide sulfur cure may increase the cured modulus and hardness of the vulcanizate. GEN: W. Hall, H. Jones, "The Effect of Zinc Oxide and Other Curatives on the Physical Properties of a Bus and Truck Tread Compound," Presented at ACS Rubber Div. Meeting, Fall, 1970.

Peroxide Cures

For peroxide cures, avoid aromatic additives such as aromatic oil, because the peroxide free radicals generated during cure will abstract labile hydrogens from the aromatic structures, rendering the free radicals "nonreactive" and thus preventing a higher crosslink density from developing. The ultimate modulus is thus retarded. RT: Chapter 17, "Peroxide Cure Systems," L. Palys, p. 430.

In a pure peroxide cure, 3,4-IR and 1,2-BR microstructures will increase the peroxide state of cure. Chapter 7, "General Purpose Elastomers and Blends," G. Day, p. 156.

For peroxide cures, select quinoline-type antioxidants that have less effect on state of cure. RT: Chapter 17, "Peroxide Cure Systems," L. Palys, p. 430.

For peroxide cures of EPDM compounds, consider selecting ZMTI as an antioxidant, which should allow good modulus development as well as impart good heat aging. RT: Chapter 17, "Peroxide Cure Systems," L. Palys, p. 430.

Using a coagent with a peroxide cure may increase cured hardness. GEN: P. Dluzneski, "Peroxide Vulcanization of Elastomers," *Rubber Chemistry and Technology*, July-August, 2001, Vol. 74, p. 451.

For peroxide cures, use a coagent as well to add unsaturation to the system, which results in higher crosslink density and cured modulus than the peroxide would have generated if used without the coagent. The reason for this is that it is more efficient and easier to crosslink with a free radical to unsaturation than by hydrogen abstraction to a saturated polymer chain. RT: Chapter 17, "Peroxide Cure Systems," L. Palys, pp. 431-432.

When choosing from the many coagents available for peroxide cures (such as MBM, TAC, TAIC, ADC, AMA, methacrylate types, acrylic types, liquid polymeric types, etc.), select one with good efficiency for the specific compound. From the selection of the appropriate coagent, it is possible to increase the modulus by 40% or more. RT: Chapter 17, "Peroxide Cure Systems," L. Palys, pp. 431-433.

For compounds cured with peroxide, avoid using acidic clays because they affect peroxide cures and reduce the state of cure much more than with sulfur cure systems. If a clay is to be used with a peroxide cure, probably it is better to select a nonacidic Burgess-type clay filler. If a process oil is to be used as well, select a paraffinic-type oil, which should have less of an effect on the peroxide cure than other types of process oils. RT: Chapter 17, "Peroxide Cure Systems," L. Palys, p. 435.

Generally there is an *order of efficiency* for crosslinking different elastomers with peroxide curatives. This efficiency is dependent on the concentration of abstractable hydrogens as well as the probability of chain scission. This reported order is as follows.

BR > NR/SBR > NBR > CR > EPDM > EPR > CPE

GEN: P. Dluzneski, "Peroxide Vulcanization of Elastomers," *Rubber Chemistry and Technology*, July-August, 2001, Vol. 74, p. 451.

N,N'-m-Phenylene dimaleimide (HVA-2) is one of the most effective coagents in a peroxide cure to increase the compound's cured modulus. GEN: P. Dluzneski, "Peroxide Vulcanization of Elastomers," *Rubber Chemistry and Technology*, July-August, 2001, Vol. 74, p. 451.

In general, avoid applying a peroxide cure to compounds that contain polyisobutylene, butyl rubber, halobutyl rubber, or polyepichlorohydrin because these polymers usually are not effectively crosslinked well by peroxide curatives. GEN: P. Dluzneski, "Peroxide Vulcanization of Elastomers," *Rubber Chemistry and Technology*, July-August, 2001, Vol. 74, p. 451.

Some peroxide-cured compounds are too soft to develop sufficient shear in order to achieve a high enough dispersion to reach cured hardness and modulus targets. Sometimes this problem can be solved by using the peroxide curative in a more dispersible form. GEN: L. H. Palys, F. Debaud, L. Keromnes, J. Brennan (Arkema Co.), "Techniques for Improving Elastomer Processing and Crosslinking Performance," Presented at the Energy Rubber Group, September 18, 2008, San Antonio, TX.

Work has been reported where the crosslink density of a peroxide cure system can be increased with more delayed scorch by using a cure system based on dicumyl peroxide, a coagent such as *N,N'*-m-phenylenedimaleimide (BMI-MP) and a sulfur "spender" such as dipentamethylenethiuram tetrasulfide (DPTT) for curing either saturated or low unsaturated rubber. GEN: M. A. Grima, J. G. Eriksson, A. G. Talma, R. N. Datta, and J. W. M. Noordermeer, "Mechanistic Studies into the New Concept of Co-agents for Scorch Delay and Property Improvement in Peroxide Vulcanization," Paper No. 86 presented at the Fall Meeting of the Rubber Division, ACS, October 10, 2006, Cincinnati, OH.

Hybrid Cures

Adding only 0.15 phr of sulfur to a dicumyl peroxide cured EPDM/DVB-based compound reportedly will significantly increase the Young's modulus. However, if the sulfur level exceeds 0.2 phr, then the modulus will drop significantly. GEN: R. Fujio, M. Kitayama, N. Kataoka, S. Anzai, "Effects of Sulfur on the Peroxide Cure of EPDM and Divinylbenzene Compounds," *Rubber Chemistry and Technology*, March–April, 1979, Vol. 52, p. 74.

Post-curing

For compounds based on specialty elastomers such as FKM, consider post-curing for a longer time and/or a higher temperature in a specially vented, environmentally safe, post-cure oven. GEN: Jim Denham, Technical Service Chemist, Dyneon Co., "Solutions for the Oil and Gas Industry," Presented at a meeting of the Energy Rubber Group, 2009, Houston, TX.

Fillers

Increasing the loading of a filler will increase the compound's hardness. GEN: R. Mastromatteo, E. Morrisey, M. Mastromatteo, H. Day, "Matching Material Properties to Application Requirements," *Rubber World*, February, 1983, p. 26.

Use a filler with higher surface area to increase the compound's hardness. GEN: R. Mastromatteo, E. Morrisey, M. Mastromatteo, H. Day, "Matching Material Properties to Application Requirements," *Rubber World*, February, 1983, p. 26.

Use a filler with higher structure to increase the compound's hardness. GEN: R. Mastromatteo, E. Morrisey, M. Mastromatteo, H. Day, "Matching Material Properties to Application Requirements," *Rubber World*, February, 1983, p. 26.

Use a filler with greater surface activity to increase the compound's hardness. GEN: R. Mastromatteo, E. Morrisey, M. Mastromatteo, H. Day, "Matching Material Properties to Application Requirements," *Rubber World*, February, 1983, p. 26.

Consider using mineral fillers with particle shapes that possess high aspect ratios (such as fibrous particles or platelets) with higher surface-to-volume ratios to increase the hardness and reinforcement of a rubber vulcanizate significantly. Some of the higher hardness values can be achieved using these fillers because the plate-like particles will line up parallel to each other during processing. RT: Chapter 13, "Precipitated Silica and Non-black Fillers," W. Waddell, L. Evans, p. 333.

Consider using higher loadings of surface-treated fillers such as stearate-treated calcium carbonate, silane-treated clays, and/or titanate-treated titanium dioxide in place of their untreated counterparts along with an appropriate compatible internal lubricant. The surface treatment often will prevent the compound viscosity from becoming too high to prevent effective processing. The higher loadings of surface-treated fillers with appropriate internal lubricants may allow higher hardness values to be achieved. GEN: R. Grossman, Q & A, *Elastomerics*, January, 1989.

The use of silane-treated talc (such as Mistron CB) reportedly can increase the cured modulus and hardness of a rubber compound. GEN: O. Noel, Education Symposium on Fillers, "Talc: A Functional Mineral for Rubber," Presented at ACS Rubber Div. Meeting, Spring, 1995. RP: O. Noel.

Carbon Black

Increasing the structure of the carbon black used in a compound will increase modulus and hardness. RT: Chapter 12, "Compounding with Carbon Black and Oil," S. Laube, S. Monthey, M-J. Wang, p. 308.

Increasing the surface area of the carbon black used in a compound will increase modulus and hardness. However, the surface area has less of an effect on the modulus of the compound than the structure does. One aspect of using finer particle size (higher surface area carbon blacks) to increase cured modulus is that it works best at low strains. When the strain is increased, this high modulus may decrease as a result of the breakdown of the filler network. The density of the filler network at low strain is related to the fineness of the particle size. RT: Chapter 12, "Compounding with Carbon Black and Oil," S. Laube, S. Monthey,

M-J. Wang, p. 308. GEN: S. Wolff, M-J. Wang, E-H. Tan, *Kaut. Gummi Kunstst.*, 1995, Vol. 48, p. 82. RP: M-J. Wang.

Increasing the loading of the carbon black used in a compound will increase modulus and hardness. As a very crude "rule of thumb," a 2-phr increase in loading will increase hardness by about one point. RT: Chapter 12, "Compounding with Carbon Black and Oil," S. Laube, S. Monthey, M-J. Wang, p. 308.

Use the hardness conversion factors for different carbon blacks to estimate the hardness increase from using one specific carbon black vs. another. GEN: R. Kirschbaum, "Approaches to Compounding Without Thermal Black," Presented at New York Rubber Group Meeting, October 14, 1976, p. 5.

Using an organosilane with a carbon black can increase the compound cured modulus. GEN: R. Swor, "Utilisation of Very High Structure Tread Blacks to Lower the Rolling Resistance of U.S. and European Radial Tires," Tire Technology International, 1994.

Using a broad aggregate carbon black (such as Vulcan 1436) in place of a typical tread-grade carbon black (such as N234) may significantly increase both durometer hardness and the cured modulus of a modern tire tread. GEN: C. Flanigan, L. Beyer, D. Klekamp, D. Rohweder (Ford Motor Co.), "Comparative Study of Silica, Carbon Black and Novel Fillers in Tread Compounds," Paper No. 34 presented at the Fall Meeting of the Rubber Division, ACS, Oct. 11, 2011, Cleveland, OH.

Chemical Promotors

Use of a carbon black-rubber coupling agent or "chemical promotor" in a carbon black loaded rubber compound may increase resilience and tensile modulus as well as reduce abrasion loss. However, even though these agents may result in an increase in tensile modulus, sometimes they can actually lower the measured hardness. In the past, coupling agents such as *N*-(2-methyl-2-nitropropyl)-4-nitrosoaniline, *N*-4-dinitroso-*N*-methylaniline, *p*-nitrosodiphenylamine, and *p*-nitroso-*N*,*N*-dimethylaniline were used. These nitroso compounds are no longer used because they give off nitrosoamines, which are suspected carcinogens. Therefore other coupling agents have been tried. A new coupling agent, *p*-aminobenzenesulfonyl azide (or amine-BSA), has recently been investigated that may impart these performance advantages to a compound. GEN: L. Gonzalez, A. Rodriguez, J. deBenito, A. Marcos, "A New Carbon Black-Rubber Coupling Agent to Improve Wet Grip and Rolling Resistance of Tires," *Rubber Chemistry and Technology*, May-June, 1996, Vol. 69, p. 266.

Silica

Increase the loading of silica in a compound. RT: Chapter 6, "Elastomer Selection," R. School, p. 134.

The use of fumed silica in silicone compounds will greatly increase the compound's cured hardness and modulus. RT: Chapter 8, "Specialty Elastomer," J. R. Halladay, p. 236.

Consider using precipitated silica in a rubber compound if an increase in vulcanizate stiffness is needed. RT: Chapter 13, "Precipitated Silica and Non-black Fillers," W. Waddell, L. Evans, p. 331.

Consider using precipitated silica with higher surface area to significantly increase the hardness of a rubber compound. RT: Chapter 13, "Precipitated Silica and Non-black Fillers," W. Waddell, L. Evans, p. 333.

The use of mercaptosilane coupling agents with precipitated silica loading can significantly increase the compound's modulus and hardness. GEN: N. Hewitt, "Compounding with Silica for Tear Strength and Low Heat Build-up," *Rubber World*, June, 1982.

Fibers

Consider using a low loading of short aramid fiber to increase the compound's modulus significantly. GEN: K. Watson, A. Frances, "Elastomer Reinforcement with Short Kevlar Aramid Fiber for Wear Applications," *Rubber World*, August, 1988, p. 20.

Consider using a low loading of short cellulose or jute fibers to significantly increase the compound's modulus. GEN: K. Watson, A. Frances, "Elastomer Reinforcement with Short Kevlar Aramid Fiber for Wear Applications," *Rubber World*, August, 1988, p. 20.

The use of 5 phr of fiber pulps such as cotton, nylon-6, and polyester will generally increase compound modulus, especially in the longitudinal direction for EPR compounds. This is true for other elastomers as well. GEN: A. Estrin, "Application of PBDMA for Enhancement of EPR Loaded with Chopped Fibers," *Rubber World*, April 2000, p. 39.

The use of 5 phr of fiber pulps such as cotton, nylon-6, and polyester with low molecular weight maleated polybutadiene (PBDMA) will significantly increase compound modulus, especially in the longitudinal direction for EPR compounds. Significantly higher modulus values are achieved using PBDMA than not using it. GEN: A. Estrin, "Application of PBDMA for Enhancement of EPR Loaded with Chopped Fibers," *Rubber World*, April 2000, p. 39.

Overall, some of the best fibers for increasing modulus are unregenerated wood cellulose fiber with an average length-to-diameter ratio of about 100 : 1 to 200 : 1

and ideally treated with a resorcinol formaldehyde vinyl pyridene latex for adhesion. GEN: L. Goettler, K. Shen, "Short Fiber Reinforced Elastomers," *Rubber Chemistry and Technology*, 1983, Vol. 56, p. 575.

Nanoclays and Carbon Nanotubes

The use of nanoclays or carbon nanotubes in rubber compounds has been reported to significantly increase the modulus and hardness of a rubber compound. GEN: Debbie Banta (Weatherford Co.), "Can Nanotechnology Provide Innovative, Affordable Elastomer Solutions to Oil and Gas Service Industry Problems?," Presented at the Energy Rubber Group, January 19, 2012, Houston, TX; David J. Lowe, Andrew V. Chapman, Stuart Cook (Tun Abdi; Razal Research Centre-TARRC), "Rubber Nanocomposites Reinforced with Organoclays," Paper No. 92 presented at the Fall Meeting of the Rubber Division, ACS, October 16, 2007, Cleveland, OH.

Oil

Reduce the level of processing oil in the recipe to increase the compound hardness. RT: Chapter 12, "Compounding with Carbon Black and Oil," S. Laube, S. Monthey, M-J. Wang, p. 311.

Plasticizers

For compounds based on polar elastomers such as NBR or CR, use lower levels of ester plasticizer. RT: Chapter 14, "Ester Plasticizers and Processing Additives," W. Whittington, p. 347.

Carbon Black/Oil Balance

Use the black/oil addition factors for different carbon blacks and oil to estimate the hardness increase from using less oil loading vs. a specific loading of carbon black. GEN: R. Kirschbaum, "Approaches to Compounding Without Thermal Black," Presented at New York Rubber Group Meeting, October 14, 1976, p. 8; C. McCormick, J. West, K. Hale, "Carbon Black Filled SBR Compounds that Achieve Maximum Treadwear, Maximum Skid Resistance, and Minimum Heat Buildup at Minimum Cost," Presented at ACS Rubber Div. Meeting, Spring, 1976, Paper No. 47, Fig. 11.

Reinforcing Resins

Use a novolak phenol-formaldehyde reinforcing resin with a methylene donor such as hexamethylenetetramine (HMT) or hexamethoxymethylo-melamine (HMMM). The reinforcing resin and the methylene donor react with each other *in situ* during the normal vulcanization of the rubber compound. The use of this reinforcing resin and methylene donor in a rubber compound can significantly increase the hardness of the vulcanizate. RT: Chapter 18, "Tackifying, Curing, and Reinforcing Resins," B. Stuck, p. 440. Also, one can consider the use of a resorcinol resin with a methylene donor such as HMT or HMMM as well. RP: J. M. Long.

Reinforcing phenol-formaldehyde novolak resins are very compatible with NBR polymers. Therefore, loadings as high as 100 phr can be used with a methylene donor to develop "ebony-type" hardnesses that are very high for the vulcanizate. RT: Chapter 18, "Tackifying, Curing, and Reinforcing Resins," B. Stuck, p. 440.

Reinforcing phenol-formaldehyde novolak resins are normally not very compatible with polychloroprene polymers. However, if 15 to 25 phr of NBR is added to such a formulation, then the hardening resin's compatibility is significantly increased, allowing greater hardness levels to be achieved. RT: Chapter 18, "Tackifying, Curing, and Reinforcing Resins," B. Stuck, p. 440.

Phenol-formaldehyde resins with 5-10% hexamethylene tetramine (as a methylene donor) can increase the hardness and modulus of an NBR compound quite well. Resin loadings in the 100 to 250 phr range are possible in order to achieve very high hardness and toughness for a molding material. Even with only 50 phr loadings of these resins, it is possible to make a 90-plus Shore A hardness compound with 300% ultimate elongation without using any carbon black or other reinforcing fillers. RP: J. R. Halladay.

High-Styrene Resins

High-styrene resins can be used to increase the compound's hardness at room temperature. RT: Chapter 18, "Tackifying, Curing, and Reinforcing Resins," B. Stuck, p. 440.

TPE

If considering the use of thermoplastic elastomers (TPEs), consider copolyesters and polyurethanes that have an upper hardness range from 80 to 90 Shore A. RT: Chapter 10, "Thermoplastic Elastomers," C. P. Rader, p. 271.

Cure Time and Temperature

Consider curing at a lower temperature for a longer time to avoid extra network sulfur and achieve a higher sulfur crosslink density so as to obtain higher cured modulus values. GEN: Flexsys Technical Bulletin, "High Temperature Curing Compounding," 1996-1997.

Electron Beam

Consider using an electron beam, especially with peroxide/promotor cures to achieve greater hardness and crosslink density. GEN: William M. Boye (Sartomer Co.), "Use of Multifunctional Crosslinking Agents in the Electron Beam Cure of Elastomers," Paper No. 84 presented at the Fall Meeting of the Rubber Division, ACS, October 13, 2009, Pittsburgh, PA.

Phase Mixing

Rubber compounds are commonly based on a blend of different elastomers. These different elastomers do not possess the same solubility parameters and establish continuous and discontinuous phases at the microscopic level from mixing. Carbon black affinity for different elastomers is reported as follows:

BR > SBR > CR > NBR > NR > EPDM > IIR

Through phase mixing techniques, the concentation and distribution of carbon black can be controlled between domains in order to maximize tensile modulus. GEN: Eric S. Castner, (Goodyear Tire and Rubber Co.), "Where's the Filler?: Morphology Control for Improved Dynamic and Mechanical Properties," Paper No. 13 presented at the Fall Meeting of the Rubber Div., ACS, October 5-6, 2004.

2.5 Reducing Set, Under Compression and/or Tension

Compression set testing probably is performed far more often and required more frequently in rubber compound specifications than tension set. As noted from the suggestions below, many aspects of rubber compounding can affect set properties. It should be mentioned that compression set and tension set are different properties. Therefore, what may work to improve compression set may not improve tension set and vice versa. In addition, compression set can be a poor predictor of sealing pressure or the sealing capacity of a rubber seal. The more difficult to run compression stress relaxation test is usually a better predictor of seal performance.

The following suggested ***experimental*** *ideas can be considered for improving a rubber compound's set properties. Also, all relevant literature sources, including the ones cited below, should be researched and read.* ***Caution:*** *these general experimental ideas may not work in all specific situations. Changes to reduce set, under compression and/or tension, will certainly affect other properties as well, for better or for worse, and this book does not purport to show how these other properties are affected. Also, this book does not purport to address safety and health issues.*

Cure System

Consider using peroxide cures to establish carbon–carbon crosslinks to help prevent a set. RT: Chapter 3, "Vulcanizate Physical Properties, Performance Characteristics, and Testing," J. S. Dick, p. 56.

Using peroxide to cure an EPDM compound will improve its compression set resistance. RT: Chapter 6, "Elastomer Selection," R. School, p. 132.

An advantage of peroxide vs. sulfur cures is that the peroxide cures can more easily be made to have better compression set resistance. RT: Chapter 17, "Peroxide Cure Systems," L. Palys, p. 434.

To significantly decrease compression set values for terpolymers of AEM (under the trade name of Vamac® from DuPont), which uses a traditional diamine cure system based on hexamethylenediamine carbamate (HMDC) and DPG, consider the addition of dicumyl peroxide and a 1,2-polybutadiene (Ricon® 152) to reportedly decrease the compression set. GEN: H. Barager, K. Kammerer, E. McBride, "Increased Cure Rates of Vamac® Dipolymers and Terpolymers Using Peroxides," Presented at ACS Rubber Div. Meeting, Fall, 2000, Paper No. 115.

Using substituted *p*-phenylenediamine AO such as *N,N'*-di-beta-naphthyl-*p*-phenylene-diamine (DNPD) or Agerite White® and zinc oxide as crosslinking agents for a bromobutyl-based compound may improve compound compression set values. GEN: D. Edwards, "A High-Pressure Curing System for Halobutyl Elastomers," *Rubber Chemistry and Technology*, March–April, 1987, Vol. 60, p. 62.

Avoid diurethane cures, which reportedly impart poor compression set properties to an NR vulcanizate. GEN: T. Kempermann, "Sulfur-Free Vulcanization Systems for Diene Rubber," *Rubber Chemistry and Technology*, July–August, 1988, Vol. 61, p. 422.

Cure Time and Temperature

Generally, better compression set resistance can be obtained by achieving a tighter state of cure, which usually can be achieved through a longer cure time, higher cure temperature, and/or a change in the cure system. GEN: "A Comparative Evaluation of Hycar Nitrile Polymers," Manual HM-1, Revised, B. F. Goodrich Chemical Co.

Crosslink Density

Increasing the crosslink density of a compound can reduce the measured compression set. GEN: R. Mastromatteo, E. Morrisey, M. Mastromatteo, H. Day, "Matching Material Properties to Application Requirements," *Rubber World*, February, 1983, p. 26.

Sulfur Cures

To improve (i.e., reduce) compression set as well as improve heat aging for EPDM compounds, one should consider a so-called "low-set" cure consisting of sulfur at 0.5 phr, ZBDC at 3.0 phr, ZMDC at 3 phr, DTDM at 2.0 phr, and TMTD at 3.0 phr. RT: Chapter 16, "Cures for Specialty Elastomers," B. H. To, p. 396.

Consider curing a W-type neoprene with thiocarbanilide (A-1) acceleration to achieve good compression set. RT: Chapter 16, "Cures for Specialty Elastomers," B. H. To, p. 401.

Avoid using *N*-(cyclohexylthio)phthalimide (CTP) as a prevulcanization inhibitor for curing a W-type neoprene with thiocarbanilide (A-1) acceleration. Even though CTP will improve the scorch safety time, it will also hurt compression set. RT: Chapter 16, "Cures for Specialty Elastomers," B. H. To, p. 401.

For curing NBR-based formulations, consider using a cure system that reduces sulfur loading and uses more sulfur donor (such as TMTD or DTDM) as a partial or total replacement for the free sulfur. The use of less elemental sulfur may improve compression set resistance. RT: Chapter 16, "Cures for Specialty Elastomers," B. H. To, pp. 398–403.

In general, for compounds based on diene elastomers, use sulfur donors and/or increase the ratio of accelerator to sulfur to reduce compression set. RP: J. M. Long.

Select a semi-EV cure system using the sulfur donor DTDM to significantly improve compression set of a butyl compound. RT: Chapter 16, "Cures for Specialty Elastomers," B. H. To, p. 403. This is reportedly true for halobutyl rubber compounds as well. RP: R. Dailey.

Cocuring a EPDM/NR blend with bis-alkylphenyl disulfide (BAPD) and dithiodicaprolactam (DTDC, Rhenocure S) in an optimized formulation can improve compression set values. GEN: A. Ahmad, "NR/EPDM Blend for Automotive Rubber Component," Rubber Research Institute of Malaysia.

Cure systems based on ultra-accelerators such as thiurams and dithio-carbamates tend to impart higher percentages of monosulfide crosslinks compared to thiazole or amine-type accelerators. Therefore these ultra-accelerators often will impart better resistance to compression set to their vulcanizates than thiazole or amine-type accelerators do. GEN: M. Studebaker, J. R. Beatty, "Vulcanization," *Elastomerics*, February, 1977, p. 41.

For a nitrile-based formulation, consider using *N,N*-phenylene dimaleimide (HVA 2) with a sulfenamide cure system for better compression set. GEN: D. Coulthard, W. Gunter, Presented at ACS Rubber Div. Meeting, Fall, 1975, Paper No. 39.

In case studies involving sulfur cures, it was reported that achieving a tighter state of cure, as measured by DSC, correlated to lower compression set. GEN: Edmee Files, "To Hell and Back," Presented at the Energy Rubber Group, September 13, 2010, San Antonio, TX.

Peroxide Cures

To achieve better compression set resistance when using BBPIB peroxide to cure a rubber compound, consider using the new proprietary "high performance" (HP) BBPIB (peroxide) formulations. RT: Chapter 17, "Peroxide Cure Systems," L. Palys, pp. 418–419.

For peroxide cures, use a coagent as well to add unsaturation to the system, which results in higher crosslink density than the peroxide would have generated if used without the coagent. The reason for this is that it is more efficient and easier to crosslink with a free radical to unsaturation than by hydrogen abstraction to a saturated backbone. Because the use of a coagent can lead to different types of crosslink networks, improved compression set resistance may be possible. RT: Chapter 17, "Peroxide Cure Systems," L. Palys, pp. 431–432.

Using a coagent with a peroxide cure may improve compression set. GEN: P. Dluzneski, "Peroxide Vulcanization of Elastomers," *Rubber Chemistry and Technology*, July-August, 2001, Vol. 74, p. 451.

Post-cure

Typically, post-curing a polyacrylate compound for about 4 to 8 hours at 170 to 190°C (338 to 374°F) may significantly lower (improve) compression set values. RT: Chapter 8, "Specialty Elastomers," P. Manley, C. Smith, p. 207.

Some types of cures generate reaction byproducts from the vulcanization process; a post-cure step at atmospheric pressure will allow these byproducts to vent from the compound and to impart better compression set properties. GEN: A. Kasner, E. Meinecke, "Porosity in Rubber: A Review," *Rubber Chemistry and Technology*, July-August, 1996, Vol. 69, p. 424.

A study of the effects of a mild post-cure procedure on an HXNBR/HNBR blended compound showed significant improvements in compression set and sealing properties of drill bit compounds. GEN: R. Pazur (Lanxess), L. Ferrari, E. Campomizzi, "HXNBR Compound Property Improvements Through the Use of Post Cure," Paper No. 70 presented at the Spring Meeting of the Rubber Division, ACS, San Antonio, TX, May 16, 2005.

FKM/Bisphenol AF Cure

Use a bisphenol cure instead of a peroxide cure for a fluoroelastomer (FKM)-based compound to achieve better resistance to compression set. RT: Chapter 6, "Elastomer Selection," R. School, p. 136.

For compression set resistance with an FKM compound, select a bisphenol AF cure system. RT: Chapter 8, "Specialty Elastomers," R. Stevens, p. 230.

Molecular Weight Effects

Increase the average molecular weight of the base rubber in a formulation to decrease the compression set. GEN: R. Mastromatteo, E. Morrisey, M. Mastromatteo, H. Day, "Matching Material Properties to Application Requirements," *Rubber World*, February, 1983, p. 25.

When selecting nitriles with concerns for lower compression set, consider NBR grades that have higher Mooney viscosity. GEN: R. Del Vecchio, E. Ferro, "Effects of NBR Polymer Variations on Compound Properties," Presented at ACS Rubber Div. Meeting, Spring, 2001, Paper No. 21.

Polychloroprene

W-type neoprenes may impact better compression set properties than G-type. RT: Chapter 6, "Elastomer Selection," R. School, p. 133.

When using neoprene in compounding, selecting W-type gives better compression set resistance. RT: Chapter 8, "Specialty Elastomers," L. Outzs, p. 208.

Thioureas, Vanax PML®, and Vulcacit CRV® cure systems reportedly give better compression set properties for CR compounds. RP: L. L. Outzs. Also see Section 4 of Appendix 2 of this book on Polychloroprene Cure Packages (A8.4).

EPDM

For the best compression set properties for a given rubber compound, avoid selecting an EPDM with high crystallinity. GEN: S. Brignac, H. Young, "EPDM with Better Low-Temperature Performance," *Rubber & Plastics News*, August 11, 1997, p. 14.

NR

Epoxidizing NR may worsen a rubber compound's compression set when ENR is used in place of NR. RT: Chapter 7, "General Purpose Elastomers and Blends," G. Day, p. 170.

DPNR

Consider selecting deproteinized natural rubber (DPNR) for engineering applications in order to achieve better creep and stress relaxation resistance as well as lower compression set values over conventional natural rubber even though the strength from the strain-induced crystallization of NR may be reduced. A similar advantage is sometimes also seen by subitituting a blend of IR/NR for NR or sometimes just replacing all NR with IR. GEN: M. Fernando, C. Forge, G. Spiller, J. Clark (Tun Abdul Razak Research Centre, Brickendonbury, Hertford, UK), "An Evaluation of Deproteinised Natural Rubber for Engineering Applications," Paper No. 41 presented at the Spring Technical Meeting of the Rubber Division, ACS, San Antonio, TX, May 16, 2005. This is also partially supported by R. Del Vecchio, E. Ferro, K. Winkler, "Fatigue Life Comparisons of NR Compounds," Paper No. 106 presented at the Fall Meeting of the Rubber Division, ACS, October 17, 2003, Cleveland, OH.

NBR

Use an emulsion NBR that is produced from a coagulating system based on calcium chloride for a lower compression set. RT: Chapter 8, "Specialty Elastomers," M. Gozdiff, p. 195.

When selecting nitriles with concerns for better compression set, consider NBR grades that have more chain branching and chain entanglements. GEN: R. Del Vecchio, E. Ferro, "Effects of NBR Polymer Variations on Compound Properties," Presented at ACS Rubber Div. Meeting, Spring, 2001, Paper No. 21.

When selecting nitriles with concerns for lower compression set, consider NBR grades that have lower ACN content. GEN: R. Del Vecchio, E. Ferro, "Effects of NBR Polymer Variations on Compound Properties," Presented at ACS Rubber Div. Meeting, Spring, 2001, Paper No. 21.

AEM

For better compression set, select a grade of ethylene-acrylic elastomer (AEM), which is a copolymer that is peroxide cured, which reportedly will provide better compression set resistance than the diamine-cured terpolymer grades. RT: Chapter 8, "Specialty Elastomers," T. Dobel, p. 223.

CPE

Selecting grades of chlorinated polyethylene with a lower percentage of chlorine will allow a better resistance to compression set. RT: Chapter 8, "Specialty Elastomers," L. Weaver, p. 213.

A four-way comparison of compounds based on CM, CSM, GECO, and NBR/PVC found that CM (chlorinated polyethylene) gave superior compression set resistance in relation to the other three elastomers. GEN: C. Hooker, R. Vara, "A Comparison of Chlorinated and Chlorosulfonated Polyethylene Elastomers with Other Materials for Automotive Fuel Hose Covers," Presented at ACS Rubber Div. Meeting, Fall, 2000, Paper No. 128.

Fluoroelastomers

Using certain types of fluoroelastomers with a specially oxidized thermal black and a 1200 molecular weight poly-(1,2-butadiene)diol prepolymer as a coupling agent with a peroxide cure reportedly can achieve an improvement in creep resistance. GEN: J. Martin, T. Braswell, H. Green, "Coupling Agents for Certain Types of Fluoroelastomers," *Rubber Chemistry and Technology*, November–December, 1978, Vol. 51, p. 897.

TPV

Select a thermoplastic vulcanizate (TPV) with a higher crosslink density in the rubbery domain from dynamic vulcanization to reduce tension set and compression set. RT: Chapter 10, "Thermoplastic Elastomers," C. P. Rader, p. 274.

Resinous Homogenizing Agents

Avoid using resinous homogenizing agents in a rubber compound because they may increase the compound's compression set. RT: Chapter 14, "Ester Plasticizers and Processing Additives," C. Stone, p. 372.

Fillers

Reducing filler loading will generally improve (i.e., reduce) compression set. GEN: R. Mastromatteo, E. Morrisey, M. Mastromatteo, H. Day, "Matching Material Properties to Application Requirements," *Rubber World*, February, 1983, p. 26.

Reducing filler structure will generally improve (i.e., reduce) compression set. GEN: R. Mastromatteo, E. Morrisey, M. Mastromatteo, H. Day, "Matching Material Properties to Application Requirements," *Rubber World*, February, 1983, p. 26.

Reducing filler surface area (increasing average particle size) will generally improve (i.e., reduce) compression set. GEN: R. Mastromatteo, E. Morrisey, M. Mastromatteo, H. Day, "Matching Material Properties to Application Requirements," *Rubber World*, February, 1983, p. 26.

Increasing filler surface activity will generally improve (i.e., reduce) compression set. GEN: R. Mastromatteo, E. Morrisey, M. Mastromatteo, H. Day, "Matching Material Properties to Application Requirements," *Rubber World*, February, 1983, p. 26.

The use of carbon nanotubes as a filler in a rubber compound has been reported to sometimes improve (reduce) compression set. GEN: Gary D. D'Abate (Pinnacle Elastomeric Technology), "Compounding Materials," Presented at the Fall Meeting of the Energy Rubber Group, September 13, 2011, Galveston, TX.

Silica

Reduce the loading of silica in a compound to improve compression set. RT: Chapter 6, "Elastomer Selection," R. School, p. 135.

In silicone compounds containing silica, compression set is positively correlated to silica surface area, oil absorption, and moisture. GEN: T. Okel, W. Waddell, "Effect of Precipitated Silica Physical Properties on Silicone Rubber Performance," *Rubber Chemistry and Technology*, March-April, 1995, Vol. 68, p. 59.

For low compression set, avoid high loadings of precipitated silica, which can impart very high compression set values to a compound. At loadings above 25 phr of silica, compression set can become extremely high. GEN: R. Tabar, P. Killgoar, R. Pett, "A Fatigue Resistant Polychloroprene Compound for High Temperature Dynamic Applications," *Rubber Chemistry and Technology*, September-October, 1979, Vol. 52, p. 781.

Silane Coupling Agents for Silica-Loaded Compounds

For rubber compounds containing a high loading of precipitated silica, consider using a mercaptosilane to improve the compound's compression set. RP: T. D. Powell.

Using silane coupling agents will improve (i.e., lower) the compression and tension set characteristics of a silica-loaded compound. Using silane coupling agents will also improve the compression set and tension set for other compounds that use siliceous fillers such as talc, clay, and so forth. RP: J. R. Halladay.

Polyurethane

Compression set for a two-part cast polyurethane system usually can be changed by adjusting the curative ratio. The curative ratio refers to the relative amounts of prepolymer and curative. The amount of curative [such as methylene-bis-orthochloroaniline (MBCA)] needed to match exactly the isocyanate groups on the prepolymer is referred to as "100% theory" or "100% stoichiometry." Thus if curative is reduced by 5%, it is referred to as "95% theory" or "95% stoichiometry." On the other hand, if the curative is increased by 5% from theory, it is referred to as "105% theory" or "105% stoichiometry." Generally, going to a lower stoichiometry, such as 95%, may decrease (i.e., improve) compression set. RT: Chapter 9, "Polyurethane Elastomers," R. W. Fuest, p. 251.

Usually, but not always, selecting a polyurethane based on a TDI prepolymer with an amine cure will provide an advantage for compression set resistance in demanding applications compared to diol-cured MDI compounds. RT: Chapter 9, "Polyurethane Elastomers," R. W. Fuest, p. 257.

To achieve good compression set resistance with polyurethanes, one should pay strict attention to recommendations from the manufacturer regarding times and temperatures for cures and post-cures. Do not underestimate the importance of post-cures. RP: R. W. Fuest.

All polyurethane rubber compounds are sometimes made more compression set resistant by blending them with conventional rubber with the appropriate curative adjustments. GEN: Thomas Jablonowski (Uniroyal Chemical), "Blends of Polyurethane Rubber with Conventional Rubbers," Paper No. 46 presented at the Spring Meeting of the Rubber Division, ACS, April 13, 1999, Chicago, IL.

PU/NBR

Adding a given quantity of NBR incrementally to a millable polyurethane compound that is sulfur cured reportedly will improve the compound's compression set. GEN: T. Jablonowski, "Blends of PU with Conventional Rubbers," *Rubber World*, October, 2000, p. 41.

Oil Loadings

Decreasing the oil loading in a compound generally will improve (i.e., reduce) the set. GEN: K. Hale, J. West, C. McCormick, "Contribution of Carbon Black Type to Skid and Treadwear Resistance," ACS Rubber Div. Meeting, Spring, 1975, Paper No. 6, Fig. 27.

Fibers

Overall, some of the best fibers to select for significantly improving creep resistance (not necessarily the compression set resistance) are unregenerated wood cellulose fibers with average length-to-diameter ratios of about 100:1 to 200:1 and ideally treated with a resorcinol formaldehyde vinyl pyridene latex for adhesion. GEN: L. Goettler, K. Shen, "Short Fiber Reinforced Elastomers," *Rubber Chemistry and Technology*, July–August, 1983, Vol. 56, p. 575.

2.6 Increasing Rebound and Decreasing Hysteresis

As is well known, no rubber compound is perfectly resilient in the cured state. Usually a given level of damping quality remains that could be significantly large or fairly small. This remaining damping or "viscous" quality of the cured rubber compound will reduce its rebound quality and increase its hysteresis or heat buildup quality from repeated deformations. Usually the higher the rebound of a compound is, the lower the hysteresis will be. However, this inverse correlation is not perfect, sometimes owing to different temperatures or different rates or amplitudes of deformation.

The reader may wish to consider the following ***experimental*** *compounding ideas for either increasing the percent rebound or decreasing the hysteresis of a compound. Also, all relevant literature sources, including the ones cited below, should be researched and read.* ***Caution:*** *these general experimental ideas may not work in all specific situations. Changes to increase rebound and decrease hysteresis will certainly affect other properties as well, for better or for worse, and this book does not purport to show how these other properties are affected. Also, this book does not purport to address safety and health issues.*

Fillers

Avoid rubber compounds containing high loadings of fully reinforcing fillers because there is an increase in hysteresis during deformation from the loss in energy resulting from internal friction or breaking and reforming of the filler/polymer contact points and the aggregate-aggregate network. RT: Chapter 7, "General Purpose Elastomers and Blends," G. Day, p. 157.

Reducing filler loading will generally improve (i.e., reduce) hysteresis heat buildup. GEN: R. Mastromatteo, E. Morrisey, M. Mastromatteo, H. Day, "Matching Material Properties to Application Requirements," *Rubber World*, February, 1983, p. 26.

Reducing filler surface area (increasing particle size) will generally improve (i.e., reduce) hysteresis heat buildup. GEN: R. Mastromatteo, E. Morrisey, M. Mastromatteo, H. Day, "Matching Material Properties to Application Requirements," *Rubber World*, February, 1983, p. 26.

Increasing filler surface activity will generally improve (i.e., reduce) hysteresis heat buildup. GEN: R. Mastromatteo, E. Morrisey, M. Mastromatteo, H. Day, "Matching Material Properties to Application Requirements," *Rubber World*, February, 1983, p. 26.

Carbon Black

Decreasing the loading of the carbon black used in a compound will increase rebound and decrease hysteresis. RT: Chapter 12, "Compounding with Carbon Black and Oil," S. Laube, S. Monthey, M-J. Wang, p. 308.

Decreasing the surface area (increasing the particle size) of the carbon black used in a compound will increase rebound and decrease hysteresis. RT: Chapter 12, "Compounding with Carbon Black and Oil," S. Laube, S. Monthey, M-J. Wang, p. 308.

Select carbon black grades with higher surface activity to reduce hysteresis and improve rebound. The higher that the surface activity of a carbon black is, the tighter the bonding will be between the carbon black particle surfaces and the elastomer medium. Thus there is less slippage between carbon black surface and the rubber chains. RT: Chapter 12, "Compounding with Carbon Black and Oil," S. Laube, S. Monthey, M-J. Wang, p. 314.

To reduce hysteresis and increase rebound, one can try to substitute a higher structure carbon black at a lower loading for an existing carbon black with the appropriate process oil adjustment in order to maintain the same compound hardness. RT: Chapter 12, "Compounding with Carbon Black and Oil," S. Laube, S. Monthey, M-J. Wang, p. 317.

Selecting a carbon black with greater fineness reportedly will significantly increase a cured compound's hysteresis under conditions of constant strain or energy. However, this fineness effect is not as pronounced at constant stress. GEN: W. Hess, W. Klamp, "The Effects of Carbon Black and Other Compounding Variables on the Tire Rolling Resistance and Traction," *Rubber Chemistry and Technology*, Vol. 56, May–June, 1983, p. 390.

Selecting a carbon black with increased structure reportedly will raise the cured compound's hysteresis at constant strain but imparts little effect at constant energy input. Under constant stress conditions this higher structure may actually lower hysteresis. GEN: W. Hess, W. Klamp, "The Effects of Carbon Black and Other Compounding Variables on the Tire Rolling Resistance and Traction," *Rubber Chemistry and Technology*, Vol. 56, May–June, 1983, p. 390.

Sometimes better carbon black dispersion from mixing can lower the cured hysteresis of a compound. On the other hand, sometimes discrete agglomerates of carbon black behave like very coarse particles, causing a decrease in cured hysteresis. This is usually more likely with low-structure carbon blacks. GEN: W. Hess, W. Klamp, "The Effects of Carbon Black and Other Compounding Variables on the Tire Rolling Resistance and Traction," *Rubber Chemistry and Technology*, Vol. 56, May–June, 1983, p. 399.

The use of a lower concentration of high-structure "LL Carbon Black" (long linkage carbon blacks that are crush resistant during mixing) in place of regular high-structure carbon black reportedly can be used to impart better rolling resistance to a tread. According to one study, high-structure carbon blacks are commonly susceptible to "crush" during mixing and convert to ordinary carbon blacks. These special high-structure LL carbon black grades are more crush resistant. GEN: H. Mouri, K. Akutagawa, "Reducing Energy Loss to Improve Tire Rolling Resistance," Presented at ACS Rubber Div. Meeting, Spring, 1997, Paper No. 14.

Lower hysteresis from a rubber compound can be achieved by using a *lower* concentration of the so-called "very high structure" carbon black in place of a conventional reinforcing carbon black. These special very high structure blacks (such as CD-2038 from Columbian) have highly branched aggregates with such high void spaces within that they are capable of absorbing a very high volume in rubber or oil. GEN: R. Swor, "Utilisation of Very High Structure Tread Blacks to Lower the Rolling Resistance of U.S. and European Radial Tires," *Tire Technology International*, 1994.

Consider using a carbon black with a broad aggregate size distribution because this promotes more efficient packing with reduced interaggregate contact and resulting lower hysteresis and higher rebound. This will result in improved tire rolling resistance. GEN: R. Swor, "Utilisation of Very High Structure Tread Blacks to Lower the Rolling Resistance of U.S. and European Radial Tires," *Tire Technology International*, 1994; M-J. Wang, S. Wolff, E-H. Tan, *Rubber Chemistry and Technology*, May–June, 1993, Vol. 66, p. 178. RP: M-J. Wang.

Consider using the new Sid Richardson's SR129 (tread grade) or SR401 (non-tread grade) from "special reactor" technology that are called "low hysteresis" carbon blacks with wider aggregate size distributions and higher structures. GEN: Leszek Nikiel, Wesley Wampler, Henry Yang, Tom Carlson (Sid Richardson Carbon and Energy Company), "Improved Carbon Blacks for Low Hysteresis Applications in Rubber," Paper No. 93 presented at the Fall Meeting of the Rubber Division, ACS, October 16, 2007, Cleveland, OH.

Carbon Black Surface Activity

Selecting carbon blacks with higher surface activities will usually reduce a compound's hysteresis. A carbon black's surface activity is related to the extent of the disorder and defects of the surface graphitic layer structure. The greater the disorder that the graphitic structure has, the higher the surface activity that the carbon black grade will have. Carbon blacks with a higher disorder in their graphitic structure also usually have a shorter heat history during their formation in the furnace reactor. Greater carbon black surface activity means greater carbon black-polymer interaction and consequently less aggregate-aggregate interaction and less compound hysteresis. GEN: W. Hess, W. Klamp, "The Effects of Carbon Black and Other Compounding Variables on the Tire Rolling Resistance and Traction," *Rubber Chemistry and Technology*, Vol. 56, May-June, 1983, p. 390.

Using nanoblacks with higher surface roughness and higher surface activity than regular carbon blacks will significantly reduce hysteresis (cured tan δ) over a wide range of applied strains or deformations. GEN: A. McNeish, "Nanoblacks for Rolling Resistance," Presented at the Fall 2000 ITEC Meeting, Paper No. 23A.

Chemical Promotors for Carbon Black

Use of a carbon black-rubber coupling agent or "chemical promotor" in a carbon black loaded rubber compound may increase resilience and modulus as well as reduce abrasion loss. In the past, coupling agents such as *N*-(2-methyl-2-nitropropyl)-4-nitrosoaniline, *N*-4-dinitroso-*N*-methylaniline, *p*-nitrosodiphenylamine, and *p*-nitroso-*N,N*-dimethylaniline were used. These nitroso compounds are no longer used because they give off nitrosoamines, which are suspected carcinogens. Therefore other coupling agents have been tried. An experimental coupling agent, *p*-aminobenzenesulfonyl azide (or amine-BSA), has recently been investigated that may impart these performance advantages to a compound and give better resiliency. GEN: L. Gonzalez, A. Rodriguez, J. deBenito, A. Marcos, "A New Carbon Black-Rubber Coupling Agent to Improve Wet Grip and Rolling Resistance of Tires," *Rubber Chemistry and Technology*, May-June, 1996, Vol. 69, p. 266.

A new highly experimental promotor, benzofurazan 1-oxide (BFO), has reportedly been used to promote greater bonding between rubber and carbon black, thus reducing cured hysteresis. To do this, mixing temperatures greater than 160°C (320°F) are usually needed. Nickel salts have been found to reduce the odoriferous benzofurazan (BFZ) byproduct that is formed at high mixing temperatures. GEN: D. Graves, "Benzofuroxans as Rubber Additives," *Rubber Chemistry and Technology*, March-April, 1993, Vol. 66, p. 61.

Recent work has shown that a chemical "coupler" for improving carbon black based rubber compounds by lowering their hysteresis has been reported. These carbon black "couplers" are an S2-S4 version of 3-thiopropionic acid and ethylene diamine diamide, both of which have been reported to lower hysteresis of a carbon black loaded rubber compound. GEN: James Burrington (Lubrizol Corp.), "Carbon Black Coupler Technology for Low Hysteresis Tire," Paper No. 108 presented at the Fall Meeting of the Rubber Division, ACS, Oct. 16, 2007, Cleveland, OH.

Consider using a special "surface modified" carbon black with in-chain functionalized solution SSBR to achieve a significant reduction on compound hysteresis. GEN: J. Douglas, S. Crossley, J. Hallett, J. Curtis, D. Hardy, T. Cross, N. Steinhauser, A. Lucassen, H. Kloppenburg (Lanxess and Columbian Chemicals), "The Use of a Surface-Modified Carbon Black with an In-Chain Functionalized Solution SSBR as an Alternative to Higher Cost Green Tire Technology," Paper No. 38 presented at the Fall Meeting of the Rubber Division ACS, Oct. 11, 2011, Cleveland, OH.

State of Mix

As reported by Hess, many times rebound will increase with increased dispersion, especially when going from 95% to greater than 99% dispersion (where good microdispersion and reduction in filler networking occur). This could be typically true for an N330 with increasing state of mix in an SBR/BR or NR/BR blend, for example. On the other hand, a carbon black such as N326 dispersed in an NR/BR compound with increasing work history may actually show the opposite effect, in which rebound decreases with improvement in state of mix. This opposite effect may be due to the fact that N326 is a lower structure black that forms "hard" agglomerates. These hard agglomerates act as large "particles." The breakdown of these "hard" agglomerates through mixing and increased work history causes a reduction in "interaggregate spacing," resulting in greater hysteresis and lower rebound. On the other hand, N347 loadings reportedly experience a rise and then fall in rebound with increasing work history and state of mix. Therefore the effects of state of mix on the cured rebound of a compound are greatly dependent on the type of carbon black and the base elastomer used. GEN: W. Hess, "Characterization of Dispersions," *Rubber Chemistry and Technology*, July–August, 1991, Vol. 64, p. 386.

For SBR-, SBR/BR-, and NR/BR-based compounds, increasing dispersion of carbon black usually results in less heat buildup and less hysteresis. However, the opposite has been observed by Gessler for compounds based on butyl rubber for the high-structure carbon blacks. This may be due to chain scission of the butyl rubber. GEN: W. Hess, "Characterization of Dispersions," *Rubber Chemistry and Technology*, July–August, 1991, Vol. 64, p. 386.

Mixing Sequence

Add the carbon black early in the mixing sequence. Avoid adding the carbon black with oil, stearic acid, or other polar ingredients such as antioxidants because these ingredients may be absorbed into the surface of the carbon black particles, which will interfere with the polymer absorption onto the carbon black surface. Thus adding oil and other ingredients with the carbon black might interfere with carbon black-rubber interaction or the formation of bound rubber. This would increase hysteresis. Therefore, it might be better to add the carbon black first before some of these other ingredients to achieve lower hysteresis. GEN: M-J. Wang, T. Wang, K. Mahmud, "Effect of Carbon Black Mixing on Rubber Reinforcement," Proceedings of the 3rd International Conference on Carbon Black, p. 205, Mulhouse, October 25–26, 2000. RP: M-J. Wang.

Heat Treatment from Mixing

For diene-based elastomers such as SBR or BR with relatively high mechanooxidation stability, increase the intensity of mixing and lengthen mixing time in order to apply "heat treatment," which may increase the bound rubber and improve the carbon black dispersion, resulting in better abrasion resistance, better microdispersion, and lower hysteresis. GEN: M-J. Wang, T. Wang, K. Mahmud, "Effect of Carbon Black Mixing on Rubber Reinforcement," Proceedings of the 3rd International Conference on Carbon Black, p. 205, Mulhouse, October 25–26, 2000. RP: M-J. Wang.

Phase Mixing

Sometimes an uneven polymer phase distribution of carbon black between different elastomer components may impart lower cured compound hysteresis. It may also be true that a lower hysteresis is more likely if the rubber phase with the higher concentration of carbon black is also the least continuous phase of the blend. GEN: W. Hess, W. Klamp, "The Effects of Carbon Black and Other Compounding Variables on the Tire Rolling Resistance and Traction," *Rubber Chemistry and Technology*, Vol. 56, 1983, May–June, p. 390.

In SBR/NR blends, a significant reduction in hysteresis and heat buildup can be achieved through phase mixing techniques (phase distribution) in which the NR phase contains approximately 75% of the carbon black loading (according to Hess and Chirico). The same principle also applies to a BR/NR blend. When the concentration of carbon black is greater in the NR phase through phase mixing techniques, hysteresis may be reduced. On the other hand, BR/SBR blends appear to be less sensitive to carbon black phase distribution differences with regard to effects on cured hysteresis. These studies were performed using primarily fully reinforcing carbon blacks. The effects from carbon black phase distribution can also be influenced by carbon black structure. GEN: W. Hess, C. Herd, P. Vegvari, "Characterization of Immiscible Elastomer Blends," *Rubber Chemistry and Technology*, July–August, 1993, Vol. 66, p. 329.

Silane Coupling Agents

Use a silane coupling agent to significantly reduce hysteresis of compounds loaded with siliceous fillers such as precipitated silica, clay, talc, and so forth. Mercaptosilanes are particularly effective when applied to sulfur-cured compounds. Vinylsilanes and methacrylsilanes work with peroxide-cured compounds. RP: J. R. Halladay.

Silica/Silane

For better rebound and less heat buildup for silica-loaded compounds, consider using low surface area silica and a silane coupling agent such as bis(3-triethoxysilypropyl)tetrasulfide. RT: Chapter 13, "Precipitated Silica and Non-black Fillers," W. Waddell, E. Evans, p. 338.

Use higher loadings of an effective organosilane coupling agent (such as TESPT) with a silica loading and allow an effective silanization reaction to occur during mixing to establish better rubber–filler interaction and reduce filler–filler interaction, which is energy absorbing, to achieve low vulcanizate hysteresis. GEN: A. McNeish, "Nanoblacks for Rolling Resistance," Presented at the ITEC Meeting, Fall, 2000, Paper No. 23A.

It is well known that using silane coupling agents with precipitated silica will generally improve the compound's abrasion resistance and lower the hysteresis of the cured compound. However, it is usually better to use an organosilane with polysulfidic sulfane functionality rather than a mercaptosilane to achieve better crosslinking. RP: T. D. Powell.

Consider using a highly dispersible grade of precipitated, hydrated silica to achieve better reduction in hysteresis. GEN: S. Daudey, L. Guy (Rhodia), "High Performance Silica Reinforced Elastomers from Standard Technology to Advanced Solutions," Paper No. 37 presented at the Fall Meeting of the Rubber Division ACS, October 11, 2011, Cleveland, OH.

Also, for silica-based compounds, consider an optimal loading of Nanoprene® BM750H VP RW from Lanxess, a BR-acrylate terpolymer additive, to reduce hysteresis. GEN: C. Flanigan, L. Beyer, D. Klekamp, D. Rohweder (Ford), B. Stuck, E. Terrill (ARDL), "Comparative Study of Silica, Carbon Black and Novel Fillers in Tread Compounds," Paper No. 34 presented at the Fall Meeting of the Rubber Division ACS, October 11, 2011, Cleveland, OH.

Consider using a special zinc soap such as ZB47 from Struktol in a TESPT-treated silica-filled natural rubber based compound in order to improve the heat buildup (HBU) and blow out (BO) times from Firestone Flexometer testing. GEN: Kwang-Jea Kim, John Vanderkooi (Struktol), "Effects of Zinc Soaps on TESPT and TESPD-Silica Mixtures in Natural Rubber Compounds," Paper No. 70

presented at the Fall Meeting of the Rubber Division, ACS, October 8, 2002, Pittsburgh, PA.

It has been reported that using "moisture-treated" silica with organosilanes can reduce the cured heat buildup of the silica-organosilane compound. GEN: Kwang-Jea Kim, John VanderKooi (Struktol), "Moisture Level Effects on Hydrolysis Reaction in TESPD/Silica/CB/S-SBR Compound," Paper No. 57 presented at the Fall Meeting of the Rubber Division ACS, October 5, 2004, Columbus, OH.

Substituting the new coupling agent AEO-MPES [alkylethoxylates-substituted mercaptopropylethoxy silane (Degussa's VP Si 363)] in place of TESPT can reduce hysteresis by as much as 13% in a silica-loaded compound. GEN: O. Klockmann, A. Blume, A. Hasse, "Fuel Efficient Silica Tread Compounds with a New Mercaptosilane: A Practical Way to Improve Its Processing," Paper No. 87 presented at the Fall Meeting of the Rubber Division, ACS, October 16, 2007, Cleveland, OH.

Carbon–Silica Dual Phase

Select the new carbon–silica dual-phase fillers with an organosilane coupling agent such as TESPT to achieve perhaps 30% reduction in hysteresis reportedly with no loss in abrasion resistance. These carbon–silica dual-phase fillers are manufactured by cofuming carbon blacks and silica. RT: Chapter 12, "Compounding with Carbon Black and Oil," S. Laube, S. Monthey, M-J. Wang, p. 314. GEN: M-J. Wang, P. Zhang, K. Mahmud, *Rubber Chemistry and Technology*, March–April, 2001, Vol. 74, p. 124. RP: M-J. Wang.

Surface Activated Fibers

Consider using a small loading level (at perhaps 2 phr) of special, chemically activated aramid fiber granules (Sulfron®) to achieve a reduction in heat buildup at constant force. GEN: N. Huntink, T. Mathew, M. Tiwari, S. Parker (Teijin Aramid), "Using Sulfron to Improve Rolling Resistance and Durability of Tires," Paper No. 1 presented at the Fall Meeting of the Rubber Division, ACS, October 11, 2011, Cleveland, OH.

Plasticizer

Select plasticizers that will lower the glass transition temperature (T_g) of the compound. RT: Chapter 3, "Vulcanizate Physical Properties, Performance Characteristics, and Testing," J. S. Dick, p. 52.

With the banning of aromatic oils for use in tires in Europe, SBR-aromatic oil masterbatches have been eliminated. The new "environmentally friendly" oil substitutes (such as medium extracted solvate or MES) used to make these new oil masterbatches have lower glass transition temperature (T_g) values than the discontinued aromatic oils. Therefore some of these new substitutes should impart lower hysteresis to the cured compound. GEN: Rudiger Engehausen (Bayer AG), "Overview of New Developments in BR and SBR and Their Influence on Tire-Related Properties," Paper No. 37 presented at the Spring Meeting of the Rubber Division ACS, April 29, 2002, Savannah, GA.

Low-Viscosity Oils

For the best compound resiliency, select a processing oil with low viscosity because high-viscosity oils can reduce a compound's resiliency. GEN: R. Tabar, P. Killgoar, R. Pett, "A Fatigue Resistant Polychloroprene Compound for High Temperature Dynamic Applications," *Rubber Chemistry and Technology*, September-October, 1979, Vol. 52, p. 781.

For the best neoprene compound resiliency, consider using rapeseed oil because it has low viscosity for good hysteresis characteristics and low volatility for good aging properties. GEN: R. Tabar, P. Killgoar, R. Pett, "A Fatigue Resistant Polychloroprene Compound for High Temperature Dynamic Applications," *Rubber Chemistry and Technology*, September-October, 1979, Vol. 52, p. 781.

Sulfur Cures

For a compound sulfur-cured with a low concentration of sulfenamide accelerator, a higher zinc oxide loading may impart lower heat buildup (from the Goodrich flexometer). GEN: W. Hall, H. Jones, "The Effect of Zinc Oxide and Other Curatives on the Physical Properties of a Bus and Truck Tread Compound," Presented at ACS Rubber Div. Meeting, Fall, 1970.

Cure Temperature

By changing the cure temperature for a gum NR/BR blend based compound, one can control the crosslink density distribution found between the NR and BR phases of this compound, which may affect rebound values. GEN: S. Groves, "Crosslink Density Distributions in NR/BR Blends: Effect of Cure Temperature and Time," Presented at ACS Rubber Div. Meeting, Fall, 1997, Paper No. 94.

Crosslink Density

Increase the ultimate crosslink density from curing to increase rebound and reduce hysteresis heat buildup. RT: Chapter 3, "Vulcanizate Physical Properties, Performance Characteristics, and Testing," J. S. Dick, p. 4. GEN: M. Studebaker, J. R. Beatty, "Vulcanization," *Elastomerics*, February, 1977, p. 41.

Rubber Glass Transition

Select base elastomers that possess lower glass transition temperatures (T_g). RT: Chapter 3, "Vulcanizate Physical Properties, Performance Characteristics, and Testing," J. S. Dick, p. 52.

The blending of elastomers with significantly different T_g values can have a great effect on the compound's T_g value and hysteresis characteristics. If two different elastomers are highly compatible, then the compound may have one broad glass transition peak. However, if the two elastomers are incompatible, then the rubber compound may have two separate T_g values. In an incompatible blend, usually the elastomer with the higher T_g will impart perhaps a disproportional effect (higher hysteresis) than if the elastomers were completely compatible. RT: Chapter 7, "General Purpose Elastomers and Blends," G. Day, p. 165.

General Comparison of Elastomers

Back in 1978, Sircar and Lamond reported that compounds based on 60 phr of N347 would impart different degrees of resiliency based on the elastomer used. All dynamic property measurements were performed at 10% double strain amplitude (DSA) in shear. The following results were obtained:

NR ($\tan \delta = 0.096$), least hysteretic
BR
CR
SBR 1500
EPDM
NBR
CIIR ($\tan \delta = 0.25$), most hysteretic

GEN: A. Medalia, "Effect of Carbon Black on Dynamic Properties of Rubber Vulcanizates," *Rubber Chemistry and Technology*, July–August, 1978, Vol. 51, p. 437.

Natural Rubber

Where possible, base a rubber compound on NR for very good resilience and high rebound (low hysteresis). RT: Chapter 6, "Elastomer Selection," R. School, p. 126.

For NR-based compounds, consider a high sulfur cure to achieve the lowest hysteresis and the highest rebound. RP: J. R. Halladay.

Avoid over-masticating or using too strong a chemical peptizer in breaking down the natural rubber "backbone" in order to retain better dynamic properties and less heat buildup. GEN C. Clarke, R. Galle-Gutbrecht, M. Hensel, K. Menting, T. Mergenhagen (Struktol), "A New 'Intelligent' Peptizer Concept, Better Dynamic Properties Plus Improved Processing," Paper No. 33 presented at the Fall Meeting of the Rubber Division, ACS, October 16, 2001, Cleveland, OH.

Avoid using epoxidized natural rubber such as ENR-20 and ENR-50, which possess a T_g higher than NR, which increases the damping quality of vulcanizates compared to regular NR. RT: Chapter 7, "General Purpose Elastomers and Blends," G. Day, p. 144.

Butyl Rubber

Avoid using butyl or halobutyl rubber because these elastomers impart a damping quality (the opposite to high rebound) to the vulcanizate. RT: Chapter 6, "Elastomer Selection," R. School, p. 134; Chapter 8, "Specialty Elastomers," G. Jones, D, Tracey, A. Tesler, p. 175.

BR

Use compounds based on monodispersed, linear *cis*-BR polymers for lower hysteresis and heat buildup in the cured state. RT: Chapter 7, "General Purpose Elastomers and Blends," G. Day, p. 145.

Use *cis*-BR with the highest *cis* content, usually based on a neodymium catalyst. RT: Chapter 7, "General Purpose Elastomers and Blends," G. Day, p. 145.

Neodymium-catalyzed high *cis*-BR can have very high linearity, very low vinyl content, and provide the highest level of crosslinking compared to other BRs, so these special high *cis*-BRs usually impart the higher rebound values to their respective vulcanizates. GEN: Lim Yew Swee (Lanxess),"Benefits of Butadiene Rubber in Natural Rubber-Based Truck and Sidewall," Paper presented at India Rubber Exposition and Conference (IRE 2011), January 19, 2011, Chennai, India.

It has been reported that Nd- and Co-catalyzed *cis*-BR usually impart the lowest hysteresis to the rubber compound. Emulsion BR (E-BR) polymers usually have a less pure microstructure, a lower molecular weight, and a wider molecular weight distribution, which results in a more hysteretic compound. The E-BR polymer's greater heat buildup may result more from greater branching and MW effects rather than differences in carbon black dispersion. RT: Chapter 7, "General Purpose Elastomers and Blends," G. Day, p. 165.

SBR

Avoid SBR with amounts of block styrene in what is normally random SBR because these blocks may increase heat buildup and hysteresis. RT: Chapter 7, "General Purpose Elastomer and Blends," G. Day, p. 148.

Use E-SBR polymerized at 5°C (41°F) (or lower) instead of 50°C (122°F) for a significant increase in percent rebound and reduction in heat buildup. RT: Chapter 7, "General Purpose Elastomers and Blends," G. Day, p. 149. RP: J. M. Long.

Consider basing a compound on solution SBR (S-SBR) instead of E-SBR for better rebound. RT: Chapter 7, "General Purpose Elastomers and Blends," G. Day, p. 149.

Use an emulsion SBR that contains a lower bound styrene (e.g., use SBR 1006 in place of SBR 1013 or SBR 1505 in place of SBR 1502) to reduce hysteresis and increase rebound of a compound. The higher bound styrene increases the T_g of the elastomer. RT: Chapter 7, "General Purpose Elastomers and Blends," G. Day, p. 165. RP: J. M. Long.

Use a cold-emulsion SBR instead of a hot-emulsion SBR to impart less hysteresis to the compound because the cold-polymerized SBR usually possesses less long-chain branching and a higher molecular weight than the hot ESBR. An example of this is the cold-emulsion SBR 1500 vs. the hot-emulsion SBR 1006. RT: Chapter 7, "General Purpose Elastomers and Blends," G. Day, p. 165.

SBR 1712 (oil-extended SBR) reportedly is better for improving compound rebound than SBR 1500 adjusted for equal oil. RP: J. M. Long.

Solution SBR

Use S-SBR polymers that have been chain end modified through the addition of tin tetrachloride, silicon tetrachloride, or carbon dioxide by the manufacturer during the polymerization process. One reason for chain end modification is to couple multiple polymer chains together. During mixing with other compounding ingredients such as carbon blacks, the tin-coupled polymer can decouple and expose chain ends that react with carbon black aggregates, thus providing lower hysteresis. RT: Chapter 7, "General Purpose Elastomers and Blends," G. Day, p. 150. GEN: V. Quiteria, C. Sierra, J. Fatou, C. Calan, L. Fraga, Presented at ACS Rubber Div. Meeting, Fall, 1995, Paper No. 78; F. Tsutsumi, M. Sakakibara, N. Oshima, "Structure and Dynamic Properties of Solution SBR Coupled with Tin Compounds," *Rubber Chemistry and Technology*, March–April, 1990, Vol. 63, p. 8.

Use S-SBR polymers in which chain ends have been modified with 4,4-bis(diethylamino)-benzophenone (EAB) by the manufacturer. This modification increases polymer reactivity with carbon black, which can lower compound hysteresis and heat buildup. RT: Chapter 7, "General Purpose Elastomers and Blends," G. Day, p. 150.

Chemically modified solution SBR grades reportedly can impart higher rebound resilience at high temperature than can tin-coupled SBR. GEN: F. Suzuki (Nippon Zeon), "Rubbers for Low Rolling Resistance," *Tire Technology International*, 1997, p. 87.

S-SBR polymers with tin–butadienyl bonds reportedly give better interaction with carbon black fillers than S-SBR polymers with tin–styryl bonds. GEN:

C. Sierra, C. Galan, J. Fatou, V. Quiteria, "Dynamic-Mechanical Properties of Tin-Coupled SBRs," *Rubber Chemistry and Technology*, May–June, 1995, Vol. 68, p. 259.

Other chain end modifications of S-SBR have been considered, such as the unique chemistry involving alkyl pyrolidinones. GEN: S. Thiele, S. Knoll (Styron Deutschland GmbH, Merseburg, Germany), "Novel Functionalized SSBR for Silica and Carbon Black Containing Tires," Presented at the Fall Meeting of the Rubber Division, ACS, October 11, 2011, Cleveland, OH.

S-SBR polymers modified with GPMOS (glycidoxypropyltrimethoxysilane) and DMI (dimethylimidazolidinone) were found to reduce hysteresis loss for silica-based compounds with TEPST coupling agent. GEN: Akira Saito, Haruo Yamada, Takaaki Matsuda, Nobuaki Kubo, Norifusa Ishimura, "Improvement of Rolling Resistance of Silica Tire Compounds by Modified S-SBR," Paper No. 39 presented at the Spring Meeting of the Rubber Division, ACS, April 29, 2002, Savannah, GA.

Functionalizing a high vinyl solution SBR with carboxylic acid along the backbone has been shown to improve filler polymer interactions for lower hysteresis for a silica-loaded compound. GEN: Thomas Gross, Judy Hannay (Lanxess), "New Solution SBRS to Meet Future Performance Demands," Paper No. 11A presented at a meeting of ITEC, September 16–18, 2008, Akron, OH.

SIBR

Consider using styrene–isoprene–butadiene rubber (SIBR) for better rebound. RT: Chapter 7, "General Purpose Elastomers and Blends," G. Day, p. 150.

Consider using SIBR polymers that have been modified with tin coupling or EAB. RT: Chapter 7, "General Purpose Elastomers and Blends," G. Day, p. 150.

The lower T_g for SIBR relates to tire rolling resistance and fuel economy. GEN: A. Halasa, B. Gross, W. Hsu (Goodyear Tire and Rubber Company), "Multiple Glass Transition Terpolymers of Isoprene, Butadiene, and Styrene," Paper No. 91 presented at the Fall Meeting of the Rubber Division, ACS, October, 2009, Cleveland, OH.

NR/BR Blend

For a rubber compound based on an NR/BR blend, generally the cured tan δ will be significantly reduced by increasing the concentration of BR. Therefore, increasing the BR content will increase the compound's resiliency and reduce its cured hysteresis. GEN: W. Hess, C. Herd, P. Vegvari, "Characterization of Immiscible Elastomer Blends," *Rubber Chemistry and Technology*, July–August, 1993, Vol. 66, p. 329.

NBR

NBR is not normally selected for compounds where heat buildup or flexing might be a problem. However, if NBR is still selected for such an application, then "soft" or large particle size carbon blacks, ester plasticizers, and a tight cure should be used for the compound for best resilience. GEN: "A Comparative Evaluation of Hycar Nitrile Polymers," Manual HM-1, Revised, B. F. Goodrich Chemical Co.

Use grades of NBR with lower ACN content to improve the compound's rebound. RT: Chapter 8, "Specialty Elastomers," M. Gozdiff, p. 194.

When selecting from nitrile elastomer grades with concerns for lower cured tan δ (more resiliency), consider NBR grades that have lower ACN content, or select NBR grades produced by the hot rather than cold polymerization process. GEN: R. Del Vecchio, E. Ferro, "Effects of NBR Polymer Variations on Compound Properties," Presented at ACS Rubber Div. Meeting, Spring, 2001, Paper No. 21.

NBR with Narrow Molecular Weight Distribution

Consider NBR grades with more narrow molecular weight distribution for higher cured compound rebound. RT: Chapter 8, "Specialty Elastomers," M. Gozdiff, p. 197.

NR/NBR Blend

In a compound based on an NR/NBR blend, decreasing the concentration of NBR will increase the rebound and decrease the hysteresis. GEN: A. Tinker, "Crosslink Distribution and Interfacial Adhesion in Vulcanized Blends of NR and NBR," J. S. Dick, H. Pawlowski, "Alternate Instrumental Methods of Measuring Scorch and Cure Characteristics," *Polymer Testing*, 1995, Vol. 14, pp. 45–84.

Triblends in Damping Applications

To achieve a better ride and less vibration in smaller cars, suspension systems must use softer rubber compounds with higher damping characteristics (lower rebound). The researchers at Ford Motor Co. reported that using compounds based on natural rubber and bromobutyl rubber blends with the addition of polyisobutylene (IM or Vistanex) as a diluent and a semi-EV cure system is useful because they are damping and have more favorable low-temperature properties. Blends of NR, BR, and NIR (Krynac 833, isoprene–acrylonitrile copolymer) were also tried. This blend also imparted very good damping qualities, which are needed in this application. However, this triblend was not found to be as good as the NR/BIIR/IM blend just discussed because the NR/BR/NIR blend possessed poor low-temperature properties. GEN: M. Lemieux, P. Killcoar, "Low Modulus, High Damping, High Fatigue Life Elastomer Compounds for Vibration Isolation," *Rubber Chemistry and Technology*, September–October, 1984, Vol. 57, p. 792.

HNBR

With HNBR compounds, it is reported that selecting a higher level of unsaturation for the HNBR base polymer used means that the dynamic heat generated from flexing probably will decrease. RT: Chapter 8, "Specialty Elastomers," M. Wood, p. 202.

At high temperatures, HNBR reportedly has better dynamic property performance than polyurethane. GEN: M. Chase, "Roll Coverings Past, Present, and Future," Paper presented at meeting of Rubber Roller Group, May 15–17, 1996, p. 4.

Polychloroprene

When using neoprene in compounding, selecting a G-type gives better rebound. RT: Chapter 8, "Specialty Elastomers," L. Outzs, p. 208.

Reactive EPDM

Substituting a 2% maleic anhydride modified EPDM in place of unmodified EPDM in a blend with natural rubber reportedly will reduce hysteresis and increase rebound of the NR/EPDM compound. GEN: A. Coran, "Blends of Dissimilar Rubbers: Cure-Rate Incompatibility," *Rubber Chemistry and Technology*, May–June, 1988, Vol. 61, p. 281.

Polynorbornene

Avoid using polynorbornene (PNR) as the base rubber in compounding because it is difficult to develop very high rebound. On the other hand, using PNR with appropriate aromatic oil loadings can be used to achieve very low cured rebound values (and very high damping). Naphthenic oils usually impart less damping and more rebound to a PNR vulcanizate. RT: Chapter 8, "Specialty Elastomers," C. Cable, p. 225.

Avoid blending PNR into NR compounds because this additive will significantly dampen the NR compound's performance (significantly lower rebound). In fact, adding PNR to NR formulations is a common method to increase damping characteristics. RT: Chapter 8, "Specialty Elastomers," C. Cable, p. 226.

Polyurethane

Usually, but not always, selecting a diol-cured ether-type polyurethane based on an MDI prepolymer will give higher rebound in demanding applications than an amine-cured TDI ester-type PU compound. In addition, ethers are better than esters for lower dynamic hysteretic heat buildup. RT: Chapter 9, "Polyurethane Elastomers," R. W. Fuest, pp. 257–258.

Polyurethane can be formulated to give very good resiliency even at high hardnesses up to 75 Shore D. GEN: M. Chase, "Roll Coverings Past, Present, and Future," Presented at Rubber Roller Group Meeting, May 15–17, 1996, p. 6.

AEM

To achieve high rebound, avoid basing the compound on ethylene-acrylic elastomer (AEM) because this polymer is a low-resilience, high damping material over a broad temperature range (broader than what butyl rubber imparts). GEN: L. Muschiatte, H. Barager, "Vamac Elastomers Serve Auto Applications," *Rubber & Plastics News*, January 10, 2000, p. 14.

Sponge Rubber

If working with cellular rubber, a closed cell structure imparts higher rebound than an open cell structure. RT: Chapter 21, "Chemical Blowing Agents," R. Annicelli, p. 477.

Molecular Weight

Increase the average molecular weight of the base rubber in a formulation to decrease the hysteresis heat buildup. GEN: R. Mastromatteo, E. Morrisey, M. Mastromatteo, H. Day, "Matching Material Properties to Application Requirements," *Rubber World*, February, 1983, p. 25.

Nanofillers

Consider experimenting with the new carbon nanotubes (CNT), organically modified layered silicates (OC), graphite with high surface area (G), or chemically reduced graphite oxide (CRGO) to achieve lower hysteresis in a rubber compound. GEN: M. Galimbeth, M. Coombs, V. Cipolletti, L. Giannini, L. Conzatti, T. Ricco, M. Mauro, G. Guerra (Politecnico di Milano, Department of Chemistry, and Pirelli Tyre Study), "Nano and Nanostructured Fillers and Their Synergistic Behavior in Rubber Composites such as Tires," Paper No. 35 presented at the Fall Meeting of the Rubber Division, ACS, October 11, 2011, in Cleveland, OH.

Double-Chain Networking

It is well known that plastics can be drawn or stretched during processing to create orientation, directional (anisotropic) properties, and improved directional physical properties on cooling below the T_m or T_g of the plastic. On the other hand, rubber compounds are quite different. Normally any orientation during the processing of a rubber compound will somewhat decay after processing.

However, "double networking" during the curing process is a unique method of introducing permanent chain orientation into a rubber compound. This is achieved first by lightly crosslinking or partially curing the rubber compound in a normal manner. This establishes the first network. This lightly cured rubber compound is stretched to an extension ratio (or set) of α_0 and crosslinked further. After this second cure, the vulcanizate is released. From the release, this second formed crosslinked network "resists the retraction of the first network." This results in a vulcanizate with a residual extension ratio (or set) of α_r. Under certain conditions, double networks might diminish the Payne effect, reduce hysteresis at lower strains, and reduce filler–filler interaction; however these points have not been verified. GEN: G. Hamed, M. Huang, "Tensile and Tear Behavior of Anisotropic Double Networks of a Black-Filled NR Vulcanizate," *Rubber Chemistry and Technology*, November–December 1998, Vol. 71, p. 846.

2.7 Increasing Tear Resistance

Tear testing methods vary. The quality of the results one obtains are greatly dependent on the shape of the test piece. Many times the standard tear tests do not actually relate to tearing that occurs with a rubber product in service. In addition, tear initiation is very different from tear propagation.

*The following **experimental** ideas may improve a compound's tear resistance. Also, all relevant literature sources, including the ones cited below, should be researched and read. **Caution:** these general experimental ideas may not work in all specific situations. Changes to increase tear resistance will certainly affect other properties as well, for better or for worse, and this book does not purport to show how these other properties are affected. Also, this book does not purport to address safety and health issues.*

Mixing

Increase dispersion of reinforcing fillers such as carbon black and/or silica through better mixing. RT: Chapter 2, "Compound Processing Characteristics and Testing," J. S. Dick, p. 42.

Adding ball-milled dispersions of reinforcing carbon black to natural rubber latex before coagulation and then compounding will yield higher tear strength than simply adding free carbon black directly into the Banbury mixer. GEN: R. Alex, K. Sasidharan, T. Kurlan, A. Kumarchandra, "Carbon Black/Silica Masterbatch from Fresh Natural Rubber Latex," Paper No. 27 presented at the IRC 11, January 19, 2011, Chennai, India.

Avoid contaminating the compounding ingredients, raw elastomers, or the mixing process with dirt, which can create "flaws" and stress points, causing poor tear strength. RT: Chapter 3, "Vulcanizate Physical Properties, Performance Characteristics, and Testing," J. S. Dick, p. 49. GEN: James Halladay, O. H. Yeoh (Lord Corp.), "Problems with Developing Visual Inspection Criteria for Replacement of Vibration Isolators and Shock Mounts," Paper No. 61 presented at the Fall Meeting of the Rubber Division, ACS, October 8, 2002, Pittsburgh, PA.

Phase Mixing

Through phase mixing techniques of an NR/BR blend, one can achieve an increase in the amount of carbon black dispersed in the BR phase, which may increase the tear strength to an optimum point. However, it may be possible to have too much of the carbon black dispersed in the BR phase, which could result in a drop-off in tear properties. GEN: E. McDonel, K. Baranwal, J. Andries, *Polymer Blends*, Vol. 2, Chapter 19, "Elastomer Blends in Tires," Academic Press,

1978, p. 282. On the other hand, Hess reported that forcing more carbon black into the natural rubber phase in an NR/BR blend will generally improve the cured tear strength of the compound. This effect is even stronger when the carbon black has a very fine particle size and lower structure. Thus if a larger amount of N110 carbon black is forced into the NR phase, it will have a much greater effect on improving the compound's tear strength than will N299. GEN: W. Hess, "Characterization of Dispersions," *Rubber Chemistry and Technology*, July-August, 1991, Vol. 64, p. 386.

Hess and Chirico reported that to optimize tear strength for compounds based on either NR/BR or NR/SBR blends, the following compounding conditions should be met: (1) Use carbon black grades with low structure and small aggregate size. (2) Through phase mixing techniques, achieve a higher concentration of carbon black in the continuous phase than in the discontinuous phase. (3) Use the elastomer with the higher strength for the continuous phase. On the other hand, blends of SBR/BR are reportedly not as sensitive to carbon black distribution differences. However, it is reported that SBR/BR blends give better tear strength when increasing the percentage of SBR and the concentration of carbon black in the SBR phase through phase mixing techniques. GEN: W. Hess, C. Herd, P. Vegvari, "Characterization of Immiscible Elastomer Blends," *Rubber Chemistry and Technology*, July-August, 1993, Vol. 66, p. 329.

Crosslink Density

Optimize the ultimate crosslink density from curing. Sometimes that optimum is lower than the optimum crosslink density for ultimate tensile strength. RT: Chapter 3, "Vulcanizate Physical Properties, Performance Characteristics, and Testing," J. S. Dick, p. 47.

Sulfur Cure

An advantage of sulfur vs. peroxide cures is that the sulfur-cured compounds can more easily be made to possess better tear resistance. RT: Chapter 17, "Peroxide Cure Systems," L. Palys, p. 434.

Cures containing higher sulfur concentrations and lower accelerator concentrations reportedly have better tear resistance than cures based on low sulfur and high accelerator concentrations. RP: J. M. Long.

Peroxide Cures

Using a coagent in a peroxide cure may help increase tear strength. GEN: P. Dluzneski, "Peroxide Vulcanization of Elastomers," *Rubber Chemistry and Technology*, July-August, 2001, Vol. 74, p. 451.

Zinc dimethacrylate metallic coagent (Saret® 634) is sometimes used in peroxide cures to improve the vulcanizate's tear strength. GEN: R. Costin, W. Nagel, "Coagents for Rubber-to-Metal Adhesion," *Rubber & Plastics News*, March 11, 1996, p. 14.

For peroxide cures, consider reducing the cure temperature and lengthening cure time to improve tear strength. GEN: Joseph Burke (Sartomer Co.), "Improving Cure Characteristics in Peroxide Cured Polyisoprene," Paper No. 28 presented at the Fall Meeting of the Rubber Division, ACS, October 14, 2003, Cleveland, OH.

Reversion Inhibitors

Reportedly, using a combination of hexamethylene-1,6-bis-thiosulfate (HTS) and 1,3-bis(citraconimidomethyl)benzene (BCI-MX) together (both reversion-resistant chemical additives) in a natural rubber compound may improve tear resistance. GEN: R. Datta, W. Helt, "New Approaches to Improve Reversion Resistance of Thick Sectioned Rubber Articles," ACS Rubber Div. Meeting, Fall, 1996, Paper No. 4.

Molecular Weight Effect

Increase the average molecular weight of the base rubber in a formulation to increase tear resistance. GEN: R. Mastromatteo, E. Morrisey, M. Mastromatteo, H. Day, "Matching Material Properties to Application Requirements," *Rubber World*, February, 1983, p. 25.

Silicone

For good tear resistance, avoid compounds based on silicone or fluorosilicone. RT: Chapter 6, "Elastomer Selection," R. School, p. 136.

Polyurethane

Polyurethane elastomers have a great advantage over conventional diene rubbers in that they can impart superior cut and tear resistance. RT: Chapter 9, "Polyurethane Elastomers," R. W. Fuest, p. 253.

Polyurethane rubber compounds can possess extremely good resistance to tear. RT: Chapter 6, "Elastomer Selection," R. School, p. 137.

Tear resistance for a two-part cast polyurethane system usually can be increased by adjusting the curative ratio. The curative ratio refers to the relative amounts of prepolymer and curative. The amount of curative [such as methylene-bis-orthochloroaniline (MBCA)] needed to match exactly the isocyanate groups on the prepolymer is referred to as "100% theory" or "100% stoichiometry."

Thus if curative is reduced by 5%, it is referred to as "95% theory" or "95% stoichiometry." On the other hand, if the curative is increased by 5% from theory, it is referred to as "105% theory" or "105% stoichiometry." Generally, going to a optimal stoichiometry usually above 100% theory, such as 105%, may increase (improve) tear strength. RT: Chapter 9, "Polyurethane Elastomers," R. W. Fuest, p. 251.

Usually, but not always, selecting an ester-type polyurethane will give an advantage for tear strength in demanding applications. RT: Chapter 9, "Polyurethane Elastomers," R. W. Fuest, p. 257.

Reportedly, castable polyurethane fabricators using MDI prepolymer systems may be able to increase their product's tear resistance by trying to use aromatic diol chain extenders such as bis-(hydroxyethyl) ether of resorcinol (HER) and hydroquinone (HQEE). The use of these chain extenders will create hard domains that have higher melting points. GEN: R. Durairaj, "Chain Extenders Increase Heat Tolerance," *Rubber & Plastics News*, November 29, 1999.

Consider blending polyurethane with conventional rubber based compound to improve tear strength. GEN: T. Jazlonowski, "Blends of Polyurethane Rubber with Conventional Rubber," Paper No. 46 presented at the Spring Meeting of the Rubber Division, ACS, April, 1999, Chicago, IL.

cis-IR

Selecting an IR with high *cis* content will improve the compound's ability to undergo strain-induced crystallization and impart higher tear strength. RT: Chapter 7, "General Purpose Elastomers and Blends," G. Day, p. 142.

XNBR

Use XNBR with the appropriate zinc oxide in place of conventional NBR to achieve a higher compound tear resistance. RT: Chapter 8, "Specialty Elastomers," M. Gozdiff, p. 199.

The hydrogenation of carboxylated nitrile (XNBR) to hydrogenated carboxylated nitrile rubber (HXNBR) may significantly improve imparted compound tear resistance in the tmperature range of 23 to 170 ºC. GEN: John E. Dato (Lanxess), "Hydrogenated Nitrile Rubber for Use in Oilfield Applications," Paper No. 57 presented at the Fall Meeting of the Rubber Division, ACS, October 10, 2006, Cincinnati, OH.

Polychloroprene

When using neoprene in compounding, selecting a G-type gives better tear strength. RT: Chapter 8, "Specialty Elastomers," L. Outzs, p. 208.

CPE

Increasing the chlorine content in chlorinated polyethylene (CM) usually results in higher compound cured tear resistance. RT: Chapter 8, "Specialty Elastomers," L. Weaver, p. 212.

cis-BR

Using BR produced from a catalyst with *cis* content greater than 96% reportedly will improve the compound's tear resistance and may create a knotty tearing characteristic because of strain crystallization from high *cis* content. GEN: J. Zhao, G. Ghebremeskel, "A Review of Some of the Factors Affecting Fracture and Fatigue in SBR and BR Vulcanizates," *Rubber Chemistry and Technology*, July-August, 2001, Vol. 74, p. 409.

SBR

Increasing the bound styrene content in an SBR used in a compound reportedly may increase the compound's flex fatigue life and tearing resistance. GEN: J. Zhao, G. Ghebremeskel, "A Review of Some of the Factors Affecting Fracture and Fatigue in SBR and BR Vulcanizates," *Rubber Chemistry and Technology*, July-August, 2001, Vol. 74, p. 409. However, this observation may apply only to gum SBR compounds. Others have found that low amounts of styrene (or no styrene, i.e., EBR) will improve flex fatigue life and tear resistance for many compounds containing carbon black or other fillers. RP: J. M. Long.

SBR/PVC Blend

The use of SBR 4503 (an SBR that is a divinylbenzene-crosslinked, hot-polymerized emulsion polymer at 30% bound styrene) with NBR/PVC blend was reported to improve the tear strength. GEN: J. Zhao, G. Ghebremeskel, J. Peasley, "SBR/PVC Blends with NBR as a Compatibilizer," *Rubber World*, December, 1998, p. 37.

Gas-Phase EPDM

Try compounding with an ultra-low Mooney viscosity gas-phase polymerized EPDM with very high ethylene content and high filler loading (the higher ethylene content reportedly increases the tear strength of the vulcanizate). GEN: A. Paeglis, "Very Low Mooney Granular Gas-Phase EPDM," Presented at ACS Rubber Div. Meeting, Fall, 2000, Paper No. 12.

Millable PU as an Additive

Adding a given quantity of millable polyurethane incrementally to an SBR-, EPDM-, or BR-based compound reportedly can improve the compound's tear resistance. GEN: T. Jablonowski, "Blends of PU with Conventional Rubbers," *Rubber World*, October, 2000, p. 41.

EPDM Nanocomposite

Consider using Buckminsterfullerene C_{60} in an EPDM with exposure to UV to develop tear strength significantly higher than that generated by free-radical crosslinking from dicumyl peroxide. This forms a "nanocomposite" from the EPDM/C_{60} combination. GEN: G. Hamed, "Reinforcement of Rubber," *Rubber Chemistry and Technology*, July–August, 2000, Vol. 73, p. 524.

SBR vs. NR/BR and SBR/BR Blends

Consider using a compound based on 100% SBR instead of a blend of SBR/BR or NR/BR or even NR/SBR for better resistance to tear. RP: J. M. Long.

Fillers

Find the optimum loading of a filler that will impart the highest tear resistance. GEN: R. Mastromatteo, E. Morrisey, M. Mastromatteo, H. Day, "Matching Material Properties to Application Requirements," *Rubber World*, February, 1983, p. 26.

Use a filler with higher surface area to increase compound tear resistance. GEN: R. Mastromatteo, E. Morrisey, M. Mastromatteo, H. Day, "Matching Material Properties to Application Requirements," *Rubber World*, February, 1983, p. 26.

Use a filler with greater surface activity to increase compound tear resistance. GEN: R. Mastromatteo, E. Morrisey, M. Mastromatteo, H. Day, "Matching Material Properties to Application Requirements," *Rubber World*, February, 1983, p. 26.

Carbon Black

Increasing the surface area (decreasing the particle size) of the carbon black used in a compound will usually increase the tear resistance of the compound. RT: Chapter 12, "Compounding with Carbon Black and Oil," S. Laube, S. Monthey, M-J. Wang, pp. 308, 317.

Increasing the loading through an optimum for the carbon black used in a compound will usually increase the tear resistance of the compound and then reduce this resistance above the optimal loading. RT: Chapter 12, "Compounding with Carbon Black and Oil," S. Laube, S. Monthey, M-J. Wang, p. 308.

Just using a carbon black with only higher structure can sometimes be detrimental to tear strength. GEN: *Carbon Black*, Chapter 9, "Carbon Black Reinforcement of Elastomers," S. Wolf, M-J. Wang, p. 334. RP: M-J. Wang.

Consider using carbon black with a low structure and moderately high surface area. The low structure will result in a lower modulus, resulting in a higher ultimate elongation. The moderately high surface area gives high tensile strength and good tear resistance. Of course, low-structure carbon blacks just described are notoriously difficult to disperse. Therefore, steps to achieve good dispersion must be performed or poor dispersion will result, which will lead to poor tear strength as well as poor cutting and chipping properties. RT: Chapter 12, "Compounding with Carbon Black and Oil," S. Laube, S. Monthey, M-J. Wang, p. 317.

According to a study by Tabar et al., when compounding a neoprene formulation *for equal hardness*, selecting a large particle size high-structure carbon black such as N765 will impart better tear resistance in a bushing compound than other carbon blacks such as N770, N330, or N220. N765 reportedly offers lower compression set as well. GEN: R. Tabar, P. Killgoar, R. Pett, "A Fatigue Resistant Polychloroprene Compound for High Temperature Dynamic Applications," *Rubber Chemistry and Technology*, September-October, 1979, Vol. 52, p. 781.

Silica

Consider using precipitated silica in a rubber compound where an improvement in tear resistance is needed. To improve tear resistance further, consider using precipitated silica treated with an organosilane. RT: Chapter 13, "Precipitated Silica and Non-black Fillers," W. Waddell, L. Evans, pp. 331, 339.

To achieve the full tear strength potential from precipitated silica, a silica loading at 30 phr or greater may be necessary. Silica can improve the tear resistance of a rubber compound significantly more than using carbon black alone. GEN: N. Hewitt, "Compounding with Silica for Tear Strength and Low Heat Build-up," *Rubber World*, June, 1982; J. Sommer, *Elastomer Molding Technology*, Elastech, Hudson, OH, 2003, p. 179.

Precipitated hydrated silica can improve tear resistance of a rubber compound without the use of silane coupling agents. GEN: Tim Okel (PPG), "Introduction to Silica," Presented at the Flexsys Compounder's Course, Akron, OH, 2006.

Caution is needed in using mercaptosilane coupling agents with precipitated silica loadings because sometimes use of the coupling agent under certain circumstances can actually degrade compound tear performance. GEN: N. Hewitt, "Compounding with Silica for Tear Strength and Low Heat Build-up," *Rubber World*, June, 1982.

Rosin derivatives (such as Staybelite Resin) and aromatic resins (such as Cumar, PICCO 100, etc.) are sometimes used in precipitated silica compounds to improve tear resistance. It is thought that a mild coupling effect may occur from the aromatic resins and silica filler. The rosin derivatives should reduce the compound's viscosity as well. GEN: N. Hewitt, "Compounding with Silica for Tear Strength and Low Heat Build-up," *Rubber World*, June, 1982.

Precipitated silica reportedly can be used in a black sidewall compound, which can include a nonstaining EPDM base to significantly improve tear strength as well as cut-growth and resistance to ozone aging. GEN: W. Waddell, L. Evans, "Use of Nonblack Fillers in Tire Compounds," *Rubber Chemistry and Technology*, July–August, 1996, Vol. 69, p. 377.

Clay

Silane coupling agents added to clays that are used as rubber fillers are reported to significantly improve the tear properties of the rubber compound. GEN: Don Askea (Polymer Valley Chemicals, Inc.), "Functional Mineral Fillers in Rubber Compounds," Presented at the Fall Meeting of the Energy Rubber Group, September 15, 2011, Galveston, TX.

Talc

Compound tear resistance may be improved by partially substituting some of the carbon black with fine particle size talc (Mistron® Vapor R). This type of substitution reportedly may improve both resistance to tearing as well as cut resistance, possibly because of the anisotropic nature of talc, which results in crack diversion and reduces the stress concentration at the crack tip. This is reportedly due as well to the imparted high cohesive fracture energy of the talc reinforcement. This effect has been observed in CSM-, NBR-, EPDM-, and SBR-based compounds. GEN: O. Noel, S. Brignac, unpublished paper, "Talc as a Reinforcing Pigment in Rubber: Synergy with Carbon Black," 1999; Roger J. Eldred, "Effect of Oriented Platy Filler on the Fracture Mechanism of Elastomers," *Rubber Chemistry and Technology*, September–October, 1988, Vol. 61, p. 619. RP: O. Noel.

Ground Rubber

It has been reported that mixing an ambient ground vulcanizate into its respective original uncured compound will result in a new composite compound that possesses better trouser tear properties than the original homogeneous compound displays (without any recycled ground rubber). However, ultimate tensile properties are exactly the opposite in that the compound containing its respective ground rubber gave poorer tensile properties than its homogeneous (no ground rubber) form. This was observed for a black-filled SBR compound. Gibala, Thomas, and Hamed concluded that the reason for the opposite results between trouser tear and ultimate tensile strength properties is that the cured sheet containing the ground cured rubber displayed tearing, taking on an "increased stick-slip character" vs. the homogeneous compound, which possessed less of this characteristic. GEN: D. Gibala, D. Thomas, G. Hamed, "Cure and Mechanical Behavior of Rubber Compounds Containing Ground Vulcanizates, Part III: Tensile and Tear Strength," Presented at ACS Rubber Div. Meeting, 1994, Paper No. 53.

Nanoclays and Carbon Nanotubes

Consider the use of nanofillers such as nanoclays or carbon nanotubes to improve tear strength. GEN: Debbie Banta (Weatherford Co.) "Can Nanotechnology Provide Innovative, Affordable Elastomer Solutions to Oil and Gas Service Industry Problems?," Presented at a meeting of the Energy Rubber Group, January 19, 2012, Houston, TX.

Wollastonite

Consider using wollastonite (calcium metasilicate) as a "nontraditional" filler (with needle-shaped particles or acicular) to directionally increase tear resistance as well as hot tear resistance. GEN: S. Robinson, M. Sheridan, A. Ferradino (R. T. Vanderbilt Co.), "The Advantages of Wollastonite, a Non-traditional Filler, in Fluorohydrocarbon (FKM) Elastomer," Paper No. 16 presented at the Spring Meeting of the Rubber Division, ACS, April 29, 2002, Savannah, GA.

Fibers

The use of 5 phr of fiber pulps such as cotton, nylon-6, and polyester will generally increase compound tear strength, especially in the transverse direction for EPR compounds. This is true for other elastomers as well. GEN: A. Estrin, "Application of PBDMA for Enhancement of EPR Loaded with Chopped Fibers," *Rubber World*, April 2000, p. 39.

The use of 5 phr of fiber pulps such as cotton, nylon-6, and polyester with low molecular weight maleated polybutadiene (PBDMA) will significantly increase compound tear strength, especially in the transverse direction for EPR compounds. The PBDMA gives significantly higher tear strength values than if it were not used. GEN: A. Estrin, "Application of PBDMA for Enhancement of EPR Loaded with Chopped Fibers," *Rubber World*, April 2000, p. 39.

Consider using a low loading of short aramid fiber to improve tear resistance. In the reference cited, this improvement was achieved with 3 and 5 phr of Kevlar pulp in an FKM compound. GEN: K. Watson, A. Frances, "Elastomer Reinforcement with Short Kevlar Aramid Fiber for Wear Applications," *Rubber World*, August, 1988, p. 20.

Ionomeric Elastomers

Using zinc oxide to crosslink a gum carboxylic-butadiene rubber that possesses one carboxyl group for approximately every 100 carbons on the backbone can produce a cured rubber that has higher tear strength than when only a normal sulfur cure is used without any carboxylic groups available. This becomes an "ionomeric elastomer" with ionic crosslink sites that form nanometer-sized domains. GEN: G. Hamed, "Reinforcement of Rubber," *Rubber Chemistry and Technology*, July-August, 2000, Vol. 73, p. 524.

Electron Beam

The use of an electron beam in the curing of EPDM compounds with peroxide/coagent cures can improve these compounds' cured tear resistance. GEN: William M. Boye (Sartomer Company), "Use of Multifunctional Crosslinking Agents in the Electron Beam Cure of Elastomers," Paper No. 84 presented at the Fall Meeting of the Rubber Division, ACS, October 13, 2009, Pittsburgh, PA.

Double Networking

It is well known that plastics can be drawn or stretched during processing to create orientation, directional (anisotropic) properties, and improved directional physical properties on cooling below the T_m or T_g of the plastic. On the other hand, rubber compounds are quite different. Normally any orientation during the processing of a rubber compound will somewhat decay after processing. However, "double networking" during the curing process is one unique method of introducing permanent chain orientation into a rubber compound. This is achieved first by lightly crosslinking or partially curing the rubber compound in a normal manner. This establishes the first network. This lightly cured rubber compound is stretched to an extension ratio (or set) of α_0 and crosslinked further. After this second cure, the vulcanizate is released. From the release, this second formed crosslinked network "resists the retraction of the first network." This results in a vulcanizate with a residual extension ratio (or set) of α_r. It has been reported that some double networks with low α_r may show better tear strength than a simple conventionally formed network. However, when α_r is high, these double-networked vulcanizates can display *very poor tear resistance.* GEN: G. Hamed, M. Huang, "Tensile and Tear Behavior of Anisotropic Double Networks of a Black-Filled NR Vulcanizate," *Rubber Chemistry and Technology*, November-December 1998, Vol. 71, p. 846.

2.8 Increasing Hot Tear Resistance

Compounds that might possess good tear resistance at room temperature may also have poor tear resistance at elevated temperatures. Often compounds are tested in a hot box to determine their hot tear resistance.

The following ***experimental*** *ideas may improve a compound's hot tear resistance. Also, all relevant literature sources, including the ones cited below, should be researched and read.* ***Caution:*** *these general experimental ideas may not work in all specific situations. Changes to increase hot tear resistance will certainly affect other properties as well, for better or for worse, and this book does not purport to show how these other properties are affected. Also, this book does not purport to address safety and health issues.*

Silica

To improve the hot tear resistance of a halobutyl compound, small amounts of silica may be added. RT: Chapter 8, "Specialty Elastomers," G. Jones, D. Tracey, A. Tisler, p. 188.

Talc

It is reported that hot tear in a peroxide-cured EPDM compound can be improved by partially substituting some of the carbon black with fine particle size talc. This was not the case reported, however, for sulfur-cured compounds. GEN: O. Noel, S. Brignac, unpublished paper, "Talc as a Reinforcing Pigment in Rubber: Synergy with Carbon Black," 1999. RP: O. Noel.

AEM

When using ethylene-acrylic elastomers (AEM), reportedly hot tear resistance can be improved by selecting terpolymer grades of AEM instead of the copolymer (dipolymer) grades, which require a peroxide cure. RT: Chapter 8, "Specialty Elastomers," T. Dobel, p. 224.

HNBR

HNBR is reported to impart good hot tear resistance to a compound. GEN: M. Chase, "Roll Coverings Past, Present, and Future," Presented at Rubber Roller Group Meeting, New Orleans, May 15–17, 1996, p. 4.

XHNBR

The hydrogenation of carboxylated nitrile (XNBR) to hydrogenated carboxylated nitrile rubber (HXNBR) will significantly improve imparted compound hot tear resistance. GEN: John E. Dato (Lanxess), "Hydrogenated Nitrile Rubber for Use in Oilfield Applications," Paper No. 57 presented at the Fall Meeting of the Rubber Division, ACS, October 10, 2006, Cincinnati, OH.

CR

Better hot tear resistance is obtained for a neoprene compound by using a sulfur-modified G-type neoprene with perhaps a 30-phr precipitated silica loading. RP: L. L. Outzs.

Fluoroelastomers with Oxidized Thermal Black

Using certain types of fluoroelastomers with a specially oxidized thermal black and a 1200 molecular weight poly-(1,2-butadiene)diol prepolymer as a coupling agent with a peroxide cure reportedly can achieve an improvement in hot tear. GEN: J. Martin, T. Braswell, H. Green, "Coupling Agents for Certain Types of Fluoroelastomers," *Rubber Chemistry and Technology*, November–December, 1978, Vol. 51, p. 897.

Surface-Modified Carbon

In hose manufacture, consider substituting at higher loading a new proprietary "surface-modified carbon" (IRX 1045 from Cabot) in place of the traditional N550 or N650 carbon blacks for better hot tear. GEN: S. Monthey, M. Lucchi, "A New Carbon Black for Peroxide-Cured EPDM Coolant Hose," Presented ACS Rubber Div. Meeting, Fall, 1999, Paper No. 34.

Fiber

The use of 5 phr of fiber pulps such as cotton, nylon-6, and polyester will generally increase compound hot tear strength, especially in the transverse direction for EPR compounds. This is true for other elastomers as well. GEN: A. Estrin, "Application of PBDMA for Enhancement of EPR Loaded with Chopped Fibers," *Rubber World*, April 2000, p. 39.

The use of 5 phr of fiber pulps such as cotton, nylon-6, and polyester with low molecular weight maleated polybutadiene (PBDMA) will significantly increase compound hot tear strength, especially in the transverse direction for EPR compounds. The PBDMA may give significantly higher hot tear strength values than if it were not used. GEN: A. Estrin, "Application of PBDMA for Enhancement of EPR Loaded with Chopped Fibers," *Rubber World*, April 2000, p. 39.

Consider using a low loading of short aramid fiber to improve hot tear resistance. GEN: K. Watson, A. Frances, "Elastomer Reinforcement with Short Kevlar Aramid Fiber for Wear Applications," *Rubber World*, August, 1988, p. 20.

Nanofillers

Nanofillers (nanoclays and/or carbon nanotubes) were reported to improve hot tear properties for an NBR compound. GEN: R. Lamba, P. Spanos, S. Meng, "Designing Elastomeric Materials for High Temperature and High Pressure Environments," Presented at the Fall Meeting of the Energy Rubber Group, September 14–16, 2010, San Antonio, TX.

Wollastonite

Consider using wollastonite (calcium metasilicate) as a "nontraditional" filler (with needle-shaped particles or acicular) to directionally increase tear resistance as well as hot tear resistance. GEN: S. Robinson, M. Sheridan, A. Ferradino (R. T. Vanderbilt Co.), "The Advantages of Wollastonite, a Non-Traditional Filler, in Fluorohydrocarbon (FKM) Elastomer," Paper No. 16 presented at the Spring Meeting of the Rubber Division, ACS, April 29, 2002, Savannah, GA.

Sulfur Cures

In general, the polysulfide crosslinks from sulfur cures may impart better hot tear resistance to a compound than peroxide cures. GEN: P. Dluzneski, "Peroxide Vulcanization of Elastomers," *Rubber Chemistry and Technology*, July–August, 2001, Vol. 74, p. 451.

Coagents

For peroxide cures, use a coagent as well to add unsaturation to the system, which results in higher crosslink density than the peroxide would have generated if used without the coagent. The reason for this is that it is more efficient and easier to crosslink with a free radical to unsaturation than by hydrogen abstraction to a saturated polymer chain. Because the use of a coagent can lead to different types of crosslink networks, improved hot tear strength may result. RT: Chapter 17, "Peroxide Cure Systems," L. Palys, pp. 431–432. GEN: A. H. Johansson (Rhein Chemie Corp.), "Peroxide Curing Trends Cure Characteristics of Peroxides," Paper No. 10 presented at the Fall Meeting of the Rubber Division, ACS, October 14, 2003, Cleveland, OH.

Silica Effects

In some situations, some precipitated silica (between 10 and 20 phr) may increase hot tear and hot tensile strength of a rubber compound. RP: J. R. Halladay.

Loading of FKMs

Reportedly, by substituting N330 in place of N990 and optimizing and adjusting concentrations of calcium and magnesium hydroxide levels as well as peroxide and coagent loadings, one can improve the hot tear resistance. GEN: J. Sommer, *Elastomer Molding Technology*, Elastech, Hudson, OH, 2003, p. 179.

2.9 Improving Low-Temperature Properties

Rubber products are used in dynamic applications over a wide temperature range. At very low temperatures, the cured rubber compound will become stiffer. At much lower temperatures, the compound becomes brittle and glass-like. It is often a challenge to the compounder to formulate so that the rubber compound will still give good performance at a specified low temperature.

The following ***experimental*** *ideas and suggestions may help impart satisfactory performance properties at low temperatures. Also, all relevant literature sources, including the ones cited below, should be researched and read.* ***Caution:*** *these general experimental ideas may not work in all specific situations. Changes to improve low-temperature properties will certainly affect other properties as well, for better or for worse, and this book does not purport to show how these other properties are affected. Also, this book does not purport to address safety and health issues.*

Selected Rubber Base

To formulate a compound to have good low-temperature properties, it is important that the compound be based on an elastomer(s) that has a low glass transition temperature (T_g). The following gives a very general comparison of the typical T_g values for different elastomers.

MVQ (dimethyl polysiloxane containing vinyl)	-120°C
BR (*cis*-1,4-polybutadiene)	-112°C
NR (natural rubber) and IR (syn. polyisoprene)	-72°C
FVMQ (fluorosilicone rubber)	-70°C
IIR (butyl rubber), CIIR (chlorobutyl), and BIIR (bromobutyl)*	-66°C
PNF (polyfluorophosphazene)	-66°C
EPDM (ethylene-propylene-terpolymer)	-55°C
EU (polyether-urethane rubber)	-55°C
SBR (styrene-butadiene rubber)	-50°C
ECO (epichlorohydrin ethylene oxide copolymer)	-45°C
NBR (nitrile), low ACN	-45°C
CR (polychloroprene)	-45°C
EAM (ethylene-acrylate rubber)	-40°C
AU (polyester urethane rubber)	-35°C
NBR (nitrile, medium ACN)	-34°C
EVM (ethylene-vinylacetate copolymer)	-30°C
HNBR (hydrogenated nitrile)	-30°C
XNBR (carboxylated NBR)	-30°C

CO (polyepichlorohydrin)	-26°C
CM (chlorinated polyethylene)	-25°C
CSM (chlorosulfonated polyethylene)	-25°C
NBR (nitrile, high ACN)	-20°C
ACM (polyacrylate)	-20 to -40°C
FKM (fluoroelastomers)	-18 to -50°C
PNR (polynorbornene)	+25°C

* Broad T_g

Of course, compounding these elastomers with other compounding ingredients will significantly affect the ultimate compound T_g and its low-temperature performance.

GEN: W. Hofmann, *Rubber Technology Handbook*, Hanser, Munich, 1989, p. 162.

Be Careful About Crystallizing Elastomers

Be careful basing a compound on crystallizing elastomers such as polychloroprene or natural rubber. These elastomers can stiffen from crystallization at lower temperature if they are not compounded properly. However, they can commonly be compounded to have good to excellent low-temperature properties. RT: Chapter 3, "Vulcanizate Physical Properties, Performance Characteristics, and Testing," J. S. Dick, p. 54.

NR/BR Blends

To improve the low-temperature properties of a natural rubber based compound, consider blending polybutadiene with the NR. Many times the improvement in low-temperature properties is directly proportional to the amount of BR added to the blend. RP: J. R. Halladay.

CR and Butyl Oleate

To meet a -40°C (-40°F) target for low-temperature properties for a polychloroprene (CR)-based compound, often a plasticizer such as butyl oleate is used as a very cost-effective plasticizer for good low-temperature properties. RP: L. L. Outzs.

Silicone

Silicone rubber (MQ) imparts extremely good low-temperature properties to a rubber compound (brittle point around -118°C). RT: Chapter 6, "Elastomer Selection," R. School, p. 128, p. 137.

Regular VMQ silicone gums will crystallize at temperatures below -45°C. However, the introduction of about 5 to 7 mol% phenyl groups during polymerization (making it a PVMQ silicone gum) will inhibit this crystallization and extend the application temperature range to −90°C. RT: Chapter 8, "Specialty Elastomers," J. R. Halladay, p. 235.

Polybutadiene

Polybutadiene imparts very good low-temperature properties. RT: Chapter 6, "Elastomer Selection," R. School, pp. 128, 129.

EAM

Ethylene-acrylic rubber, for an oil-resistant elastomer, imparts fairly good low-temperature properties compared to some other oil-resistant rubbers such as polyacrylates. RT: Chapter 6, "Elastomer Selection," R. School, p. 137.

EPDM

Use amorphous EPDM grades (with lower ethylene content) to impart better low-temperature properties. RT: Chapter 8, "Specialty Elastomers," R. Vara, J. Laird, p. 191.

Single site constrained geometry metallocene catalyst technology for commercial polymerization makes it possible to produce efficiently commercial grades of EPDM with high ethylene content. However, through this technology, it is possible to manipulate the ethylene distribution in such a way as to affect the distribution of the melting endotherm. Thus this technology can provide commercial EPDM grades with improved low-temperature properties. GEN: D. Parikh, M. Hughes, M. Laughner, L. Meiske, R. Vara, "Next Generation of Ethylene Elastomers," Presented at ACS Rubber Div. Meeting, Fall, 2000, Paper No. 158.

Selecting an EPDM with higher diene content may improve low-temperature properties because the diene disrupts the crystallization of the ethylene segments. GEN: John Dewar, Don Tsou (Lanxess), "Factors Influencing Low Temperature Performance of EPDM Compounds," Paper No. 66 presented at the Fall Meeting of the Rubber Division, ACS, October 5, 2004, Columbus, OH.

NBR

Select grades of NBR with lower levels of ACN to improve a compound's low-temperature flexibility. RT: Chapter 8, "Specialty Elastomers," M. Gozdiff, p. 194.

SBR

Select grades of SBR with lower levels of bound styrene to improve a compound's low-temperature flexibility. RP: J. M. Long, J. R. Halladay.

NBR/PVC Blend

Adding a given quantity of millable polyurethane incrementally to an NBR/PVC-based compound reportedly will improve the compound's low-temperature properties. GEN: T. Jablonowski, "Blends of PU with Conventional Rubbers," *Rubber World*, October, 2000, p. 41.

CPE

Selecting grades of chlorinated polyethylene with a lower percentage of chlorine will allow for better low-temperature properties. RT: Chapter 8, "Specialty Elastomers," L. Weaver, p. 213.

CSM

Selecting grades of chlorosulfonated polyethylene with a lower percentage of chlorine will allow for better low-temperature properties. RT: Chapter 8, "Specialty Elastomers," C. Baddorf, p. 215.

Alkylated CSM

Alkylated chlorosulfonated polyethylene (Acsium®) is an alkylated form of CSM that possesses better low-temperature properties than CSM (Hypalon®). The alkyl side chains break up crystallinity, which helps low-temperature properties. GEN: R. Fuller, "Alkylated Chlorosulfonated Polyethylene," DuPont Dow Elastomers, September, 1997.

Polynorbornene

For compounds based on polynorbornene, naphthenic oils impart much better low-temperature properties than do aromatic process oils. In addition, low-viscosity naphthenic oils impart better low-temperature properties than higher viscosity naphthenic oils. RT: Chapter 8, "Specialty Elastomers," C. Cable, pp. 225, 226.

FKM

When using fluoroelastomer polymers, selecting ones that were manufactured with perfluoromethylvinyl ether (PMVE) in place of hexafluoropropylene (HFP) may improve the compound's low-temperature flexibility. Examples of these types of polymers are Viton GLT and GFLT, which require peroxide cures. RT: Chapter 8, "Specialty Elastomers," R. D. Stevens, pp. 229–231.

Generally, FKM may give better low-temperature properties compared to FEPM-based compounds. GEN: R. Campbell (Greene Tweed Co.), "History of Sealing

Products and Future Challenges in the Oil Field," Presented at a meeting of the Energy Rubber Group, September 13, 2011, Galveston, TX.

Polyurethane

Usually, but not always, selecting an ether-type polyurethane based on an MDI prepolymer will give better low-temperature properties in demanding applications than an ester-type PU. RT: Chapter 9, "Polyurethane Elastomers," R. W. Fuest, p. 257.

TPV

Low-temperature limits for thermoplastic vulcanizates (TPVs) are determined by the glass transition temperature (T_g) of the rubber phase. Therefore, selecting an EPDM/PP in place of an NBR/PP will provide better low-temperature properties. RT: Chapter 10, "Thermoplastic Elastomers," C. P. Rader, p. 276.

HNBR

Select a special low-temperature grade of HNBR (LT-HNBR), which contains a low acrylonitrile level and an additional third monomer possessing a bulky, flexible side group that will disrupt crystallization, to improve the HNBR's low-temperature properties. GEN: J. N. Gamlin, S. X. Guo, D. Achten (Bayer AG), "Stretching the Temperature Limits of HNBR Elastomers," Paper No. 54 presented at the Fall Meeting of the Rubber Division, ACS, October 8, 2002, Pittsburgh, PA.

Oils

Select oils that are not highly aromatic for a compound. Generally, aromatic process oils tend to impart poor low-temperature properties to a given compound. RT: Chapter 12, "Compounding with Carbon Black and Oil," S. Laube, S. Monthey, M-J. Wang, p. 312.

Plasticizer

Consider the use of adipate and sebacate plasticizers with halobutyl rubber to improve low-temperature properties. RT: Chapter 8, "Specialty Elastomers," G. Jones, D. Tracey, A. Tisler, p. 180.

Azelate polymeric plasticizers are reported to impart "excellent" low-temperature properties. GEN: Stephen O'Rourke (C. P. Hall Co.), "High Performance Ester Plasticizers," Paper No. 61 presented at the Fall Meeting of the Rubber Division, ACS, October 16, 2001, Cleveland, OH.

For NBR and CR compounds, consider using dioctyl adipate (DOA) as the selected plasticizer in a rubber compound because reportedly it imparts excellent low-temperature properties to a rubber compound. RT: Chapter 14, "Ester Plasticizers and Processing Additives," W. Whittington, pp. 350, 351.

Consider using dinormal 7,9,11 carbon phthalate (C7C11P) as the selected plasticizer in a rubber compound because reportedly in some cases it can impart even better low-temperature properties to an NBR-based compound than DOA. RT: Chapter 14, "Ester Plasticizers and Processing Additives," W. Whittington, pp. 354, 378.

Avoid polymeric synthetic plasticizers or synthetic plasticizers with higher viscosity and use monomeric plasticizers with lower viscosity instead for better low-temperature properties. RT: Chapter 14, "Ester Plasticizers and Processing Additives," W. Whittington, p. 362.

Reportedly, DOA (dioctyl adipate), DMBTG (dibutylmethylene bis-thioglycolate), and DBEEA [*di*(butoxyethoxy ethyl)adipate] impart better low-temperature properties for HNBR compounds than many other plasticizers. In addition, DBEEA reportedly is the "most recommended" for HNBR-based, sulfur-cured compounds because it imparts a good balance between low-temperature properties and heat resistance. GEN: S. Hayashi, H. Sakakida, M. Oyama, T. Nakagawa, "Low-Temperature Properties of Hydrogenated Nitrile Rubber (HNBR)," *Rubber Chemistry and Technology*, September–October, 1991, Vol. 64, p. 534.

Adding only a small amount of ester plasticizer to a natural rubber based compound can significantly improve the low-temperature properties of the compound (small loadings of synthetic ester remain compatible with NR provided the loading is below a critical concentration). RP: J. R. Halladay.

Consider using "low-polarity polymer modifiers" (LPPM) to improve low-temperature properties for rubber compounds based on EPDM, SBR, NR, or CR. LPPMs are special high molecular weight esters with low oxygen-to-carbon ratios with reportedly low solubility parameters that behave as "efficient" plasticizers. The resulting compound possesses better low-temperature properties without the LPPM bleeding to the surface of the compound as a conventional plasticizer might. LPPM reportedly can plasticize semicrystalline polymers and also provide very good low-temperature properties. In plasticizing the amorphous regions but allowing the regions of crystallinity to remain intact, these LPPMs impart to these low-polarity elastomers strength, good high-temperature properties, and good low-temperature properties as well. GEN: Stephen O'Rourke (CPH Innovation Corp., Chicago, IL), "New Line of Modifiers for Low Polarity Elastomers and Plastics," Paper No. 4 presented at the Fall Meeting of the Rubber Division, ACS, October, 2004.

Crosslink Density

Increasing the chemical crosslink density of a natural rubber vulcanizate will decrease the rate of low-temperature crystallization. GEN: D. Campbell, A. Chapman, "Relationships Between Vulcanizate Structure and Vulcanizate Performance," Malaysian Rubber Producers Research Association, Brickendonbury, Hertford, UK.

Coagent with Peroxide Cure

Use low molecular weight (liquid) high vinyl 1,2-polybutadiene resins (such as Ricon® grades) as a coagent in a peroxide cure of EPDM compounds to obtain better low-temperature properties than some other traditional coagents might impart. GEN: R. Drake, "Using Liquid Polybutadiene Resin to Modify Elastomeric Properties," *Rubber & Plastics News*, February 28 and March 14, 1983.

2.10 Increasing Electrical Conductivity

This property is difficult to measure for a cured compound. Sample preparation is very critical. Different test methods will give different results.

The following ***experimental*** *ideas may increase the electrical conductivity of a rubber compound. Also, all relevant literature sources, including the ones cited below, should be researched and read.* ***Caution:*** *these general experimental ideas may not work in all specific situations. Changes to increase electrical conductivity will certainly affect other properties as well, for better or for worse, and this book does not purport to show how these other properties are affected. Also, this book does not purport to address safety and health issues.*

Hydrin Rubber

Epichlorohydrin rubber probably imparts a little more electrical conductance to a rubber compound than many other synthetic elastomers. Loading an epichlorohydrin rubber with N472 high-structured carbon black can help increase somewhat the electrical conductivity of the compound. RT: Chapter 8, "Specialty Elastomer," C. Cable, p. 217.

CR Blends

Sircar reported the use of heterogeneous elastomer blends based on polychoroprene (CR) to achieve better electrical conductivity. The carbon black theoretically forms a "conductive path" at the phase boundaries. CIIR/CR blends reportedly give the best electrical conductance followed by NBR/CR blends followed by NR/CR blends, which may give the weakest electrical conductance of these blends. Usually the optimal electrical conductance is achieved at close to a 50:50 blend ratio. Usually N472 at 60 phr with 20 phr oil is a good starting position. GEN: W. Hess, C. Herd, P. Vegvari, "Characterization of Immiscible Elastomer Blends," *Rubber Chemistry and Technology*, July–August, 1993, Vol. 66, p. 329.

Mixing a natural rubber black masterbatch with a polychloroprene causes the carbon black particles to move to the boundary phases, which reportedly increases electrical conductivity. GEN: J. Pyne, "Processing Report, Conductive Rubbers Advance with New Blacks," *European Rubber Journal*, November, 1981, p. 17.

Low Molecular Weight Rubber Base

Using a lower viscosity base rubber with a given carbon black may actually impart higher electrical conductivity to the final vulcanizate because the lower viscosity causes less breakdown of the carbon black's structure during mixing as a result of lower shear forces. GEN: J. Pyne, "Processing Report, Conductive Rubbers Advance with New Blacks," *European Rubber Journal,* November, 1981, p. 17.

Conductive Silicones

Special grades of "electrically conductive silicones" are commercially available. GEN: J. Pyne, "Processing Report, Conductive Rubbers Advance with New Blacks," *European Rubber Journal,* November, 1981, p. 17; R. Norman, "Conductive Rubber's Growing Applications," *European Rubber Journal,* November, 1981; SWS Silicone Corporation Brochure, "Electrically Conductive Silicones for EMI/RFI Shielding."

Carbon Black Loading

Increasing the loading of the carbon black used in a compound will usually increase electrical conductivity. RT: Chapter 12, "Compounding with Carbon Black and Oil," S. Laube, S. Monthey, M-J. Wang, p. 308.

Increasing the loading level of a conductive carbon black will usually decrease the electrical resistivity (improve electrical conductance) of the rubber compound. GEN: R. Juengel, "Carbon Black Selection for Conductive Rubber Compounds," *Rubber World,* September, 1985, p. 30.

With an increase in carbon black loading, there is a minimum loading level called the "percolation threshold" at which the rubber compound's electrical resistivity begins to decrease rapidly with a corresponding increase in electrical conductance. The phr concentration for the occurrence of this percolation threshold changes with different grades of carbon black. Carbon black grades that have higher surface areas will most likely have percolation thresholds at lower phr (concentration) than carbon black grades with lower surface area. However, this same inverse correlation does not appear to exist for grades of carbon black with different structure. This percolation threshold is also somewhat dependent on the type of rubber used and its wettability in relation to the carbon black. GEN: C. O'Farrell, M. Gerspacher, L. Nikiel, "Electrical Resistivity of Rubber Compounds Role of Carbon Black," ITEC '98 Select, p. 71; *Carbon Black,* Chapter 8, "Conducting Carbon Black," N. Probst, p. 271.

Carbon Black Surface Area

Increasing the surface area of the carbon black used in a compound will usually increase electrical conductivity, presumably because smaller carbon black particle size means smaller size aggregates. With smaller aggregates dispersed in the rubber matrix, there is less distance among the aggregates and a better opportunity for electron transfer from one aggregate to another. RT: Chapter 12, "Compounding with Carbon Black and Oil," S. Laube, S. Monthey, M-J. Wang, p. 308. RP: M-J. Wang.

Carbon Black Structure

Using a carbon black with higher structure may increase electrical conductivity. It is thought that this is true because the higher the structure is, the more irregular will be the shape of the aggregates on average, which will also afford more opportunities for potential paths for electron transfer in the rubber compound matrix. GEN: R. Mastromatteo, E. Morrisey, M. Mastromatteo, H. Day, "Matching Material Properties to Application Requirements," *Rubber World*, February, 1983, p. 26; *Carbon Black*, Chapter 8, "Conducting Carbon Blacks," N. Probst, p. 276. RP: M-J. Wang.

It is reported that some developed conductive carbon blacks tend to have higher structures and sometimes wider particle size distributions. GEN: J. Pyne, "Processing Report," *European Rubber Journal*, November, 1981.

Carbon Black Particle Surface

Carbon blacks with high levels of extractables or oxygenated surfaces may not impart as high electrical conductivity characteristics. It is reported that high volatile content may insulate carbon black aggregates from electron transfer. Thus higher carbon black surface oxidation may lower the electrical conductivity of the compound. Therefore, carbon blacks with lower volatile content may be considered for better electrical conductivity. GEN: N. Probst, "Conductive Carbon Blacks," *Rubber Technology* '97; J. Accorsi, E. Romero, *Plastics Engineering*, April, 1995. RP: M-J. Wang.

Avoid using a carbon black–rubber coupling agent or "chemical promotor" in a carbon black loaded rubber compound. These promotors may actually reduce the electrical conductivity of the rubber compound, probably because they will reduce the aggregate–aggregate interaction of the carbon black loading. GEN: L. Gonzalez, A. Rodriguez, J. deBenito, A. Marcos, "A New Carbon Black–Rubber Coupling Agent to Improve Wet Grip and Rolling Resistance of Tires," *Rubber Chemistry and Technology*, May–June, 1996, Vol. 69, p. 266.

Types of Carbon Black

Generally speaking, a specially made carbon black for improving electrical conductivity would generally possess a finer particle size, a higher structure, a lower volatile content, and a higher porosity. GEN: *Carbon Black*, Chapter 8, "Conducting Carbon Black," N. Probst, p. 271. RP: M-J. Wang.

In one compound comparison, it was reported that SCF black was the most conductive, followed by N472 and N293, followed by N294, followed by acetylene black as the least conductive. GEN: R. Juengel, "Carbon Black Selection for Conductive Rubber Compounds," *Rubber World*, September, 1985, p. 30.

Graphite

While graphitized carbon black imparts poorer reinforcement to a vulcanizate, it also has reduced interaction with the rubber hydrocarbon medium and thus stronger particle-particle interactions, which might enhance to some extent electrical conductivity. Higher carbon black porosity may also help improve compound electrical conductivity. GEN: Joel G. Neilsen, Sed Richardson (Carbon Black Co.), "How Carbon Black Affects Electrical Properties of Rubber Compounds," Presented at a meeting of the Southern Rubber Group, July 13, 2011, Charleston, SC.

Consider expanded graphite blended with other conventional fillers to improve electrical conductivity of a rubber compound. GEN: Thomas Gruenberger, Nicolas Probst (Timcal Belgium SA), "Graphite in Rubber Compounds: New Opportunities with Graphite Products in the Development of Rubber Compounds," Paper No. 102 presented at the Fall Meeting of the Rubber Division, ACS, November 1-3, 2005, Pittsburgh, PA.

Nanotechnology

With the proper use of nanotechnology, graphene and carbon nanotubes can be used to increase electrical conductivity of a rubber compound. GEN: Edmee Files, "The Good, The Bad and the Ugly: Challenges Down Under," Presented at the Winter Meeting of the Energy Rubber Group, January 15, 2009, Houston, TX.

N472 Carbon Black

The historic popularity of N472 carbon black selection for electrical conductance is probably attributable to its fairly good balance between possible electrical conductance and its imparted physical properties. GEN: R. Juengel, "Carbon Black Selection for Conductive Rubber Compounds," *Rubber World*, September, 1985, p. 30.

"Conductive" Carbon Blacks

Some so-called conductive carbon blacks are described as blacks that are able to form a conductive carbon network at basically low concentrations. In some cases, the structure of these special blacks is very important. GEN: N. Probst, "Conductive Carbon Blacks," *Rubber Technology* 1997.

Carbon Black Particle Size Distribution

Selecting carbon black grades that have broader particle size distributions may increase a rubber compound's electrical conductance because the smaller particle fraction coalesces to smaller agglomerates and their greater number causes "smaller gap width" among agglomerates. GEN: A. Sircar, T. Lamond, "Effect of Carbon Black Particle Size Distribution on Electrical Conductivity," *Rubber Chemistry and Technology*, March-April, 1978, No. 51, p. 126.

Ketjenblack

Ketjenblack EC is reported to be one of the blacks that imparts higher electrical conductance to the vulcanizate than many other so-called conductive carbon blacks. Ketjenblack is reported to possess a DBP absorption number of 320 ml/100 g, a surface area of 950 m^2/g, and a particle size of 30 nm. This higher surface area is not necessarily a result of smaller particle size. Sometimes these very conductive blacks with ultrahigh surface area actually have particles that are closer to "hollow" spheres, not truly solid spheres. GEN: J. Pyne, "Processing Report," *European Rubber Journal*, November, 1981.

Other Conductive Blacks

Reportedly using approximately 50 phr of XC-72 or acetylene black may improve electrical conductivity of the compound to the 10 to 1000 Ohm-cm range (semiconductivity). Also, this loading might possibly not result in major impairments of other physical properties. RP: R. J. Del Vecchio.

Silica-Loaded Compound

Rubber compounds loaded with silica as the only filler will generally be subject to static electricity buildup in dynamic applications (such as tires or belts) because the silica is a poor conductor of electricity. However, adding a few parts of a conductive carbon black to this silica-filled compound may help reduce the occurrence of this electrostatic charge. RP: T. D. Powell.

Mixing

The degree of carbon black dispersion (the compound state-of-mix) will affect compound electrical conductance for better or worse. GEN: R. Juengel, "Carbon Black Selection for Conductive Rubber Compounds," *Rubber World*, September, 1985, p. 30.

Increasing the carbon black dispersion by applying more work history or mixing for a longer time can either increase or decrease the electrical conductance of the compound depending on what type of carbon black is being used and what its loading level is in relation to its "percolation threshold" level. GEN: C. O'Farrell, M. Gerspacher, L. Nikiel, "Electrical Resistivity of Rubber Compounds Role of Carbon Black," *ITEC '98 Select*, p. 71.

To achieve good electrical conductance for a rubber compound, reportedly you need to achieve a homogeneous distribution of the carbon black in the rubber compound. On the other hand, when the carbon black is submitted to high mechanical shearing in the mixer, the secondary structure, which consists of the agglomerates, is destroyed, which may cause the electrical conductivity to decrease. Therefore, there may be an optimum during a rubber mixing cycle to achieve the highest electrical conductivity (the optimal balance between homogenization that usually increases electrical conductivity, and destruction of the agglomeration or secondary structure, which may decrease electrical conductivity). GEN: *Carbon Black*, Chapter 8, "Conducting Carbon Black," N. Probst, p. 279. RP: M-J. Wang.

Flexing Vulcanizate

Distortion of the vulcanizate will affect electrical conductance. In addition, flexing of the vulcanizate may increase its electrical resistivity compared to its previous unstressed state. GEN: R. Juengel, "Carbon Black Selection for Conductive Rubber Compounds," *Rubber World*, September, 1985, p. 30.

Deformations, that is, straining and flexing of a rubber compound, will affect its electrical conductance. GEN: J. Pyne, "Processing Report, Conductive Rubbers Advance with New Blacks," *European Rubber Journal*, November, 1981, p. 17.

One study that stretched and relaxed carbon black loaded compounds based on different elastomers ranked butyl rubber with the least loss in conductivity, followed by CR, NR, EPDM, and SBR, with NBR ranking the worst for loss in electrical conductivity. GEN: J. Pyne, "Processing Report, Conductive Rubbers Advance with New Blacks," *European Rubber Journal*, November, 1981, p. 17.

Compression

Electrical conductivity may increase or decrease with an increase in compression depending on such factors as the type of base elastomer, its viscosity, and so forth. For example, if the compound is based on a high-viscosity, stiff EPDM, the electrical conductance of the compound may decrease with a rise in applied pressure. On the other hand, if the compound is based on a low-viscosity NBR, the electrical conductance might increase with a rise in applied pressure. Thus changes in electrical conductivity, whether up or down from increasing pressure applied, are very dependent on many compound factors such as applied strain amplitude, amount of filler loading, type of filler loading, type of rubber, and so forth, making generalizations very difficult. GEN: K. Sau, T. Chaki, D. Khastgir,

"The Effect of Compressive Strain and Stress on Electrical Conductivity of Conductive Rubber Composites," *Rubber Chemistry and Technology*, May-June, 2000, Vol. 73, p. 310.

According to Thompson and Allen (1994), molding a rubber part at higher pressures reportedly can reduce the resistivity (increase electrical conductivity) for certain types of rubber compounds filled with carbon black. However, Thompson, et al. (1988) have reported just the opposite effect for different compounds. Therefore, molding pressure may affect the ultimate electrical conductivity or resistivity; however, it is not always easy to predict the direction of change. GEN: C. Thompson, J. Allen, "The Effect of Mold Pressure on the Electrical Resistivity of Elastomers," *Rubber Chemistry and Technology*, March-April, 1994, Vol. 67, p. 107; C. Thompson, T. Besuden, L. Beumel, "Resistivity of Rubber as a Function of Mold Pressure," *Rubber Chemistry and Technology*, November-December, 1988, Vol. 61, p. 828.

Metal Powder Loadings

Metal powder loadings can be used in a rubber compound to significantly increase electrical conductance even though these fillers may be detrimental to other compound physical properties. GEN: J. Pyne, "Processing Report, Conductive Rubbers Advance with New Blacks," *European Rubber Journal*, November, 1981, p. 17.

Consider powdered metals as a compounding ingredient to increase both a compound's thermal conductivity and electrical conductivity. Among the metal powders to select from, silver, copper, and aluminum are usually the best conductors. Silver is usually too expensive. Copper is a very strong prooxidant and degradant to elastomers such as natural rubber. Aluminum powder is used to improve both thermal and electrical conductivity; however, the use of this material can pose special safety problems such as fire hazards and so forth. In fact, the use of aluminum powder or some other metallic powders in rubber can make the rubber compound highly flammable. In addition, reportedly steel wool has been used in polyurethane foams to improve antistatic properties. GEN: V. Vinod, S. Varghese, R. Alex, B. Kuriakose, "Effect of Aluminum Powder on Filled Natural Rubber Composites," *Rubber Chemistry and Technology*, May-June, 2001, Vol. 74, p. 236.

Silver-Coated Glass Microspheres

Loading a rubber compound with silver-coated microspheres should increase electrical conductivity. RP: R. J. Del Vecchio.

Direction of Vulcanizate Grain

The direction of a vulcanizate's "grain" can also affect its electrical conductance. GEN: J. Pyne, "Processing Report, Conductive Rubbers Advance with New Blacks," *European Rubber Journal*, November, 1981, p. 17.

Antistatic Agents and Humidity

Some "antistatic agents" are reported to be very sensitive to changes in humidity. GEN: R. Norman, "Conductive Rubber's Growing Applications," *European Rubber Journal*, November 1981, p. 21.

Oxidized Carbon Black Loaded EPDM

Controlled thermo-oxidation and degradation of a carbon black loaded EPDM compound with the development of oxidized surface layers can result in higher electrical conductivity. GEN: B. Mattson, B. Stenberg, "Electrical Conductivity of Thermo-oxidative Degraded EPDM Rubber," *Rubber Chemistry and Technology*, May-June, 1992, Vol. 65, p. 315.

Temperature

The temperature of a rubber compound has an effect on its electrical conductivity. However, different rubber compounds show different changes in electrical conductivity to a given temperature change. There is no established theory yet to explain these different responses. GEN: A. Voet, "Temperature Effect of Electrical Resistivity of Carbon Black Filled Polymers," *Rubber Chemistry and Technology*, March-April, 1981, Vol. 54, p. 42.

2.11 Increasing Thermal Conductivity

Thermal conductivity of a rubber compound can be very important, especially for products performing dynamically with thick cross-sections. This property partially determines how fast a given rubber product can dissipate heat generated internally from dynamic deformations while in service. It is also important in determining the performance efficiency of such products as rubber curing bladders. Thermal conductivity can even affect the heat history of a rubber compound during processing or curing in the factory.

*The following **experimental** ideas can be considered for increasing thermal conductivity of a cured rubber. Also, all relevant literature sources, including the ones cited below, should be researched and read. **Caution:** these general experimental ideas may not work in all specific situations. Changes to increase thermal conductivity will certainly affect other properties as well, for better or for worse, and this book does not purport to show how these other properties are affected. Also, this book does not purport to address safety and health issues.*

High Loading of Low-Structure Carbon Blacks

Reportedly, thermal conductivity usually increases proportionally to an increase in carbon black loading. GEN: *Carbon Black*, Chapter 8, N. Probst, p. 285. RP: M-J. Wang.

Consider using one of the new ultra-low-structure, semireinforcing carbon blacks, which can be used at a higher loading to increase thermal conductivity. GEN: S. Bussolari, S. Laube, "A New Cabot Carbon Black for Improved Performance in Peroxide Cured Injection Molded Compounds," Presented at ACS Rubber Div. Meeting, Fall, 2000, Paper No. 98.

Types of Carbon Black

Acetylene black reportedly imparts higher thermal conductivity to a rubber compound than furnace carbon blacks because of the graphiticity of acetylene blacks. GEN: *Carbon Black*, Chapter 8, N. Probst, p. 285. RP: M-J. Wang.

Carbon Black/Silica Ratio

Some rubber compounds use both carbon black and silica loadings to achieve a given balance of dynamic properties. If the concentration of carbon black is increased relative to the silica concentration, thermal conductivity should improve. GEN: J. Dick, H. Pawlowski, "Applications of the Rubber Process Analyzer in Characterizing the Effects of Silica on Uncured and Cured

Compound Properties," Paper No. 34 presented at the Spring Meeting of the Rubber Division, ACS, May 4–8, 1996, Montreal, Canada.

New Class of Thermally Modified Carbon Blacks

Consider a new series of thermally modified carbon blacks that possess different levels of graphitization over conventional carbon black in order to increase thermal conductivity. These carbon blacks were introduced to butyl tire curing bladders in order to improve their thermal conductivity performance. GEN: W. Wang, R. Lamba, C. Herd, D. Tandon, C. Edwards (Columbian Chemical), "A New Class of High Performance Carbons for Improved Thermal Conductivity and Service Life in Bladder Compounds," Paper No. 15 presented at the Spring Meeting of the Rubber Division, ACS, May 16–18, 2005, San Antonio, TX.

Powdered Metals

Consider powdered metals as a compounding ingredient to improve a compound's thermal conductivity. Among the metal powders to select from, silver, copper, and aluminum are usually the best conductors. Silver is usually too expensive. Copper is a very strong prooxidant and degradant to elastomers such as natural rubber. Aluminum powder is used to improve both thermal and electrical conductivity; however, the use of this material can pose special safety problems such as fire hazards and so forth. In fact, the use of aluminum powder or some other metallic powders in rubber can make the rubber compound highly flammable. GEN: V. Vinod, S. Varghese, R. Alex, B. Kuriakose, "Effect of Aluminum Powder on Filled Natural Rubber Composites," *Rubber Chemistry and Technology*, May–June, 2001, Vol. 74, p. 236.

Graphite

Consider expanded graphite blended with other conventional fillers to improve thermal conductivity of a rubber compound. GEN: Thomas Gruenberger, Nicolas Probst (Timcal Belgium SA), "Graphite in Rubber Compounds: New Opportunities with Graphite Products in the Development of Rubber Compounds," Paper No. 102 presented at the Fall Meeting of the Rubber Division, ACS, November 1–3, 2005, Pittsburgh, PA.

Pyrolytic Graphite Fibers

Consider using fibers of pyrolytic graphite to increase the thermal conductivity of a rubber compound. GEN: C. Ettles, J. Shen, "The Influence of Frictional Heating on the Sliding Friction of Elastomers and Polymers," *Rubber Chemistry and Technology*, March–April, 1988, Vol. 61, p. 119.

PTFE Additive

Use a PTFE additive such as Alphaflex® to try to increase the thermal conductivity of a rubber compound. GEN: J. Menough, "A Special Additive," *Rubber World*, May, 1987, p. 12.

Nanotechnology

With the proper use of nanotechnology, graphene and carbon nanotubes can be used to increase thermal conductivity of a rubber compound. GEN: Edmee Files, "The Good, The Bad and the Ugly: Challenges Down Under," Presented at the Winter Meeting of the Energy Rubber Group, January 15, 2009, Houston, TX.

2.12 Decreasing Coefficient of Friction

In some product applications the coefficient of friction can be important.

*The following **experimental** ideas may decrease the coefficient of friction for a rubber product. Also, all relevant literature sources, including the ones cited below, should be researched and read. **Caution:** these general experimental ideas may not work in all specific situations. Changes to decrease the coefficient of friction will certainly affect other properties as well, for better or for worse, and this book does not purport to show how these other properties are affected. Also, this book does not purport to address safety and health issues.*

Molybdenum Disulfide, Teflon, or Silicone Additives to PU Formulations

In polyurethane formulations, molybdenum disulfide, fluorocarbons (Teflon®), or special silicone oils are sometimes used to lower friction properties and wear characteristics. RT: Chapter 9, "Polyurethane Elastomers," R. W. Fuest, p. 252.

TPU

If considering the use of thermoplastic elastomers (TPEs), consider thermoplastic polyurethanes (TPUs), which possess very low coefficients of friction against other surfaces. RT: Chapter 10, "Thermoplastic Elastomers," C. P. Rader, p. 271.

Chlorination of NR

Consider chlorinating natural rubber products. The process of chlorinating the surface of a NR vulcanizate will significantly reduce its surface friction and tackiness. This chlorination can also be applied to the surfaces of some other articles made of certain synthetic elastomers. Often this chlorination is achieved from immersion in dilute sodium hypochlorite solution. All appropriate safety procedures should be followed for any chlorination procedure. GEN: C. Extrand, A. Gent, "Contact Angle and Spectroscopic Studies of Chlorinated and Unchlorinated Natural Rubber Surfaces," *Rubber Chemistry and Technology*, September-October, 1988, Vol. 61, p. 688; J. Sommer, *Elastomer Molding Technology*, Elastech, Hudson, OH, 2003.

Filler Loadings

Increasing the hardness of a rubber compound by using more reinforcing filler will generally reduce the cured coefficient of friction of the compound. RP: J. R. Halladay.

2.13 Decreasing Air Permeability

A rubber compound's resistance to air permeability can be very important in certain applications such as pneumatic tires. In such applications, the ability to prevent contained air from escaping can be very important.

The following ***experimental*** *ideas may decrease the rate of air permeability through a rubber compound. Also, all relevant literature sources, including the ones cited below, should be researched and read.* ***Caution:*** *these general experimental ideas may not work in all specific situations. Changes to decrease air permeability will certainly affect other properties as well, for better or for worse, and this book does not purport to show how these other properties are affected. Also, this book does not purport to address safety and health issues.*

Butyl Rubber

Base a compound on either butyl or halobutyl elastomer for greatly improved air permeability resistance compared to other general purpose elastomer bases. RT: Chapter 3, "Vulcanizate Physical Properties, Performance Characteristics, and Testing," J. S. Dick, p. 57; Chapter 6, "Elastomer Selection," R. School, p. 132; Chapter 8, "Specialty Elastomers," G. Jones, D, Tracey, A. Tesler, p. 175.

ENR

Use more epoxidized natural rubber to increase polarity and lower permeability to gases. RT: Chapter 7, "General Purpose Elastomers and Blends," G. Day, p. 144.

NBR

Use a grade of NBR with a high level of ACN to achieve high air permeability resistance for a compound. RT: Chapter 8, "Specialty Elastomers," M. Gozdiff, p. 194.

Compounds based on NBR grades with 40% ACN reportedly have about the same permeability rates as some butyl rubber formulations for air and other gases. NBR grades with even higher ACN content can impart superior permeability resistance compared with some butyl compounds. GEN: "A Comparative Evaluation of Hycar Nitrile Polymers," Manual HM-1, B. F. Goodrich Chemical Co.

ECO

Compounds based on polyepichlorohydrin (because of their high polarity) have very good resistance to air permeability. Hydrin 100 reportedly has air permeability resistance that is better than that of butyl rubber compounds.

RT: Chapter 8, "Specialty Elastomers," C. Cable, p. 216. GEN: "Designing with Elastomers," B. F. Goodrich Chemical Co.

BIMSM

Brominated isobutylene-co-*para*-methylstyrene (BIMSM) is the latest isobutylene copolymer to be used to improve air permeability resistance. GEN: W. Waddell, R. Napier (ExxonMobil Chemical Co.), "Polymers for Innerliner Applications: New Developments," Paper No. 122 presented at the Fall Meeting of the Rubber Division, ACS, October 13–15, 2009, Pittsburgh, PA.

New TPE for Tire Innerliner

ExxonMobil has developed a new thermoplastic vulcanizate (TPV) based on a new alloy of BIMSM (brominated isobutylene-co-*para*-methylstyrene) and nylon that can be blow molded directly into the tire to form a more efficient innerliner that is lighter weight with better air permeability resistance. GEN: D. Tracey, A. Tsou (ExxonMobil Chemical Co.), "DVA Innerliners," Paper No. 41 presented at the Fall Meeting of the Rubber Division, ACS, October 10–12, 2006, Cinncinati, OH.

Millable Polyurethane

It is reported that some polyester grades of millable polyurethane can give gas permeation resistance close to that of IIR (butyl rubber). GEN: Thomas Jablonowski (TSE), "Millable Polyurethane Rubber," Presented at the Southern Rubber Group, February 28, 2012, Asheville, NC.

Talc

Consider using talc as a filler in a rubber compound where an improvement in air permeability resistance is needed. RT: Chapter 13, "Precipitated Silica and Non-black Fillers," W. Waddell, L. Evans, p. 328. RP: O. Noel.

Special Glass Flakes

Consider evaluating a low loading of special glass flakes with only a moderate amount of work history during mixing (to avoid excessive breakdown of the glass flakes while in the mixer) and aligned in the calendering direction in downstream processing to increase orientation of the flakes and increase the diffusion path to inhibit gas permeability. (It is uncertain how this affects dynamic durability.) GEN: S. Fulton, D. Mason (NGF Europe, Ltd.), "Improved Barrier Properties of Tires by the Incorporation of Glass Platelets," Presented as Paper No. 9B at the International Tire Expo and Conference (ITEC), September 16, 2008, Akron, OH.

Expandable Graphite

Very special mixing of graphite and expanded graphite with some rubbers might help improve permeability resistance. GEN: R. Faulkner, K. Mumby, "Comparison of Graphite and Graphene Precursors in HNBR," Paper No. 33 presented at the Fall Meeting of the Rubber Division, ACS, October 13–15, 2009, Pittsburgh, PA.

Brominated isobutylene-co-*para*-methylstyrene (BIMSM)/nanoclay composites have been shown to have better air permeability resistance than traditional tire innerliner compounds. GEN: J. Soisson, B. Rodgers, W. Weng, R. Webb, S. Jacob (ExxonMobil Chemical Company), "Vulcanization of Nanocomposite Tire Innerliner Compounds and Permeability," Paper No. 112 presented at the Fall Meeting of the Rubber Division, ACS, October 13–15, 2009, Pittsburgh, PA.

Carbon Black

Increasing the loading of the carbon black used in a compound will usually increase permeability resistance. RT: Chapter 12, "Compounding with Carbon Black and Oil," S. Laube, S. Monthey, M-J. Wang, p. 308. GEN: R. R. Juengel, D. C. Novakoski, S. G. Laube, ITEC, October, 1994. RP: M-J. Wang.

Sometimes, thermal black, with very large particle size and very low structure, can be used to reduce air permeability mainly because of its higher loading capability in a rubber compound. GEN: R. R. Juengel, D. C. Novakoski, S. G. Laube, ITEC, October, 1994. RP: M-J. Wang.

Decreasing the surface area (increasing the particle size) of the carbon black used in a compound will usually increase permeability resistance. RT: Chapter 12, "Compounding with Carbon Black and Oil," S. Laube, S. Monthey, M-J. Wang, p. 308.

Use the new ultra-low-structure semireinforcing carbon blacks to achieve higher carbon black loadings, which will improve air permeability resistance. GEN: S. Bussolari, S. Laube, "A New Cabot Carbon Black for Improved Performance in Peroxide Cured Injection Molded Compounds," Presented at ACS Rubber Div. Meeting, Fall, 2000, Paper No. 98. Fine particle size, low-structure carbon blacks that may be very expensive to mix in a natural rubber based compound may be mixed faster and more economically with QDI from Flexsys. GEN: F. Ignatz-Hoover, Presented at ACS Rubber Div. Meeting, Fall, 2002, Paper No. 106; ACS Rubber Div. Meeting, Fall, 2003, Paper No. 98. RP: F. Ignatz-Hoover.

Oil Loading

Avoid high oil loadings. Often higher oil loadings will hurt air permeability resistance. RP: R. Schaefer.

2.14 Increasing Rubber-to-Metal Adhesion

The improvement of cured rubber-to-metal adhesion is a separate science in and of itself. It is sometimes very difficult to achieve good rubber-to-metal adhesion. Even when good "original" adhesion values are achieved, the aged adhesion, the corrosion resistance, the humidity aging results, and so forth may still be poor. The original adhesion values cannot predict what the aged adhesion values will be. In addition, standardized lab adhesion tests do not necessarily correlate to actual rubber-to-metal adhesion achieved in the manufacture of the rubber product (RP: R. J. Del Vecchio).

The most common rubber-to-metal adhesion is "rubber-to-brass" adhesion in which usually the "metal" is "brass-plated steel." However, there are other types of rubber-to-metal adhesion as well that are also mentioned below.

The following ***experimental*** *ideas might help in achieving better rubber-to-metal adhesion values. Also, all relevant literature sources, including the ones cited below, should be researched and read.* ***Caution:*** *these general experimental ideas may not work in all specific situations. Changes to increase rubber-to-metal adhesion will certainly affect other properties as well, for better or for worse, and this book does not purport to show how these other properties are affected. Also, this book does not purport to address safety and health issues.*

Natural Rubber

NR is commonly used as the elastomer of choice to achieve good brass wire adhesion. RT: Chapter 20, "Compounding for Brass Wire Adhesion," A. Peterson, p. 473; Chapter 12, "Compounding with Carbon Black and Oil," S. Laube, S. Monthey, M-J. Wang, p. 319.

Cobalt Salts

Cobalt salts are commonly used in rubber formulations to achieve good rubber-to-metal adhesion to brass wire. Cobalt salts reportedly affect the growth of copper sulfide at the wire surface. This helps "anchor" the rubber to the wire. The cobalt salts reportedly can improve both the initial and aged wire adhesion. But increasing cobalt levels can reduce humidity-aged adhesion and affect cure acceleration. In fact, data show that increasing cobalt levels may increase unaged adhesion but decrease humidity-aged adhesion values. Therefore, optimal levels of cobalt, sulfur, and accelerator have to be established for the best balance of properties. The cobalt salts that are commonly used as rubber compounding ingredients in tire wire coat formulations include cobalt naphthenate, cobalt neodecanoate, cobalt stearate, and special organocobalt salts.

RT: *Rubber Technology, Compounding and Testing for Performance.* 2nd ed. Edited by J. S. Dick, Hanser Publications, 2009 ■ GEN: General References from the Literature ■ RP: Review Panel (Advisors for this book) ■ IMPORTANT: The information contained herein is experimental in nature and is meant for individuals with advanced scientific training and rubber compounding experience. All safety precautions should be taken when replicating or applying this information. The author makes no representation or warranty, either expressed or implied.

Comparative evaluations of these salts should be performed on an equal cobalt basis. RT: Chapter 20, "Compounding for Brass Wire Adhesion," A. Peterson, pp. 464-471.

Consider using a small quantity of cobalt boroacylate (Manobond 680C) to further improve rubber to brass plated steel tire cord adhesion. GEN: W. Stephen Fulton (OMG UK Ltd.), "Source of Zinc Can Influence the Structure of the Bonding Interface," Paper No. 84 presented at the Fall Meeting of the Rubber Division, ACS, October 5-8, 2004, Columbus, OH.

RF Resin and HMMM

Resorcinol-formaldehyde (RF) resins with HMMM (for *in situ* crosslinking with the RF resin during vulcanization) are usually used in combination with cobalt salt to improve initial adhesion and preserve aged adhesion by protecting the system from moisture attack over time. RT: Chapter 20, "Compounding for Brass Wire Adhesion," A. Peterson, pp. 466, 471-472.

Higher Sulfur to Accelerator Ratio

When compounding for good brass wire adhesion, generally a higher than normal loading of insoluble sulfur with a lower level of accelerator is used. This usually ensures good formation of copper sulfide (Cu_xS) at the surface of the brass wire. RT: Chapter 20, "Compounding for Brass Wire Adhesion," A. Peterson, p. 464.

DCBS

One of the most popular accelerators for a brass wire adhesion compound is selecting N,N-dicyclohexyl-2-benzothiazyl sulfenamide (DCBS) accelerator, which is slower curing than many other sulfenamides. Increasing the concentration of sulfur/DCBS (in phr) reportedly will also increase unaged and humidity-aged adhesion values. RT: Chapter 20, "Compounding for Brass Wire Adhesion," A. Peterson, pp. 469-470. GEN: T. Kleiner, L. Ruetz, "DCBS: An Accelerator for Adhesion Compounds and Other Tire Applications," *Rubber World*, November, 1996, p. 34.

Silica

Precipitated silica is commonly used as a partial replacement for carbon black in a brass adhesion formulation because it reportedly enhances the formation of ZnO at the interface and improves aged adhesion. When some of the carbon black is replaced with precipitated silica, an increase in unaged and humidity-aged adhesion values is observed. RT: Chapter 20, "Compounding for Brass Wire Adhesion," A. Peterson, pp. 466, 472.

N326 Carbon Black

N326 is reported to be almost the universally accepted carbon black of choice for wire coat compounds. Supposedly this carbon black is chosen by many because it imparts good green strength and good reinforcement at a low compound viscosity level (and phr level) to achieve good penetration into the brass-plated cord. RT: Chapter 12, "Compounding with Carbon Black and Oil," S. Laube, S. Monthey, M-J. Wang, p. 319; Chapter 20, "Compounding for Brass Wire Adhesion," A. Peterson, p. 473.

Effect of Stearic Acid/Zinc Oxide

An excessive level of stearic acid in a wire coat compound has been reported to decrease humidity-aged adhesion, especially in compounds that contain a large level of cobalt naphthenate. GEN: Y. Ishikawa, "Effects of Compound Formulation on the Adhesion of Rubber to Brass-Plated Steel Cord," *Rubber Chemistry and Technology*, Vol. 57, 1984, pp. 855–878. RP: M. A. Lawrence.

Excess stearic acid reportedly has a negative effect on wire coat adhesion because it is corrosive to brass. Thus the zinc oxide film that normally forms on the brass surface is actually attacked and dissolved by excess stearic acid. To avoid this, the stearic acid should be consumed very quickly in the vulcanization process. Therefore, the zinc oxide grade that is selected should be a highly reactive grade that will react quickly with the stearic acid. In addition, the ratio of ZnO to stearic acid should be somewhat high. GEN: W. J. van Ooij, "Mechanism and Theories of Rubber Adhesion to Steel Tire Cords: An Overview," *Rubber Chemistry and Technology*, Vol. 57, 1984, p. 451.

Effect of Cure Conditions

It has been observed that by increasing cure temperature from 130 °C (266 °F) to 190 °C (374 °F), the pull-out force after cure decreases linearly. GEN: G. S. Jeon, G. Seo, "Influence of Cure Conditions on the Adhesion of Rubber Compound to Brass-Plated Steel Cord: Part I, Cure Temperature," *Journal of Adhesion*, Vol. 76, pp. 201–221. RP: M. A. Lawrence.

Copper/Zinc Plating Composition and Thickness

Typically for good brass wire adhesion to rubber, brass-plated steel cords are used with copper/zinc plating 0.1 to 0.5 µm in thickness and a chemical composition of 60 to 70% copper. RT: Chapter 20, "Compounding for Brass Wire Adhesion," A. Peterson, pp. 464–465.

Optimizing the Levels of Sulfur, Cobalt, Resorcinol Resin, HMMM, Silica, and Accelerator

One can optimize the levels of sulfur, cobalt, and resorcinol resin in a wire coat stock to maximize the unaged and aged rubber adhesion to brass-plated steel cord. From the reported completion of a design of experiment (DOE), an optimized rubber formulation was found that maximized adhesion. In this one situation, an optimal combination was reported to be 4.0 parts Penacolite® Resin B20-S (resorcinol resin), 0.45 parts Manobond® 680C (0.1 parts active cobalt), 7.1 parts sulfur, and 2.66 parts active hexamethoxymethylmelamine (HMMM). GEN: M. A. Lawrence, J. de Almeida, "Maximize Steel Cord Adhesion with Resorcinol Formaldehyde Resins," *Tire Technology International*, 2001, pp. 58–61. RP: Mark A. Lawrence.

In addition, replacing some of the carbon black with silica in a wire coat compound can reportedly increase adhesion to brass-plated steel tire cord. From another DOE, adhesion was found to be maximized with a combination of 25 parts silica, 42.6 parts carbon black, 1.2 parts cobalt neodecanoate, 4 parts sulfur, and 0.5 parts TBBS accelerator. GEN: L. R. Evans, W. H. Waddell, *Rubber World*, June 1997, pp. 22–28. RP: M. A. Lawrence.

Air Diffusion

It is known that through normal use of a steel belted radial tire, the inflating air can permeate the crown region, thus promoting an oxidative condition with the belt-skim rubber compound. The diffused air and higher operating temperatures can change the physical properties of a belt-skim compound. By selecting the proper compounding formulations and tire construction design, these degrading conditions may be minimized. GEN: H. Kaidou, A. Ahagon, "Aging of Tire Parts During Service. II. Aging of Belt-Skim Rubbers in Passenger Tires," *Rubber Chemistry and Technology*, November–December, 1990, Vol. 63, No. 5, p. 698.

Coagent with Peroxide Cure for Rubber-to-Metal Adhesion

Using a coagent with a peroxide cure may improve rubber-to-metal adhesion. GEN: P. Dluzneski, "Peroxide Vulcanization of Elastomers," *Rubber Chemistry and Technology*, July–August, 2001, Vol. 74, p. 451.

In some cases, increasing the concentration of zinc diacrylate metallic coagent (Saret® 633) will increase a given peroxide-cured rubber compound's rubber-to-metal adhesion strength for aluminum, zinc, brass, or steel surfaces. GEN: R. Costin, W. Nagel, "Coagents for Rubber-to-Metal Adhesion," *Rubber & Plastics News*, March 11, 1996, p. 14.

Polychloroprene

Good adhesion of a polychloroprene (CR) compound to brass is achievable when the CR compound contains as little as only 0.5 phr of sulfur (unlike NR compounds that might require a minimum of 3 phr to achieve good rubber-to-metal adhesion to brass). GEN: G. Hamed, F. Liu, "The Bonding of Polychloroprene to Brass: Rate and Temperature Effects," *Rubber Chemistry and Technology*, November–December, 1984, Vol. 57, p. 1036.

Level of Antiozonants Used

Sometimes too much antiozonant, or the selection of antiozonant, such as IPPD, 77PD, or 6PPD, may affect adhesion for better or for worse. GEN: S. Chaudhuri, J. Halladay, P. Warren (Lord Corp.), "The Effect of Antiozonants on Rubber-to-Metal Adhesion," Presented at the India Rubber Expo, January 19–22, 2011, Chennai, India; J. Halladay, P. Warren, (Lord Corp.), "The Impact of Antiozonants on Rubber-to-Metal Adhesion, Part 2," Paper No. 50 presented at the Fall Meeting of the Rubber Division, ACS, October 11–13, 2011, Cleveland, OH.

Precoating Steel Tire Cord

Consider precoating steel cord with 2-amino-4,6-dimethoxypyrimidine to help improve adhesion strength. GEN: M. Kim, Y. Kim, B. Sohn, M. Han (Hyosung Co. and Kumho Industrial Co.), "8C: Adhesion Improvement between Brass Plated Steel Cord and Rubber Compound by Coating the Organic Adhesion Promoter," Paper No. 8C presented at ITEC 2002, September, 2002, Akron, OH.

Surface Preparation of Metal Substrate

To achieve good rubber-to-metal adhesion, the metal surface should first be cleaned and properly prepared. RP: R. L. Del Vecchio.

2.15 Improving Rubber-to-Fabric Adhesion

Rubber-to-fabric adhesion is very important to prevent product failure while in service.

The following ***experimental*** *ideas from the literature may help the rubber technologist establish and maintain good rubber-to-fabric adhesion. Also, all relevant literature sources, including the ones cited below, should be researched and read.* ***Caution:*** *these general experimental ideas may not work in all specific situations. Changes to improve rubber-to-fabric adhesion will certainly affect other properties as well, for better or for worse, and this book does not purport to show how these other properties are affected. Also, this book does not purport to address safety and health issues.*

Polyester Adhesion to Rubber

If one wishes to maintain good polyester adhesion, it is important that the rubber compound not contain curatives or other compounding ingredients that will promote aminolysis or hydrolysis of the ester bonds in the PET tire cord. GEN: Y. Shindada, D. Hazelton, "Polyester in Reinforced EPDM: Factors Affecting Thermal Degradation," *Rubber Chemistry and Technology*, May–June, 1978, Vol. 51, p. 253.

RF Resin/Methylene Donor

The addition of a resorcinol-formaldehyde resin with a methylene donor to a rubber compound may significantly improve its adhesion to nylon or polyester. GEN: "Penacolite Resins Dry Bonding Systems for Adhesion to Organic Fibers," Technical Literature, INDSPEC Chemical Corp. RP: M. A. Lawrence.

RFL-Treated Fabric

To maintain good rubber-to-fabric adhesion, minimize the exposure of the RFL-treated fabric to oxygen, ozone, nitrogen oxides, and ultraviolet light. Exposure to these environmental factors during storage can significantly degrade fabric adhesion. GEN: T. S. Solomon, "An Overview of Tire Cord Adhesion," Presented at ACS Rubber Div. Meeting, Fall, 1983, Paper No. 57, p. 10.

Reportedly, better rubber-to-nylon adhesion can be achieved by using carboxylated nitrile (XNBR) that has been hydrogenated into HXNBR for oil field applications. GEN: John E. Dato (Lanxess), "Hydrogenated Nitrile Rubber for Use in Oilfield Applications," Paper No. 57 presented at the Fall Meeting of the Rubber Division, ACS, October 10–12, 2006, Pittsburgh, PA.

2.16 Improving Flame Retardance

There are several standardized methods for measuring the flame retardancy of a cured rubber compound, and they all give different results. Sometimes they do not even correlate very well with each other. Therefore the compounder should be very careful with his or her assumptions regarding the results of these tests and the actual flame retardancy that a given rubber product might actually possess while in service. Many of these standard lab tests are not perfect predictors of the flame retardancy of a product in service.

*The following **experimental** ideas may improve a compound's flame retardancy. Also, all relevant literature sources, including the ones cited below, should be researched and read. **Caution:** these general experimental ideas may not work in all specific situations. Changes to improve flame retardancy will certainly affect other properties as well, for better or for worse, and this book does not purport to show how these other properties are affected. Also, this book does not purport to address safety and health issues.*

Silicone

Compounds based on silicone or fluorosilicone rubber can be made to have some degree of flame retardancy. Improvements in flame retardancy for a silicone compound may be achieved by substituting the methyl groups in the polymer with phenyl or vinyl groups, by using platinum cures to promote crosslinking, and using zinc borate and/or alumina trihydrate additives in the formulation. RT: Chapter 6, "Elastomer Selection," R. School, p. 136; Chapter 22, "Flame Retardants," K. K. Shen, D. Schultz, p. 500.

CSM

Generally, compounds based on chlorosulfonated polyethylene (CSM) possess some flame resistance. Selecting grades of CSM with higher chlorine content or the addition of antimony oxide should increase the compound's flame retardancy. RT: Chapter 8, "Specialty Elastomers," C. Baddorf, p. 213. RP: K. K. Shen.

CPE

For compounds based on chlorinated polyethylene (CM), select a polymer with higher chlorine content or add antimony oxide/zinc borate as a synergist. RT: Chapter 22, "Flame Retardants," K. K. Shen, D. Schultz, p. 495. RP: K. K. Shen.

NBR/PVC

When NBR/PVC blends are used, a certain degree of flame retardancy may be provided by the presence of the PVC (polyvinyl chloride). Of course other flame retardants are also used. RT: Chapter 22, "Flame Retardants," K. K. Shen, D. Schultz, p. 500.

CR

Compounds based on polychloroprene (CR) and highly loaded with more than 40 phr of hard clay can possess some degree of flame retardancy. Chlorinated plasticizers can also improve flame retardance but cause heavy smoke. RP: L. L. Outzs.

Chlorine Donors

Chlorinated paraffin (from 40 to 70% chlorine) sometimes can function as a plasticizer and serve as a "chlorine donor" to help impart some degree of flame retardancy. RT: Chapter 22, "Flame Retardants," K. K. Shen, D. Schultz, p. 491.

Chlorinated paraffins are thermally less stable than many chlorinated aromatic materials. Therefore, if a compound will be exposed to very high processing temperatures, then a chlorinated alicyclic material such as Dechlorane Plus® may be used instead. RT: Chapter 22, "Flame Retardants," K. K. Shen, D. Schultz, p. 491.

A chlorine donor such as chlorinated paraffin is commonly used in conjunction with antimony oxide, zinc borate, and/or alumina trihydrate for more effective flame retardancy. Sometimes a true synergy can be achieved. RT: Chapter 22, "Flame Retardants," K. K. Shen, D. Schultz, p. 491.

Combinations of chlorinated paraffin with a bromine donor (DBDPO) reportedly have been used together in the same compound. Using these chlorine and bromine donors together might produce a synergistic effect. RT: Chapter 22, "Flame Retardants," K. K. Shen, D. Schultz, p. 499.

Bromine Donors

Bromine donors are sometimes more effective than chlorine donors as flame retardants. In the past, decabromodiphenyl oxide (DBDPO) with 83% bromine had been used as a bromine source for this purpose, typically at 3 phr for every 1 phr of antimony oxide used. Because it is an aromatic halogen material, it is thermally very stable. However, more recently new environmental concerns have arisen regarding the use of DBDPO and its toxicity. Therefore decabromodiphenylethane is being used more and more to replace DBDPO. Also consider using ethylene bis-tetrabromophthalimide when good UV stability and/or less tendency to bloom is needed. RT: Chapter 22, "Flame Retardants," K. K. Shen, D. Schultz, p. 491.

Additives

Magnesium hydroxide or alumina trihydrate can also be used as a flame retardant; however, without halogen sources, very high loadings (100–250 phr) are required to be effective. RT: Chapter 22, "Flame Retardants," K. K. Shen, D. Schultz, pp. 492–493. Coadditives such as polydimethylsiloxane, zinc borate, melamine polyphosphate, and so forth can be added to possibly increase "char strength" and overall fire test performance. RP: K. K. Shen.

Often carbon black is required for reinforcement. However, fine particle carbon blacks do present a special problem because they tend to prolong afterglow combustion. This effect has to be compensated for by the proper use of flame retardant additives such as zinc borate or phosphate. RT: Chapter 22, "Flame Retardants," K. K. Shen, D. Schultz, p. 493.

Phosphate ester plasticizers or chlorinated paraffins are the plasticizers of choice in many rubber compounds where a higher degree of flame retardancy is needed. Most other conventional processing oils and synthetic plasticizers are simply "fuel sources" that obviously are detrimental to flame retardancy. RT: Chapter 22, "Flame Retardants," K. K. Shen, D. Schultz, pp. 492–493.

Inorganic ammonium polyphosphate reportedly has been used in compounds that are based on halogen-free elastomers, but it was not reported how effective this additive is as a flame retardant. RT: Chapter 22, "Flame Retardants," K. K. Shen, D. Schultz, p. 492.

Fillers

Other nonblack fillers, such as calcium carbonate, clay, talc, and silica, may display weak or low levels of flame retardancy at high loadings mainly because they are displacing combustible elastomers; however, caution should be exercised because some of these mineral fillers, such as calcium carbonate, can also reduce the effectiveness of any halogenated flame retardants used because they scavenge hydrogen halides. Although these fillers are inexpensive, they generally are not very effective as flame retardants. Thus if any of these mineral fillers are used with halogen sources, an upward adjustment in the concentration of a halogen donor and other flame retardants might be required to maintain flame retardancy. The use of calcium carbonate fillers with halogen sources should be avoided for this reason. RT: Chapter 22, "Flame Retardants," K. K. Shen, D. Schultz, p. 493.

Mixing

To achieve good flame retardancy properties, it is very important that all of the flame retardant ingredients be properly and uniformly dispersed through the mixing process. Sometimes with halogen-free elastomers, fillers such as alumina trihydrate or magnesium hydroxide, which are commonly used at high loadings, are treated with stearate, silane, or titanate coupling agents to try to achieve good dispersion. RT: Chapter 22, "Flame Retardants," K. K. Shen, D. Schultz, pp. 493–494. One can also consider adding a compatibilizer such as polyolefin grafted with maleic anhydride. RP: K. K. Shen.

Crosslink Density

It is reported that increasing the crosslink density of a vulcanizate containing flame retardants may sometimes possibly improve the flame retardancy of the compound. GEN: H. Kato, H. Adachi, H. Fujita, "Innovation in Flame and Heat Resistant EPDM Formulations," *Rubber Chemistry and Technology*, May–June, 1983, Vol. 56, p. 287.

Some fluoroelastomer-based compounds are considered "self-extinguishing or nonburning in air." GEN: Jim Denham (Dyneon 3M), "Solutions for the Oil and Gas Industry," Presented at a meeting of the Energy Rubber Group, May 27, 2009, Dallas, TX.

2.17 Reducing Compound Cost

In the competitive economic environment of the rubber industry, the cost of a compound can be extremely important for the economic success of a given product line. It may be possible to develop a rubber compound that meets all the customer specifications; however, if it is too expensive, the compound may not be accepted. Also, most rubber products are sold by volume rather than weight (sold by dimensions usually from a mold). Therefore it makes more sense usually to compare the cost of a rubber compound as a "pound-volume cost" rather than a "cost per pound."

The following ***experimental*** *ideas might reduce the economic cost of a rubber compound. Also, all relevant literature sources, including the ones cited below, should be researched and read.* ***Caution:*** *these general experimental ideas may not work in all specific situations. Changes to reduce compounding cost will certainly affect other properties as well, for better or for worse, and this book does not purport to show how these other properties are affected. Also, this book does not purport to address safety and health issues.*

Carbon Black/Oil

Use higher structure carbon blacks with increased oil (provided the oil is compatible with the rubber compound) to maintain modulus levels and reduce cost. RT: Chapter 4, "Rubber Compound Economics," J. Long, p. 76.

Carbon Black Loadability

Consider using a carbon black with a low structure and a low surface area. This carbon black most likely will not only be a less expensive grade, but it will also have a "higher loadability" character, which can also reduce the compound cost (assuming it can be dispersed during the mixing). GEN: S. Monthey, "The Influence of Carbon Blacks on the Extrusion Operation for Hose Production," *Rubber World*, May, 2000, p. 38.

Consider using one of the new ultra-low-structure, semireinforcing carbon blacks that possess higher "loadability." By being able to increase carbon black levels, one can achieve a significant cost savings. GEN: S. Bussolari, S. Laube, "A New Cabot Carbon Black for Improved Performance in Peroxide Cured Injection Molded Compounds," Presented at ACS Rubber Div. Meeting, Fall, 2000, Paper No. 98.

Consider using low surface area and low-structure carbon blacks to load up an expensive polymer without raising the compound viscosity too much. By keeping the compound viscosity reasonably low, the compound can still be injection molded or cured in other ways. RT: Chapter 12, "Compounding with Carbon Black and Oil," S. Laube, S. Monthey, M-J. Wang, p. 321.

Silica

To achieve various compound cured properties, such as low tire rolling resistance and good wet traction, many times it is necessary to use an organosilane coupling agent with a silica loading. Organosilanes are quite expensive. If a lower concentration of organosilane coupling agent can be used and the physical property improvements still achieved, then a significant cost savings can be gained. One way to use a lower concentration of organosilane is to use a special silica with an increased relative amount of hydroxyls on the particle surface and specifically a higher concentration of "geminal hydroxyls," which are reported to be more "couple-able." With a higher concentration of hydroxyls or "more efficient" geminal hydroxyls, less organosilane coupling agent is required to achieve the same enhancement in cured physical properties. GEN: L. Gatti, "Reduced Silane Usage in Wet Traction Oriented Compounds Through High Surface Activity-Reduced Surface Area Highly Dispersible Silica or Through High Density Filler Blends," Presented at ACS Rubber Div. Meeting, Spring, 2001, Paper No. 57.

Reportedly, a 25% reduction in organosilane loading in a low rolling resistance tire tread was achieved by using an emulsion SBR that was hydroxy functionalized with hydroxy propyl methacrylate (HPMA) during polymerization. Organosilane is the more expensive compounding additive. Reducing its concentration results in a significant cost reduction. GEN: V. Monroy, S. Hofmann, R. Tietz (Dow Chemical), "Effects of Chemical Functionalization of Polymers on Tire Silica Compounds: Emulsion Polymers," Paper No. 63A presented at the Fall Meeting of the Rubber Division, ACS, October 16–19, 2001, Cleveland, OH.

Fillers

In white compounds that use titanium dioxide, it may be possible to partially substitute some of the TiO_2 with other less-expensive white fillers such as water-washed clays, calcium carbonate, and optical whiteners and still have sufficient hiding power and whiteness. RT: Chapter 4, "Rubber Compound Economics," J. Long, p. 76.

Using a carbon–silica dual-phase filler system in place of a traditional silica-loaded tread compound could possibly have some economic advantages such as a lower amount of expensive organosilane being required and less heat treatment being required during mixing. On the other hand, the expense of these new dual-phase fillers may possibly offset these cited cost savings advantages. RT: Chapter 12, "Compounding with Carbon Black and Oil," S. Laube, S. Monthey, M-J. Wang, p. 318.

Consider using calcium carbonate as a filler in a rubber compound to significantly reduce compound cost. RT: Chapter 13, "Precipitated Silica and Non-black Fillers," W. Waddell, L. Evans, p. 326.

Consider using clay as a filler in a rubber compound to significantly reduce compound cost. Clays are commonly used at loadings of 20 to 150 phr. RT: Chapter 13, "Precipitated Silica and Non-black Fillers," W. Waddell, L. Evans, p. 327.

In general, air-floated clay and whiting (ground limestone) are two of the most common and least expensive "economic diluents" in rubber compounding that are used to reduce the cost of the compound. GEN: D. Askea (Polymer Valley Co.), "Functional Mineral Fillers in Rubber Compounds," presented at the Winter Meeting of the Southern Rubber Group, February 24, 2009, Asheville, NC.

Sepiolite, a mineral filler that is a form of magnesium silicate from Spain, reportedly can be used as a partial substitute for carbon black (up to 30%) in tire treads to reduce compound costs. It can reportedly improve wet traction. If used with silane coupling agents, it can also possibly improve tire rolling resistance. GEN: L. Hernandez, L. Rueda, C. Anton, "Magnesium Silicate Filler in Rubber Tread Compounds," *Rubber Chemistry and Technology*, September-October, 1987, Vol. 60, p. 606.

Consider partially substituting some of the carbon black with talc. Although talc has a higher density (2.7) than carbon black (1.8), it is usually much less expensive. Therefore, one can substitute one part of carbon black with 1.5 parts talc and still reduce the compound cost. In addition, talc usually provides faster extrusion capability, which can result in a higher output and a further reduction in cost. RP: O. Noel.

Consider using rice husk charcoal as an economic diluent in rubber compounds. GEN: Y. Yamashita, A. Tanaka (Univ. of Shiga Prefecture Hassaka, Hikone, Japan), "Mechanical Property of Rubber Reinforced with Rice Husk Charcoal," Paper No. 18 presented at the Spring Meeting of the Rubber Division, ACS, April 29-May 1, 2002, Savannah, GA.

Consider directionally oriented layered silicate minerals and/or directionally oriented treated short cellulose fibers in rubber tube construction products so as to replace traditional continuous filament constructions in order to achieve a cost savings. GEN: L. Goettler, M. Benes, F. Al-Yamani (University of Akron), "Discontinuous Reinforcement in Rubber Hose Construction and Performance," Presented at the Hose Manufacturer's Conference, August 19, 2008, Independence, OH.

DBP Weighted Averaging Technique

Try substituting thermal blacks with furnace blacks and selected inexpensive nonblack fillers by using what is called "DBP weighted averaging technique." Some of these nonblack fillers to consider are hard and soft clays. GEN: B. Topcik, "MT Black Replacement Compounding," *Rubber World*, November 1973, p. 42; B. H. Topcik, "Compounding with Substitute Materials," *Rubber World*, March, 1977, p. 51; B. H. Topcik, "Compounding Neoprene to Replace Medium Thermal Black Low Compression Set Development," *Rubber World*, May, 1974, p. 84.

Lower Density

Because most rubber products are sold on a constant volume basis (not a constant weight basis), if compound changes can be implemented to lower the density of the rubber compound but still have the same raw material price profile (on a per weight basis), a cost savings can be realized. RT: Chapter 4, "Rubber Compound Economics," J. Long, p. 77.

If a lower density base elastomer (or elastomer blend) can be used in place of a high-density elastomer without a per pound cost penalty, then a lower pound-volume cost may be realized. For example, substituting NBR in place of CR may result in a lower pound-volume cost provided the other needed compound changes do not offset these savings. RT: Chapter 4, "Rubber Compound Economics," J. Long, p. 77.

One-Pass Mix in Place of Two-Pass (Reduction of Multipass Mixes)

If possible, convert two-step mixes to one-step mixes by using energy-controlled mixing techniques and effective processability testing. GEN: J. S. Dick, M. Ferraco, K. Immel, T. Mlinar, M. Senskey, J. Sezna, "Utilization of the Rubber Process Analyzer in Six Sigma Programs to Improve Quality and Reduce Production Costs," Presented at ACS Rubber Div. Meeting, Fall, 2001, Paper No. 15; F. Myers, S. Newell, "Use of Power Integrator and Dynamic Stress Relaxometer to Shorten Mixing Cycles and Establish Scale-up Criteria for Internal Mixers," *Rubber Chemistry and Technology*, May–June, 1978, Vol. 51, p. 180.

Careful use of the prevulcanization inhibitor cyclohexylthiophthalimide (CTP) will enable a compounder to specifically adjust the scorch safety needed for a given compound when using a sulfenamide cure system. Through wise use of this inhibitor, one might be able to convert a two-pass mix into a one-pass mix in which all the ingredients, including curatives, are placed in an internal mixer during a single mix cycle to reduce overall mixing cost. Without an effective use of CTP, the compound might reach scorch prematurely because of the longer mix cycle time, possibly at higher temperatures, which may be needed in the single pass to achieve adequate dispersion. RT: Chapter 15, "Sulfur Cure Systems," B. H. To, p. 387.

Use an internal mixer that has an effective cooling system so that the batch temperature does not rise too quickly due to viscous heating during the mixing process. Rising batch temperature adds to the batch's heat history (contributing to scorch for a final or one-pass mix) and causes the compound viscosity to decrease, which causes lower shear mixing forces and less effective dispersion at the higher temperature range. Therefore keeping the batch at a lower temperature range for a longer time period should improve the ultimate dispersion and state of mix. Besides making sure that the optimum batch size has been selected and working with an effective cooling system, one can also gain some additional control of mixing temperature by adjusting either ram pressure or adjusting rotor speed. If temperature control of the batches is not achieved during mixing, it is very possible that poor quality of mix will result, which will increase quality costs (such as a great deal of downstream scrap), higher rework costs, and/or higher external failure costs. RT: Chapter 23, "Rubber Mixing," W. Hacker, p. 514.

Masterbatches

Use an oil-extended polymer with a higher molecular weight in place of a lower molecular weight equivalent clear polymer (not oil extended) with the appropriate compound adjustments. An example of this is substituting SBR 1712 in place of SBR 1500 with the appropriate oil adjustments. RT: Chapter 4, "Rubber Compound Economics," J. Long, p. 79.

Using carbon black/oil masterbatches (such as SBR 1606) in place of free black mixing with clear SBR polymer (such as SBR 1500) may eliminate the need for multiple Banbury passes and require only one pass for a cost savings. RT: Chapter 4, "Rubber Compound Economics," J. Long, p. 80; Chapter 23, "Rubber Mixing," W. Hacker, pp. 520–521.

Process Aids

Using a processing aid as a compound additive to extrude or calender a compound at a faster rate can result in a cost savings. RT: Chapter 4, "Rubber Compound Economics," J. Long, p. 81.

Cure Time

The intelligent use of secondary accelerators (such as TMTM) with primary accelerators can serve to reduce cure times in the factory. RT: Chapter 4, "Rubber Compound Economics," J. Long, p. 82.

Consider using EV cures (more accelerator, less elemental sulfur) to cure at higher cure temperatures, reducing the cure time (and mold residence time) without greatly hurting cured physical properties. GEN: A. Bhowmick, S. De, "Dithiodimorpholine-Based Accelerator System in Tire Tread Compound for High-Temperature Vulcanization," *Rubber Chemistry and Technology*, November–December, 1979, Vol. 52, p. 985.

Consider using new, experimental, silica-coated iron oxide from Evonik to shorten the cure time for a thick rubber article by using an electromagnetic alternating field process by induction heating. GEN: O. Taikum, A. Korch, R. Friehmelt, F. Minister, M. Schotz, H. Herzog, S. Katusic (Evonik Degussa), "Novel Silica Coated Iron Oxide, 'Magsilica' to Speed Up Crosslinking in Rubber," Paper No. 16 presented at the Fall Meeting of the Rubber Division, ACS, October 11–13, 2011, Cleveland, OH.

Cure System

Consider selecting a less expensive cure system for vulcanizing EPDM. Such a less expensive cure system could be sulfur at 1.5 phr, TMTD at 1.5 phr, and MBT at 0.5 phr. RT: Chapter 16, "Cures for Specialty Elastomers," B. H. To, p. 395.

Base Rubber

When possible, base the rubber compound on less expensive base elastomers such as NR or SBR. RT: Chapter 6, "Elastomer Selection," R. School, pp. 126–129.

EPDM can accept very high loadings of inexpensive fillers such as clay along with carbon black, silica, and talc. Generally, the higher the total phr, the less expensive the compound. Chapter 6, "Elastomer Selection," R. School, p. 132.

A higher viscosity grade of CSM, such as Hypalon 4085, allows greater "dilutability" for preparing extremely extended formulations, which could result in a cost savings. RT: Chapter 8, "Specialty Elastomers," C. Baddorf, p. 215.

Consider the use of "lower cost" polyurethanes based on polypropylene glycol (PPG) ethers where highly engineered parts and applications are not required, rather than compounds based on premium-grade ethers or esters. RT: Chapter 9, "Polyurethane Elastomers," R. W. Fuest, p. 248, p. 259. RP: R. W. Fuest.

When possible, select an ordinary emulsion SBR for the rubber base of a compound for a significant cost savings. This might be feasible if no specific specifications are required. GEN: M. Chase, "Roll Coverings Past, Present, and Future," Presented at Rubber Roller Group Meeting, New Orleans, LA, May 15–17, 1996, p. 7.

Gas-Phase EPDM

Try compounding with a low Mooney viscosity, powdered, gas-phase polymerized EPDM with carbon black (usually N650) to possibly achieve a faster mix cycle and higher throughput. GEN: A. Paeglis, "Very Low Mooney Granular Gas-Phase EPDM," Presented at ACS Rubber Division Meeting, Fall, 2000, Paper No. 12.

Liquid BR/Silicone Rubber

Consider using low molecular weight (liquid) high-vinyl 1,2-polybutadiene resins (such as special Ricon® grades that contain certain antioxidants for improved heat resistance of the BR) as an additive at low part levels in a peroxide-cured silicone compound to allow this compound to accept more carbon black loading and reduce the compound cost. R. Drake, "Using Liquid Polybutadiene Resin to Modify Elastomeric Properties," *Rubber & Plastics News*, February 28 and March 14, 1983.

Millable PU/SBR

Adding a given quantity of SBR incrementally to a millable polyurethane-based compound reportedly can reduce the compound's cost. GEN: T. Jablonowski, "Blends of PU with Conventional Rubbers," *Rubber World*, October, 2000, p. 41.

SBR with NBR/PVC Blend

The use of SBR with NBR/PVC blend was reportedly used to reduce compound cost without greatly deteriorating the physical properties of the compound. GEN: J. Zhao, G. Ghebremeskel, J. Peasley, "SBR/PVC Blends with NBR as a Compatibilizer," *Rubber World*, December, 1998, p. 37.

FKM/ACM Alloy

Consider using a compound based on a special proprietary alloy of FKM and ACM (Dai-El Alloy AG-1530) using a special peroxide cure to serve as a less expensive substitute for FKM. This alloy reportedly will provide a compound with oil and heat resistance between that of FKM and ACM compounds. GEN: M. Kishine, T. Noguchi, "New Heat-Resistance Elastomers," *Rubber World*, February, 1999, p. 40.

AEM

In some cases, AEM might be a less expensive alternative to using FKM in certain automotive applications. GEN: E. McBride, K. Kammerer, L. Lefebvre (DuPont), "Testing of AEM and FKM Compounds in Acid Condensates for Turbocharger Systems," Paper No. 7 presented at the Fall Meeting of the Rubber Division, ACS, October 11–13, 2011, Cleveland, OH.

TPE

If considering the use of thermoplastic elastomers (TPEs), consider styrenics, which are typically the lowest cost TPEs. RT: Chapter 10, "Thermoplastic Elastomers," C. P. Rader, p. 270.

Even though precompounded TPVs may have a higher material cost than conventional rubber compounds, it is quite possible that a production cost savings can be realized by (1) significantly shortening the mold cycle (no curing required), (2) easy recycling of scrap (it's a thermoplastic, not a thermoset), and (3) fabrication to much tighter dimensional tolerances (lower total quality costs). RT: Chapter 10, "Thermoplastic Elastomers," C. P. Rader, p. 276

Consider using scrap tire rubber in making TPVs for less critical applications. GEN: H. Chandra, K. Chandra, C. Pillai (Quantum Polymer Composites LLC, Cleveland), "Advances inThermoplastic Rubbers and Vulcanizates made with Recycled Polymers," Paper presented at India Rubber Expo 2011, January 19–21, Chennai, India.

Ground Rubber

Consider the use of ambient ground rubber from scrap tires or other rubber products. In fact, consider having your cured scrap ground by an outside grinder and then working it away in small percentages where possible in noncritical rubber compounds. RT: Chapter 11, "Recycled Rubber." K. Baranwal, W. Klingensmith, p. 286. GEN: R. Swor, L. Jensen, M. Budzol, "Ultrafine Recycled Rubber," *Rubber Chemistry and Technology*, November–December, 1980, Vol. 53, p. 1215.

Antioxidant Combinations

It is possible to replace perhaps 4 phr of 6PPD antiozonant with a combination of a lower concentration of 6PPD and TMQ and wax and still achieve similar antidegradation protection. RT: Chapter 4, "Rubber Compound Economics," J. Long, p. 76.

6-Ethoxy-2,2,4-trimethyl-1,2-dihydroquinoline (ETMQ) can be used as a relatively inexpensive antiozonant. Usually ETMQ is used in combination with PPDs. ETMQ is sometimes partially substituted for PPDs to achieve a cost savings provided that the protection against ozone attack is still satisfactory. RT: Chapter 19, "Antidegradants," F. Ignatz-Hoover, p. 457.

Extruding

Consider adding a gear pump to the extruder in order to increase productivity and reduce costs. GEN: Olaf Skibba (VMI), "Extruder Gear Pump Systems for Improved Extrusion Output and Precision," Paper presented at ITEC 2008, Akron, OH.

Continuous Mixing

Consider going to a continuous mixing operation, such as with a ring extruder, to improve productivity and reduce manufacturing costs. GEN: Gerard Nijman (Vredestein Banden BV), "Continuous Mixing, a Challenging Opportunity?," Paper No. 72 presented at the Fall Meeting of the Rubber Division, ACS, October 8–11, 2002, Pittsburgh, PA.

Blending

For EPDM/NR blends, one might consider using less EPDM and more NR if possible to reduce the cost of the compound. GEN: A. Ahmad, "NR/EPDM Blend for Automotive Rubber Component," Rubber Research Institute of Malaysia.

Note: For additional ideas on reducing manufacturing costs, refer to Section 5.3 (Ways to Reduce Mixing Time) and Section 5.9 (Ways to Increase Extrusion Speed Without Loss in Quality).

3 Improving Degradation Resistance of Cured Rubber Compounds

3.1 Improving Hot Air and/or Heat Aging Resistance

Hot air aging and heat aging resistance are becoming very important today, especially in automotive applications where "under-the-hood" operating temperatures have been increasing and there is more pressure from automotive manufacturers to have rubber parts that last longer (part of the move toward a 150,000-mile car warranty!). Anaerobic heat aging resistance can be quite different from hot air aging resistance. For example, a rubber compound might possess good reversion resistance (anaerobic), yet still be subject to oxidative attack (or vice versa). The type of aging properties that a cured rubber compound possesses is very important in determining how that compound will be classified in accordance with SAE J200/ASTM D2000 for automotive applications.

The following ***experimental*** *ideas might improve a rubber compound's hot air aging resistance and/or heat aging resistance (but not always both). Also, all relevant literature sources, including the ones cited below, should be researched and read.* ***Caution:*** *these general experimental ideas may not work in all specific situations. Changes to improve hot air aging or heat aging resistance will certainly affect other properties as well, for better or for worse, and this book does not purport to show how these other properties are affected. Also, this book does not purport to address safety and health issues.*

Subjective Ranking of Rubber

One overall subjective ranking that can be applied to different base elastomers for heat resistance is as follows:

FKM (Best) > VMQ > Fluorosilicone > ACM > AEM > EPDM > CO/ECO ≈ CM > CSM ≈ NBR > CR > IIR ≈ NBR/PVC > SBR > NR

RT: *Rubber Technology, Compounding and Testing for Performance*. 2nd ed. Edited by J. S. Dick, Hanser Publications, 2009 ■ GEN: General References from the Literature ■ RP: Review Panel (Advisors for this book) ■ IMPORTANT: The information contained herein is experimental in nature and is meant for individuals with advanced scientific training and rubber compounding experience. All safety precautions should be taken when replicating or applying this information. The author makes no representation or warranty, either expressed or implied.

GEN: J. Horvath, "Selection of Polymers for Automotive Hose and Tubing Applications," *Rubber World*, December 1987, p. 21.

Another subjective ranking that can be applied to base elastomers for heat resistance is the following:

FKM (Best) ≈ TFE > VMQ/PVMQ/MQ > FVMQ > AEM > EPM > EPDM ≈ ACM > CSM > CR > NR ≈ NBR > AU/EU > SBR/BR

GEN: R. Mastromatteo, E. Morrisey, M. Mastromatteo, H. Day, "Matching Material Properties to Application Requirements," *Rubber World*, February, 1983, p. 28.

The so-called "continuous use" temperature rankings for specialty elastomers were reported as follows:

Type of rubber	Continuous use temperature
HNBR	145°C (293°F)
ACM	150°C (302°F)
AEM	160°C (320°F)
AG (DAI-EL Alloy AG-1530)	175°C (347°F)
FKM	230°C (446°F)

GEN: M. Kishine, T. Noguchi, "New Heat-Resistance Elastomers," *Rubber World*, February, 1999, p. 40.

Perfluoroelastomers

Often perfluoroelastomers (FFKMs) are selected as the base rubber for a compound when the ultimate heat resistance is needed. FFKM polymers reportedly can impart useful service life at temperatures as high as 316°C (601°F). GEN: M. Coughlin, R. Schnell, S. Wang, "Perfluoroelastomers in Severe Environments: Properties, Chemistry, and Applications," Presented at ACS Rubber Div. Meeting, Spring, 2001, Paper No. 24.

Fluoroelastomers

Compounds based on FKM possess extremely good heat aging resistance. RT: Chapter 6, "Elastomer Selection," R. School, p. 135.

Fluoroelastomer (FKM) compounds can achieve heat resistance up to 260°C (500°F). GEN: M. Chase, "Roll Coverings Past, Present, and Future," Presented at Rubber Roller Group Meeting, New Orleans, May 15–17, 1996, p. 7.

To maximize compound high-temperature performance, the best acceptor for the chosen FKM must be selected. Common types of acceptors for FKM compounds are low-activity magnesium oxide, high-activity magnesium oxide, calcium oxide, calcium hydroxide, zinc oxide, and lead oxide (litharge), which is toxic. RT: Chapter 6, "Elastomer Selection," R. School, p. 136.

Fluoroelastomer-based compounds provide extremely good protection against air aging and can provide a "useful service life" at temperatures greater than 200°C (392°F). RT: Chapter 8, "Specialty Elastomers," R. Stevens, p. 229.

For even better heat aging resistance of a fluoroelastomer compound, select a bisphenol AF cure system. RT: Chapter 8, "Specialty Elastomers," R. Stevens, p. 230.

New fluoroelastomers made with vinylidene fluoride, tetrafluoroethylene, and propylene reportedly possess better aging properties for many "aggressive engine oils" at high temperatures compared to conventional FKM elastomers. This is reported to be attributable to the use of propylene in place of hexafluoropropylene in the polymerization process. GEN: W. Grootaert, R. Kolb, A. Worm, "A Novel Fluorocarbon Elastomer for High-Temperature Sealing Applications in Aggressive Motor-Oil Environments," *Rubber Chemistry and Technology*, September–October, 1990, Vol. 63, p. 516.

The following is a general ordinal ranking of different types of fluoroelastomers in regards to the heat resistance they impart to a rubber compound:

FFKM (72% F) Best > FKM Type 2 (70% F) > FKM Type 1 (65% F)

GEN: Jim Denham (3M), "Basic Fluoroelastomer Technology," Presented at the Energy Rubber Group, September 13, 2011, Galveston, TX.

Under severe conditions of very high pressure and very high temperature (HPHT) in "down-hole" oil drilling operations using fluoroelastomer compounds, to help avoid explosive decompression (ED) as a mode of failure, consider using the new ultra-high-viscosity (UHV) fluoroelastomers [ML (1 + 10 @ 121°C) = 150], 66% fluorine content, bisphenol AF curable. GEN: J. Denham (Dyneon 3M), "Solutions for the Oil and Gas Industry," Presented at the Spring Meeting of the Energy Rubber Group, May 28, 2009, Arlington, TX.

HNBR

Compounds based on HNBR possess very good heat aging resistance. The higher the level of hydrogenation is, the better the heat aging will be because there are fewer unsaturated sites on the polymer backbone that can be attacked. RT: Chapter 8, "Specialty Elastomers," M. Wood, p. 202.

Some grades of HNBR can be sulfur cured because they still possess some unsaturation. However, an HNBR compound will usually possess better heat aging resistance if it is cured with peroxide instead of sulfur. RT: Chapter 8, "Specialty Elastomers," M. Wood, p. 202.

TOTM and TINTM plasticizers reportedly impart better heat resistance for HNBR compounds than plasticizers such as DOP and DBEEA because these trimellitate plasticizers have lower volatility and higher molecular weights. DBEEA reportedly is the "most recommended" for HNBR-based, sulfur-cured compounds because it imparts a good balance between low-temperature properties and heat resistance. GEN: S. Hayashi, H. Sakakida, M. Oyama, T. Nakagawa, "Low-Temperature Properties of Hydrogenated Nitrile Rubber (HNBR)," *Rubber Chemistry and Technology*, September-October, 1991, Vol. 64, p. 534.

NBR

High-temperature applications for NBR-based compounds often use peroxides and thiurams as curatives. RT: Chapter 6, "Elastomer Selection," R. School, p. 131.

Using NBR grades with higher ACN content may improve compound heat aging resistance. RT: Chapter 8, "Specialty Elastomers," M. Gozdiff, p. 194.

Use an emulsion NBR that is produced from a coagulating system based on magnesium sulfate for better heat resistance. RT: Chapter 8, "Specialty Elastomers," M. Gozdiff, p. 196.

To impart better heat resistance to a nitrile formulation, consider ether thioether plasticizers, in which the sulfur functions as a secondary antioxidant. GEN: J. Dunn, "Compounding Elastomers for Tomorrow's Automotive Market, Part II," *Elastomerics*, February, 1989, p. 29.

To impart better heat resistance to a nitrile formulation, consider using precipitated silica, magnesium oxide, and zinc oxide in place of carbon black. GEN: J. Dunn, "Compounding Elastomers for Tomorrow's Automotive Market, Part II," *Elastomerics*, February, 1989, p. 29.

Natural Rubber

For NR compounds, cure at a lower temperature for a longer period of time to increase the preponderance of monosulfide crosslinks for better aging properties. GEN: M. Studebaker, J. Beatty, "Vulcanization," *Elastomerics*, February, 1977, p. 41.

For NR formulations, selecting diphenylguanidine (DPG) as the secondary accelerator ("kicker") instead of a dithiocarbamate or a thiuram "kicker" for a primary accelerator may improve the compound's heat aging characteristics. GEN: B. Ashworth, K. Crawford, "Effect of Secondary Accelerator Selection on the Aging Characteristics of Natural Rubber Vulcanizates," *Rubber World*, December, 1982, p. 20.

Never use nickel dibutyldithiocarbamate (NBC) in a natural rubber compound because NBC is a strong pro-oxidant and degradant to NR. RT: Chapter 19, "Antidegradants," F. Ignatz-Hoover, p. 453.

Cardanol-formaldehyde (CF) novolak curing resin can be used with HMTA to cure a natural rubber compound so as to impart good aging properties. CF novolak curing resins are more compatible with NR than phenol-formaldehyde novolak curing resins. In addition, the CF resins have some antioxidant properties. The CF resins are manufactured using cardanol as a raw material. Cardanol is usually extracted from CNSL or "cashew nut shell liquid," which is relatively inexpensive and a natural product. GEN: Y. Vu, J. Mark, L. Pham, "Blends of Natural Rubber with Cardanol-Formaldehyde Resins," Presented at ACS Rubber Div. Meeting, Spring, 1998, Paper No. 58.

Consider using a diurethane cure (Novor® crosslinking agents from Hughson Chemical Co.), which reportedly impart better heat aging resistance to a NR vulcanizate. GEN: T. Kempermann, "Sulfur-Free Vulcanization Systems for Diene Rubber," *Rubber Chemistry and Technology*, July-August, 1988, Vol. 61, p. 422.

For an all-natural rubber motor mount, consider applying a curable formulated chloroprene/polyoctenamer undercoat (drying the solvent) followed by the application of a formulated curable halobutyl/polyoctenamer top coating followed by drying. When properly applied and cured on the surface of the NR motor mount, this two-layer coating serves as an effective barrier against ozone attack as well as thermooxidative attack of the natural rubber. It significantly slows down the diffusion of ozone and oxygen into the NR and prevents the cleavage of the backbone of the NR molecules. Reportedly, fatigue cycling testing of NR motor mounts showed that the two-layer coating improved the cycles-to-failure results significantly. GEN: H. Graf, E. Sayej, "Reversion Resistance of Engine Mounts," *Rubber World*, February, 2000, p. 55.

Consider coating products of natural rubber with a thin layer of HXNBR to help prolong high-temperature service life. GEN: Rani Joseph (Cochin University, India), "HXNBR for Improving Ageing Resistance of Natural Rubber Products," Paper No. 1 presented at the Fall Meeting of the Rubber Division, ACS, November 1-3, 2005, Pittsburgh, PA.

Polychloroprene

W-type neoprenes generally impact better heat aging properties to a vulcanizate than G-type neoprene. RT: Chapter 6, "Elastomer Selection," R. School, p. 133; Chapter 8, "Specialty Elastomers," L. L. Outzs, p. 208.

Octylated diphenylamine is reported to be one of the better antioxidants for polychloroprene-based compounds. GEN: R. Tabar, P. Killgoar, R. Pett, "A Fatigue

Resistant Polychloroprene Compound for High Temperature Dynamic Applications," *Rubber Chemistry and Technology*, September-October, 1979, Vol. 52, p. 781.

Reportedly, for better AO protection for a CR compound, try Stalite S (octylated diphenylamine) at 4 phr, Aranox (*N*-phenyl-*N'*-Cp-toluenesulfonyl) at 1 phr, zinc oxide at up to 10 phr, and magnesium oxide up to 6 phr. GEN: L. Outzs, "Neoprene," Presented at the DuPont Compounders Course, May, 2006, Fairlawn, OH.

EVM

Ethylene vinyl acetate (EVM), sometimes called the "forgotten rubber," possesses good resistance to heat aging when properly compounded because it possesses a saturated backbone. Because of this fact, EVM is being used more in automotive seal and gasket applications where good heat aging resistance is needed. The EVM grades that contain a vinyl acetate content between 40 and 80% are good elastomers for rubber applications. GEN: R. Pazur, L. Ferrari, H. Meisenheimer (Bayer Inc.), "Ethylene Vinyl Acetate Copolymers: The Forgotten Rubber," Paper No. XVI presented at the Spring Meeting of the Rubber Division, ACS, May 17-19, 2004, Grand Rapids, MI.

EPDM

EPDM reportedly can be compounded to have heat resistance up to 125°C (257°F). GEN: K. Dominic, V. Kothari, "Overview of Automotive Wire and Cable and Recent Advances," ACS Rubber Division Meeting, Spring, 1998, Paper No. 32, p. 13.

Using peroxide to cure an EPDM compound will improve its heat aging resistance. RT: Chapter 6, "Elastomer Selection," R. School, p. 132.

Low-Viscosity, Gas-Phase EPDM Needs Less Oil

Try compounding with an ultra-low Mooney viscosity gas-phase polymerized EPDM with very high ethylene content and filler loading. Higher ethylene content reportedly improves the processability of the compound without dependency on the use of traditional processing oils and plasticizers, which can sometimes impart poor hot air aging properties. GEN: A. Paeglis, "Very Low Mooney Granular Gas-Phase EPDM," Presented at ACS Rubber Div. Meeting, Fall, 2000, Paper No. 12.

EPDM/CR

Consider adding a small quantity of CR (or other chlorinated polymer) to a peroxide-cured EPDM compound to improve the compound's heat resistance. GEN: R. Ohm, R. Annicelli, T. Jablonowski, C. Lahiri, R. Mazzeo, "Optimizing the Heat Resistance of EPDM and NBR," ACS Rubber Div. Meeting, Fall, 2000, Paper No. 99, p. 5.

EPDM/CR Alloy

Consider developing a rubber compound based on a blend of 70 phr of Neoprene GRT, 30 phr of EPDM, with a compatibilizer of escor acid terpolymer at 10 phr (ethylene-methacrylate-acrylic acid terpolymer), which displays superior DeMattia cut growth resistance, heat resistance, and ozone resistance, which is claimed to be good for many power transmission belt applications. GEN: P. Arjunan, R. Kusznir, A. Dekmezian, "Compatibilization of CR/EPM Blends for Power Transmission Belt Applications," *Rubber World*, February, 1997, p. 21.

EPDM/POE

EPDM compound heat aging resistance can be improved by blending with polyolefin elastomer (POE). RT: Chapter 6, "Elastomer Selection," R. School, p. 139.

Butyl and Halobutyl

Consider using compounds based on butyl and halobutyl rubber instead of general purpose rubber. RT: Chapter 8, "Specialty Elastomers," G. Jones, D. Tracey, A. Tesler, p. 174.

ExxonMobil's Exxpro (BIMSM, brominated copolymer of isobutylene and paramethylstyrene) has no unsaturation in its backbone. Therefore, BIMSM has more heat resistance than halobutyl rubbers, which do possess a small level of unsaturation in their backbone. Therefore, using BIMSM in place of halobutyl rubber in the tire innerliner may result in the innerliner being more heat resistant, thus improving the service life of the tire innerliner under severe service conditions. GEN: G. Jones, "Exxon's Exxpro Innerliner for Severe Service Tire Applications," Presented at the 1996 ITEC Meeting, September, 1996, Akron, OH.

HIIR

Halobutyl rubber may provide better resistance to heat aging than regular butyl rubber cured with a conventional sulfur vulcanization system. RT: Chapter 8, "Specialty Elastomers," G. Jones, D. Tracey, A. Tisler, p. 185.

Higher heat resistance is obtained with halobutyl rubber when cured with bismaleimide (HVA-2) either with or without peroxide. RT: Chapter 8, "Specialty Elastomers," G. Jones, D. Tracey, A. Tisler, p. 186.

Cure bromobutyl-based compound with a proprietary curative called Vanax 189 from R. T. Vanderbilt for significantly improved heat aging resistance. Vanax 189 is a derivative of 1,2-dimercapto-1,3,4-thiadiazole. GEN: R. Ohm, "New Developments in Curing Halogen-Containing Polymers," Presented at ACS Rubber Div. Education Symposium No. 45, "Automotive Applications II," Spring, 1998, p. 4.

Consider curing a halobutyl compound with zinc oxide and certain diamine AOs such as di-beta-naphthyl-*p*-phenylenediamine (Agerite White®). This may improve the compound's heat resistance. (The mechanism may involve a Friedel-Crafts alkylation reaction.) GEN: J. Dunn, "Compounding Elastomers for Tomorrow's Automotive Market, Part II," *Elastomerics*, 1989, p. 28.

Using a substituted *p*-phenylenediamine AO such as *N,N'*-di-beta-naphthyl-*p*-phenylene-diamine (DNPD or Agerite White®) and zinc oxide as crosslinking agents for a bromobutyl-based compound may improve air aging resistance. GEN: D. Edwards, "A High-Pressure Curing System for Halobutyl Elastomers," *Rubber Chemistry and Technology*, March-April, 1987, Vol. 60, p. 62.

In tire innerliner compounds, use less natural rubber and more halobutyl rubber to improve heat resistance. GEN: W. Waddell, R. Napler, D. Tracey, "Nitrogen Inflation of Tires," Paper No. 45 presented at the Fall Meeting of the Rubber Divison, ACS, October 14-16, 2008, Louisville, KY.

BIIR vs. CIIR

Compounds based on BIIR may give slightly better heat resistance than those based on CIIR. GEN: J. Fusco, "New Isobutylene Polymers for Improved Tire Processing," Presented at the Akron Rubber Group Meeting, January 24, 1991, Akron, OH.

BIMS

Use Exxon's EXXPRO® (polyisobutylene with brominated *p*-methylstyrene cure site) for further heat resistance over regular halobutyl rubber, which relies on unsaturation sites as cure sites. GEN: R. Ohm, "New Developments in Curing Halogen-Containing Polymers," Presented at ACS Rubber Div. Education Symposium No. 45, "Automotive Applications II," Spring, 1998, p. 5.

Brominated isobutylene-co-*p*-methylstyrene (BIMS)-based compounds may have better heat aging properties than a conventional halobutyl rubber compound under more severe temperatures such as 125 °C (257 °F) or 150 °C (302 °F). GEN: G. Jones, "Exxpro Innerliners for Severe Service Tire Applications," Presented at ITEC, 1998, Paper No. 7A.

Avoid Zinc for CM and CSM

For compounds based on chlorinated polyethylene (CM) or chlorosulfonated polyethylene (CSM), *avoid* incorporating *any* zinc-containing materials into the formulations or in any way using any zinc-containing coatings or dusting agents that might come in contact with these formulations because they will seriously degrade heat resistance and state of cure. RT: Chapter 8, "Specialty Elastomers," L. Weaver, C. Baddorf, pp. 212, 215.

CR

Generally, mercaptan-modified neoprene has better heat resistance than the sulfur-modified or xanthate-modified neoprene. GEN: Nobuhiko Fujii, Denki Kagaku Kogyo K. K. (Denka), "Recent Technical Improvements of CR and ER in Industrial Applications," Denka Literature, 2011.

CSM

Epoxidized soybean oil (ESO) and sometimes epoxidized glycerol dioleate (EpGdO) reportedly impart better stability from exposure to air oven aging of chlorosulfonated polyethylene (CSM)-based compounds. RT: Chapter 14, "Ester Plasticizers and Processing Additives," W. Whittington, p. 358.

Nickel dibutyldithiocarbamate (NBC) reportedly may improve the heat resistance of Hypalon®. RT: Chapter 19, "Antidegradants," F. Ignatz-Hoover, p. 453.

CPE

Selecting grades of chlorinated polyethylene with lower percentage of chlorine may allow better resistance to heat aging. RT: Chapter 8, "Specialty Elastomers," L. Weaver, p. 213.

Epoxidized ester plasticizers reportedly impart better stability from exposure to air oven aging of CPE-based compounds. RT: Chapter 14, "Ester Plasticizers and Processing Additives," W. Whittington, p. 359.

A four-way comparison of compounds based on CM, CSM, GECO, and NBR/PVC found that CM gave superior heat resistance in relation to the other three elastomers. GEN: C. Hooker, R. Vara, "A Comparison of Chlorinated and Chlorosulfonated Polyethylene Elastomers with Other Materials for Automotive Fuel Hose Covers," Presented at ACS Rubber Div. Meeting, Fall, 2000, Paper No. 128.

Avoid High-Styrene Resins

High-styrene resins should be avoided for compounds that will be used in service at high temperatures. RT: Chapter 18, "Tackifying, Curing, and Reinforcing Resins," B. Stuck, p. 446.

Silicone Rubber

PVMQ silicone gums impart to a rubber compound somewhat better resistance to heat than do VMQ silicone gums. RT: Chapter 8, "Specialty Elastomers," J. R. Halladay, p. 235.

In silicone rubber compounding, precipitated silica can impart superior heat aging properties above 200°C (392°F). RT: Chapter 8, J. R. Halladay, "Specialty Elastomers," p. 236.

In silicone compounding, traditional antioxidants are not used; however, heat stabilizers are used. Red iron oxide (at 1 to 2 phr) is an inexpensive heat stabilizer that is sometimes used in silicone rubber compounding. Barium zirconate at about 4 phr is also an important stabilizer for light-colored compounds. Many commercial silicone bases already have heat stabilizers and may not require additional stabilizers. RT: Chapter 8, J. R. Halladay, "Specialty Elastomers," p. 236.

Reportedly, some silicones can be compounded to display heat resistance up to 260°C (500°F). GEN: M. Chase, "Roll Coverings Past, Present, and Future," Presented at Rubber Roller Group Meeting, New Orleans, May 15–17, 1996, p. 7.

Polyurethane

Polyurethanes are in general not considered "high-temperature materials." However, usually, but not always, selecting a polyurethane based on a TDI prepolymer will give better high-temperature properties. In addition, selecting an ester-type polyurethane will usually give better heat aging resistance in demanding applications. RT: Chapter 9, "Polyurethane Elastomers," R. W. Fuest, p. 257.

Castable polyurethane fabricators using MDI prepolymer systems may be able to increase their product's tolerance to high-temperature exposure by trying to use aromatic diol chain extenders such as bis(hydroxyethyl) ether of resorcinol (HER) and hydroquinone (HQEE). The use of these chain extenders will create hard domains that have higher melting points. GEN: R. Durairaj, "Chain Extenders Increase Heat Tolerance," *Rubber & Plastics News*, November 29, 1999.

A novel polyurethane elastomer based on *p*-phenylene diisocyanate (PPDI) has been reported to give very good aging property retention at 150°C. GEN: Z. Zhu, R. Rosenberg, V. Gajewski, G. Nybakken, M. Ferrandino (Chemtura Corp.), "High Performance Polyurethane Elastomers," Paper No. 31 presented at the Fall Meeting of the Rubber Division, ACS, November 1–3, 2005, Pittsburgh, PA.

XLPE

Reportedly, XLPE can be compounded to have heat resistance up to 125 to 150°C (257 to 302°F). GEN: K. Dominic, V. Kothari, "Overview of Automotive Wire and Cable and Recent Advances," Presented at ACS Rubber Div. Meeting, Spring, 1998, Paper No. 32, p. 13.

Silicone/EPDM

Using compounds based on silicone-EPDM in place of straight EPDM compounds can enable heat resistance up to 204°C (399°F). GEN: M. Chase, "Roll Coverings Past, Present, and Future," Presented at Rubber Roller Group Meeting, New Orleans, May 15-17, 1996, p. 7.

TPV

Thermoplastic vulcanizates (TPVs) based on EPDM and polypropylene (PP) compositions provide excellent retention of physical properties at elevated temperatures with aging, provided the service temperatures are sufficiently below the melt transition temperature of the TPV. RT: Chapter 10, "Thermoplastic Elastomers," C. P. Rader, p. 274.

High-temperature limits for TPVs are determined by the susceptibility to oxidative attack by the thermoplastic phase. For polypropylene (PP), this oxidative limit may be approximately 125 to 135°C (257 to 275°F). RT: Chapter 10, "Thermoplastic Elastomers," C. P. Rader, p. 276.

Thermoplastic vulcanizate alloys consisting of dynamically vulcanized polyacrylate rubber (ACM) with polyamide (PA) provide material that can withstand long-term exposure to temperatures around 150°C. GEN: Jiri G. Drobay, "High Performance Thermoplastic Elastomers: A Review," Paper No. 69 presented at the Fall Meeting of the Rubber Division ACS, October 14-16, 2008, Louisville, KY.

Also, consider using Dupont's thermoplastic elastomers called the ETPV family consisting of a copolyester matrix with highly crosslinked AEM (modified ethylene-acrylate rubber), which reportedly may handle temperatures of 135°C continuously and peaks that are as high as 180°C. GEN: J. Drobny (Drobny Polymer Associates), "High Performance Thermoplastic Elastomers: A Review," Paper No. 69 presented at the Fall Meeting of the Rubber Division, ACS, October 14-16, 2008, Louisville, KY.

Silica

Consider using precipitated silica in a rubber compound where an improvement in heat aging resistance is needed. RT: Chapter 13, "Precipitated Silica and Non-black Fillers," W. Waddell, L. Evans, p. 331.

Carbon Nanotubes

Reportedly, from the use of multiwalled carbon nanotubes (MWCNT) in an NBR formulation, the percent retention in ultimate tensile strength from high-temperature exposure was significantly improved. GEN: Steve Driscoll, Status Report titled "U. Mass. Lowell Today"; N. Warasitthinon, L. Schulte, A. Erley, A. Hope, et al., "Nanocomposites of Nitrile Rubber with Multi-walled Carbon Nanotubes," Presented at the India Rubber Expo 2011, January 2011, Chennai, India.

CPE Nanoclay composites

Consider using organically modified clay montmorillonite (such as natural montmorillonite modified with quaternary ammonium salts) dispersed in a CPE compound. Reportedly this improves heat resistance. GEN: S. Kar, K. Bhowmick, "Thermal Mechanical Behavior of Chlorinated Polyethylene Nanocomposites," Paper No. 56 presented at the Fall Meeting of the Rubber Division, ACS, October 14–16, 2008, Louisville, KY.

Talc

It has been reported that substituting 40% of the carbon black in an EPDM hose compound with talc resulted in an improvement in heat aging resistance. Certain grades of talc may show a significant advantage over treated and untreated clay as well. GEN: O. Noel, unpublished draft, "Talc Synergy with Carbon Black in Sulfur Donor Cured EPDM," June, 2003; H. Bertram, "Influence of Light Colored Fillers on the Aging Behavior of NBR Vulcanizates," Bayer TIB 17. RP: O. Noel.

Plasticizers

Polymeric plasticizers should be considered instead of monomeric plasticizers for better heat aging and permanence. RT: Chapter 14, "Ester Plasticizers and Processing Additives," W. Whittington, pp. 356, 362.

Synthetic plasticizers with higher viscosity, such as polymeric plasticizers, impart better permanence and aging properties to a rubber compound than monomeric synthetic plasticizers generally do. RT: Chapter 14, "Ester Plasticizers and Processing Additives," W. Whittington, p. 362.

Consider substituting hydrocarbon resins in place of the more volatile plasticizers and processing oils to retain better permanence and impart better heat aging resistance. GEN: "Hydrocarbon Resins for Rubber Compounding," Neville Literature, 2008, Neville Chemical Co.

High-Viscosity Oils

For good heat aging resistance, use processing oils with high-viscosity properties because they usually have fewer low molecular weight components and better permanence (less volatility) in a compound than a low-viscosity oil. GEN: R. Tabar, P. Killgoar, R. Pett, "A Fatigue Resistant Polychloroprene Compound for High Temperature Dynamic Applications," *Rubber Chemistry and Technology*, September-October, 1979, Vol. 52, p. 781.

Rapeseed Oil for CR

For the best neoprene compound resiliency, consider using rapeseed oil because it has low viscosity for hysteresis and low volatility for good aging properties. GEN: R. Tabar, P. Killgoar, R. Pett, "A Fatigue Resistant Polychloroprene Compound for High Temperature Dynamic Applications," *Rubber Chemistry and Technology*, September-October, 1979, Vol. 52, p. 781.

Vulcanized Vegetable Oil

If a brown vulcanized vegetable oil (VVO) is being used in a natural rubber compound, it may be better to use a VVO based on soybean oil rather than on boiled linseed oil for the better retention of compound air aging properties. GEN: S. Botros, F. El-Mohsen, E. Meinecke, "Effect of Brown Vulcanized Vegetable Oil on Ozone Resistance, Aging, and Flow Properties of Rubber Compounds," *Rubber Chemistry and Technology*, March-April, 1987, p. 159.

If an NR compound contains a brown vulcanized vegetable oil (VVO), the compound's heat resistance will increase if VVOs are used that contain lower percent free sulfur levels or lower percent acetone extract values. GEN: S. Botros, F. El-Mohsen, E. Meinecke, "Effect of Brown Vulcanized Vegetable Oil on Ozone Resistance, Aging, and Flow Properties of Rubber Compounds," *Rubber Chemistry and Technology*, March-April, 1987, p. 159.

Ultra-accelerators

Cure systems based on ultra-accelerators such as thiurams and dithiocarbamates tend to impart higher percentages of monosulfidic crosslinks compared to thiazole or amine-type accelerators. Therefore these ultra-accelerators often will impart better aging properties to their vulcanizates. GEN: M. Studebaker, J. R. Beatty, "Vulcanization," *Elastomerics*, February, 1977, p. 41.

EV/Semi-EV Cures

Use compounds based on an efficient vulcanization (EV) or semi-efficient (semi-EV) cure system where the ratio of accelerator to sulfur is high and sulfur donors are sometimes used in place of free sulfur. These cure systems generate a higher preponderance of monosulfidic and disulfidic crosslinks, which are thermodynamically more stable than polysulfidic crosslinks. Therefore, semi-EV and EV cures impart better heat aging properties to a rubber compound and better thermal stability than conventional sulfur cures. RT: Chapter 15, "Sulfur Cure Systems," B. H. To, p. 387.

HTS

Hexamethylene-1,6-bisthiosulfate (HTS), a post-vulcanization stabilizer, reportedly forms special hybrid crosslinks when it is used as a supplement to a sulfenamide cure to impart better heat aging resistance to a natural rubber compound and good flex fatigue resistance. RT: Chapter 15, "Sulfur Cure Systems," B. H. To, p. 391. Similar effects may be achieved if Perkalink 900® (from Flexsys) is considered for use. RP: B. H. To.

Thiuram Cure

Consider using a "thiuram cure" for improved heat aging resistance provided that environmental problems with nitrosamines can be avoided through good ventilation and so forth. GEN: T. Kempermann, "Sulfur-Free Vulcanization Systems for Diene Rubber," *Rubber Chemistry and Technology*, July–August, 1988, Vol. 61, p. 422.

Dithiophosphates vs. Dithiocarbamates

Functionally, dithiophosphate rubber accelerators are similar to dithiocarbamate accelerators except the phosphorus atoms replace the nitrogen atoms for these traditional dithiocarbamate cure packages. The dithiophosphate cures reportedly give potentially better heat aging resistance and reversion resistance as well compared to a dithiocarbamate. GEN: S. Monthey, M. Saewe, V. Meenenga (Rhein Chemie), "Using Dithiophosphate Accelerators to Improve Dynamic Properties in Vibration Isolation Applications," Presented at the Spring Southern Rubber Group, June 11–14, 2012, Myrtle Beach, SC.

DIPDIS Cure

Consider the use of an experimental accelerator, bis(diisopropyl)thiophosphoryl disulfide (DIPDIS), with synergistic combinations of conventional thiazole-type accelerators to produce NR vulcanizates with more stable crosslinks and better aging properties. GEN: S. Mandal, R. Datta, D. Basu, "Studies of Cure Synergism, Part 1: Effect of Bis(diisopropyl)thiophosphoryl Disulfide and Thiazole-Based Accelerators in the Vulcanization of NR," *Rubber Chemistry and Technology*, September–October, 1989, Vol. 62, p. 569.

New Crosslinking Agent

Reportedly a new crosslinking agent [1,6-bis(*N*,*N*-dibenzylthiodicarbamyldithio) hexane] gives stable hybrid crosslinks in truck tread compounds for better tread wear properties as well as better heat aging properies. GEN: T. Kleiner (Bayer AG), "Improvements in Abrasion and Heat Resistance by Using a New Crosslink Agent," Paper No. 12A presented at the ITEC 2002 Meeting, September, 2002, Akron, OH.

High Zinc Oxide Loading

Sulfur/sulfenamide-cured compounds that contain higher than usual levels of zinc oxide may possess better heat aging characteristics and better resistance to overcure. GEN: W. Hall, H. Jones, "The Effect of Zinc Oxide and Other Curatives on the Physical Properties of a Bus and Truck Tread Compound," Presented at ACS Rubber Div. Meeting, Fall, 1970.

NBR Cures

For curing NBR-based formulations, consider using a cure system that uses a lower amount of sulfur and a larger quantity of sulfur donor (such as TMTD or DTDM) as a partial or total replacement for the free sulfur. The use of less elemental sulfur may improve aging resistance in the nitrile compound. RT: Chapter 16, "Cures for Specialty Elastomers," B. H. To, pp. 398–403.

Butyl Cure

Select a semi-EV cure system using the sulfur donor DTDM to significantly improve heat aging properties of a butyl compound. RT: Chapter 16, "Cures for Specialty Elastomers," B. H. To, p. 403.

EPDM Cure

To improve heat aging as well as improve (i.e., reduce) compression set for EPDM compounds, one should consider a so-called "low-set" cure consisting of sulfur at 0.5 phr, ZBDC at 3.0 phr, ZMDC at 3 phr, DTDM at 2.0 phr, and TMTD at 3.0 phr. RT: Chapter 16, "Cures for Specialty Elastomers," B. H. To, p. 396.

ECO Cures

Zinc oxide as an activator in a Hydrin/ECO terpolymer compound has been reported to cause poor hot air oven aging. When service temperatures are near 125°C, perhaps one should consider also using calcium oxide as an activator, perhaps with a reduced quantity of zinc oxide in order to achive better heat aging resistance. GEN: S. Harber (Zeon Chemical), "Metal Oxide Activation Systems for Improved Aging in Epichlorohydrin Terpolymer Compounds," Paper No. 55 presented at the Fall Meeting of the Rubber Division, ACS, October 10–12, 2006, Cincinnati, OH.

NR/EPDM Blend

Consider covulcanizing an NR/EPDM blend with a sulfur/peroxide cure. The EPDM will impart better heat aging properties to the compound. GEN: S. Tobing, "Co-vulcanization in NR/EPDM Blends," *Rubber World*, February, 1988, p. 33.

Peroxide Cure for EPDM Compounds

For peroxide cures of EPDM compounds, consider selecting ZMTI as an antioxidant, which should allow good modulus development as well as impart good heat aging. RT: Chapter 17, "Peroxide Cure Systems," L. Palys, p. 430.

Peroxide/Coagent Use

Using a coagent with a peroxide cure may improve heat aging. GEN: P. Dluzneski, "Peroxide Vulcanization of Elastomers," *Rubber Chemistry and Technology*, July–August, 2001, Vol. 74, p. 451.

For peroxide cures, use a coagent as well to add unsaturation to the system, which results in a higher crosslink density than the peroxide would have generated if used without the coagent. The reason is that it is more efficient and easier to crosslink with a free radical to unsaturation than by hydrogen abstraction to a saturated polymer chain. Because the use of a coagent can lead to different types of crosslink networks, improved heat aging resistance may result. RT: Chapter 17, "Peroxide Cure Systems," L. Palys, pp. 431–432.

Peroxide vs. Sulfur Cure

An advantage of peroxide vs. sulfur cures is that the peroxide cures can more easily be made to impart better heat aging resistance. The reason is that the carbon–carbon bond strength is about 350 kJ/mol vs. sulfur–sulfur bond strength of only 115–270 kJ/mol and carbon–sulfur bond strength of only 285 kJ/mol. RT: Chapter 17, "Peroxide Cure Systems," L. Palys, p. 434.

DBU/MMBI (Heat Stabilizer) Substitute for MgO in CR and HIIR Cures

For halobutyl (HIIR) and polychloroprene (CR) compounds, it has been reported that the presence of magnesium oxide as an acid acceptor actually promotes more C–C bond links than ether and thioether links. By substituting a new heat stabilizer called "DBU/MMBI" in place of magnesium oxide, significantly improved hot air aging properties are reported for polychloroprene-, chlorobutyl-, and bromobutyl-based compounds. The chemical identity of this "DBU/MMBI" heat stabilizer is only partially disclosed as "DBU: 1,5-diazabicyclo-(4.5.0)-undecen-(11)" or $C_9H_{16}N_2$. GEN: R. Musch, R. Schubart, A. Sumner, "Heat Resistant Curing System for Halogen-Containing Polymers," Presented at ACS Rubber Div. Meeting, Spring, 1999, Chicago, IL.

Use Higher Molecular Weight TMQ

Polymerized 1,2-dihydro-2,2,4-trimethylquinoline (TMQ) is a very cost-effective and widely used general purpose antioxidant. Using a higher molecular weight version of TMQ, such as Flectol H, will provide more antioxidant persistence over time. RT: Chapter 19, "Antidegradants," F. Ignatz-Hoover, p. 454.

TMQ/6PPD

Even though TMQ by itself does not offer ozone or fatigue protection, it is commonly used to extend the effectiveness of 6PPD itself. This is particularly important if the compound is exposed to oxidative aging before exposure to ozone. Therefore it is quite common to see a synergistic combination of TMQ and 6PPD used together to achieve the best combination of both oxidative aging protection as well as protection against ozone attack. RT: Chapter 19, "Antidegradants," F. Ignatz-Hoover, pp. 454–460.

TMQ/BLE

Consider a blend of TMQ and BLE if both good heat aging and flex fatigue properties are needed. GEN: S. Hong, C. Lin, "Improved Flex Fatigue and Dynamic Ozone Crack Resistance Through the Use of Antidegradants or Their Blends in Tire Compounds," Presented at ACS Rubber Div. Meeting, Fall, 1999, Paper No. 27.

Metal "Poisons"

In general, avoid rubber compounding ingredients and raw rubbers themselves that may contain any copper, manganese, nickel, or cobalt. These transition elements can function as pro-oxidants and promote the degradation of the vulcanizate. To some extent, alkyl-aryl PPDs and dihydroquinoline-type AOs in black compounds will react with these "metal poisons." For nonblack compounds, dinaphthyl-PPD and TMQ can also be used for partial protection. RT: Chapter 19, "Antidegradants," F. Ignatz-Hoover, pp. 454, 461–462.

Bound Antioxidants

Antioxidants can be chemically bound to the backbone of the raw NBR polymer before it is added to the mixing process. This improves the compound's resistance to heat while also providing resistance to extraction. RT: Chapter 6, "Elastomer Selection," R. School, p. 131.

Some improvement in air oven aging for nitrile-based compounds can be achieved if an NBR is substituted that contains a bound antioxidant; however, the amount of improvement in air oven aging is not even comparable to the much greater improvement achieved by using HNBR instead. RT: Chapter 8, "Specialty Elastomers," M. Gozdiff, p. 199.

Use raw synthetic elastomers in which a hindered phenol or an amine antioxidant is "bound in" to the polymer backbone and can thus impart antioxidant protection to a rubber compound even under harsh environmental conditions. RT: Chapter 19, "Antidegradants," F. Ignatz-Hoover, pp. 454–455.

Use *N*-phenyl-*N'*-1,3-dimethylbutyl-*p*-quinone diimine (6-QDI) in a rubber compound that uses sulfur-vulcanization of a diene rubber base. After vulcanization, a portion of the 6QDI is chemically bonded to the polymer backbone and/or the carbon black and cannot be extracted. Thus this bound antioxidant is reported to perform better than a traditional 6PPD/TMQ combination. RT: Chapter 19, "Antidegradants," F. Ignatz-Hoover, p. 454.

Resin Cures

The use of a methylol phenol–formaldehyde resin cure in place of a sulfur cure for a butyl compound should significantly improve the compound's heat aging resistance. These resins can be used to cure some other elastomers as well, not just butyl rubber. RT: Chapter 8, "Specialty Elastomer," G. Jones, D. Tracey, A. Tisler, p. 178; Chapter 18, "Tackifying, Curing and Reinforcing Resins," B. Stuck, p. 443.

Bisphenol Cures for FKM

In some situations, for better heat resistance, consider using a bisphenol cure system in place of a traditional peroxide cure system for vulcanizing FKM compounds. GEN: Daikin America Inc., "New DAI-EL Fluoroelastomers for Extreme Environments," Presented at the Fall Meeting of the Energy Rubber Group, September 15–18, 2008, San Antonio, TX.

Fibers to Retain Physical Properties

Consider using a low loading of short aramid fiber to improve cured physical property retention at high temperatures such as 150 °C (302 °F). GEN: K. Watson, A. Frances, "Elastomer Reinforcement with Short Kevlar Aramid Fiber for Wear Applications," *Rubber World*, August, 1988, p. 20.

Nitrogen Inflation

Tires inflated with nitrogen instead of air may impart some protection against oxidative aging. GEN: J. MacIsaac, L. Evans, J. Harris, E. Terrill, "The Effects of Inflation Gas on Tire Laboratory Test Performance," Paper No. 18C-1 presented at ITEC 2008, September 16, 2008, Akron, OH; U. Karmarkar, A. Pannikottu (ARDL), "Role of Materials Research in Laboratory Tire Aging and Durability Test Development," Paper No. 17 presented at the Fall Meeting of the Rubber Division ACS, 1999, Cincinnati, OH.

3.2 Improving Ozone Resistance

Ozone attack on a rubber product is usually a surface phenomenon. The average ambient concentration of ozone in the air we breathe has been rising. Ozone attack on certain strained rubber products usually results in cracking. This can severely damage or ruin a rubber product.

The following ***experimental*** *ideas might improve a cured rubber compound's resistance to ozone attack. Also, all relevant literature sources, including the ones cited below, should be researched and read.* ***Caution:*** *these general experimental ideas may not work in all specific situations. Changes to improve ozone resistance will certainly affect other properties as well, for better or for worse, and this book does not purport to show how these other properties are affected. Also, this book does not purport to address safety and health issues.*

Waxes

For static ozone protection, it can be important to use a special blend of protective waxes, usually consisting of low molecular weight paraffin, high molecular weight paraffin, and microcrystalline wax. These blends are usually necessary to achieve the proper performance. Wax is used mostly for static protection. In fact if too much wax is used in the formulation, dynamic ozone protection could be harmed. RT: Chapter 3, "Vulcanizate Physical Properties, Performance Characteristics, and Testing," J. S. Dick, p. 64; Chapter 19, "Antidegradants," F. Ignatz-Hoover, p. 456. GEN: R. Layer, R. Lattimer, "Protection of Rubber Against Ozone," *Rubber Chemistry and Technology*, July-August, 1990, Vol. 63, p. 426.

PPDs

Paraphenylene diamines (PPDs) are very potent antiozonants providing good dynamic protection. RT: Chapter 19, "Antidegradants," F. Ignatz-Hoover, p. 454.

Many times, for good dynamic ozone protection, it is important to use an effective *p*-phenylenediamine antiozonant such as 6PPD. RT: Chapter 3, "Vulcanizate Physical Properties, Performance Characteristics, and Testing," J. S. Dick, p. 64.

A combination of 6PPD (for long-term applications) and 77PPD (for short-term applications) is commonly used for better protection from ozone. GEN: L. Walker, J. Luecken, "Antidegradants for Ozone and Fatigue Resistance: Laboratory and Tire Tests," *Elastomerics*, May, 1980, p. 36.

The new high silica loaded tire compounds for lower rolling resistance may require higher loadings of PPD antiozonants because the silica particles absorb some of the PPD antiozonant itself. GEN: F. Ignatz-Hoover, D. Killmeyer,

B. To (Flexsys), "Aging Characteristics of Carbon Black Filled vs. Silica Filled Compound," Paper No. 73 presented at the Fall Meeting of the Rubber Division, ACS, October 11–13, 2011, Cleveland, OH.

PPD Microencapsulation

Consider using a microencapsulation process for an antiozonant such as 6PPD. This microencapsulation agent can be cellulose acetate. Using microencapsulated 6PPD as a compounding ingredient can provide better controlled release for more effective compound protection from ozone attack over a longer period of time. GEN: L. Evans, D. Benko, J. Gillick, W. Waddell, "Microencapsulated Antidegradants for Extending Rubber Lifetime," *Rubber Chemistry and Technology*, March–April, 1992, Vol. 65, p. 201.

Wax/PPDs

Usually, for the best combination of both static and dynamic protection against ozone attack, a combination of a special wax blend and an effective *p*-phenylene diamine antiozonant such as 6PPD is commonly used. RT: Chapter 19, "Antidegradants," F. Ignatz-Hoover, p. 456; Chapter 3, "Vulcanizate Physical Properties, Performance Characteristics, and Testing," J. S. Dick, p. 64.

NBC AO

Nickel dibutyldithiocarbamate (NBC) reportedly is sometimes used as an antiozonant to impart some static ozone protection for only NBR, CR, and SBR compounds. It is not used for dynamic protection. RT: Chapter 19, "Antidegradants," F. Ignatz-Hoover, p. 453.

AO Molecular Weight and Solubility

An effective antiozonant must be soluble in the rubber compound and it must migrate to the cured compound's surface while in service to protect against ozone attack. Generally this rate of migration is dependent on the molecular weight as well as the solubility of the AO in the rubber compound. RT: Chapter 19, "Antidegradants," F. Ignatz-Hoover, p. 450.

6QDI

Use *N*-phenyl-*N'*-1,3-dimethylbutyl-*p*-quinone diimine (6QDI) in a rubber compound that uses sulfur vulcanization of a diene rubber base. After vulcanization, a portion of the 6QDI is chemically bonded to either the polymer backbone and/or the carbon black and cannot be extracted. Thus this bound AO is reported in some cases to perform better than a traditional 6PPD/TMQ combination in functioning as an antioxidant. It has been also reported that another portion of the 6QDI is converted to 6PPD, which also functions as a very effective antiozonant. RT: Chapter 19, "Antidegradants," F. Ignatz-Hoover, p. 454.

77PD

77PD (a dialkyl PPD) is used primarily when static ozone protection is needed. Its use is mostly for short-term protection and it is normally not used alone when dynamic protection is also needed. The 77PD is commonly used with alkyl-aryl PPDs for better overall ozone protection under both static and dynamic conditions. Commonly, compounders will use a blend of 6PPD and 77PD (2:1 blend) for the best protection, 6PPD for dynamic protection, and 77PD for static protection against ozone attack. RT: Chapter 19, "Antidegradants," F. Ignatz-Hoover, pp. 454–460.

TMQ/6PPD

Even though TMQ by itself does not usually afford ozone or fatigue protection, it is commonly used to extend the effectiveness of 6PPD itself. This is particularly important if the compound is exposed to higher temperature oxidative aging before exposure to ozone. Therefore it is quite common to see a synergistic combination of TMQ and 6PPD together. RT: Chapter 19, "Antidegradants," F. Ignatz-Hoover, pp. 454–460.

Nonstaining Acetal AO

For white or colored rubber compounds in which no staining can be tolerated, consider using cyclic acetal antiozonant (Vulkazon AFS from Bayer AG) for effective protection against ozone attack across a wide range of different types of elastomers. GEN: W. Jeske, "How to Avoid Ozone Cracking: A Solution for White and Colored Rubber Goods," Presented at ACS Rubber Div. Meeting, Fall, 1999, Paper No. 65.

It was reported by Warrach and Tsou that bis(1,2,3,6-tetrahydrobenzaldehyde)-pentaerythrityl acetal imparted better ozone protection to compounds based on CR, IIR, CIIR, or BIIR than *p*-phenylenediamine AOs, without discoloring or staining. GEN: W. Waddell, "Tire Black Sidewall Surface Discoloration and Non-staining Technology: A Review," *Rubber Chemistry and Technology*, July–August, 1998, Vol. 71, p. 590.

Butyl and Halobutyl Rubber

Butyl and halobutyl elastomers possess good resistance to ozone attack. RT: Chapter 6, "Elastomer Selection," R. School, p. 133.

Select a butyl rubber with the lowest level of unsaturation (0.8 mol%) for better resistance against ozone attack because this will provide the lowest number of unsaturated sites or opportunities for ozone attack. RT: Chapter 8, "Specialty Elastomers," G. Jones, D. Tracey, A. Tesler, p. 176.

DNPD/ZnO Cure for BIIR

Using a substituted *p*-phenylenediamine AO such as *N,N'*-di-beta-naphthyl-*p*-phenylene diamine (DNPD or Agerite White®) and zinc oxide as crosslinking agents for a bromobutyl-based compound may help improve ozone resistance. GEN: D. Edwards, "A High-Pressure Curing System for Halobutyl Elastomers," *Rubber Chemistry and Technology*, March-April, 1987, Vol. 60, p. 62.

BIMS

Brominated isobutylene-co-*p*-methylstyrene (BIMS)-based compounds may have better ozone resistance than some other butyl or halobutyl rubber compounds because the BIMS compounds have completely saturated backbones. GEN: A. Tisler, K. McElrath, D. Tracey, M. Tse, "New Grades of BIMS for Non-stain Tire Sidewalls," Presented at ACS Rubber Div. Meeting, Fall, 1997, Paper No. 66.

Flowers and Fusco reported that substituting BIMS in place of halobutyl rubber or EPDM in a triblend with NR in a tire sidewall compound caused a significant improvement in resistance to fatigue crack growth and ozone resistance. GEN: W. Waddell, "Tire Black Sidewall Surface Discoloration and Non-staining Technology: A Review," *Rubber Chemistry and Technology*, July-August, 1998, Vol. 71, p. 590.

The use of BIMS as a substitute for one or more of the EPDM/HIIR/NR triblend polymers used in tire sidewall compounds combines the excellent ozone resistance and flex fatigue resistance of the BIMS for better compound performance. GEN: W. Waddell, "Tire Black Sidewall Surface Discoloration and Non-staining Technology: A Review," *Rubber Chemistry and Technology*, July-August, 1998, Vol. 71, p. 590.

Butyl Rubber Dioxime Cure

The ozone resistance of a butyl rubber compound may be improved by using a dioxime cure. RT: Chapter 8, "Specialty Elastomers," G. Jones, D. Tracey, A. Tesler, p. 177.

Alkyl Phenol Disulfide Cure of Halobutyl Rubber

Alkyl phenol disulfides are usually very effective in covulcanizing halobutyls with general purpose elastomers (with higher unsaturation) to significantly improve the ozone resistance beyond what the higher unsaturated polymers would otherwise display by themselves. RT: Chapter 16, "Cures for Specialty Elastomers," B. H. To, p. 409.

EPDM and EPM

Compounds based on EPDM and EPM usually possess superior ozone resistance. RT: Chapter 6, "Elastomer Selection," R. School, p. 132.

High-Gloss Ozone-Resistant Compounds

One of the best ways to achieve a "high-gloss" appearance over the rubber product's service life is to use a rubber blend that contains a sufficient quantity of rubber with a saturated (or highly saturated) polymer backbone. By using enough of this "saturated" elastomer in the blend, the resulting rubber compound will be inherently resistant to ozone attack and without the need to use antiozonants. These antiozonants, if used, bloom to the compound's surface and destroy the "high-gloss" appearance. Examples of ozone-resistant elastomers that could be used in such a blend include EPDM, BIIR, CIIR, and the new BIMS elastomer. GEN: W. Waddell, "Tire Black Sidewall Surface Discoloration and Non-staining Technology: A Review," *Rubber Chemistry and Technology*, July–August, 1998, Vol. 71, p. 90.

Avoid Natural Rubber

Avoid basing a rubber compound on natural rubber because NR will impart poor resistance to ozone attack unless strong antiozonants are used. Sometimes NR compounds are blended with certain synthetic elastomers to help improve resistance to ozone. RT: Chapter 6, "Elastomer Selection," R. School, p. 127.

Flexible Coating Over NR Mount

For an all-natural rubber motor mount, consider applying a curable formulated polychloroprene/polyoctenamer undercoat (drying off the solvent) followed by the application of a formulated curable halobutyl/polyoctenamer top coating followed by drying. When properly applied and cured on the surface of the NR motor mount, this two-layer coating serves as an effective barrier against ozone attack as well as thermooxidative attack of the NR. It significantly slows down the diffusion of ozone and oxygen into the NR and prevents the cleavage of the backbone of the NR molecules. Fatigue cycling testing of NR motor mounts showed that the two-layer coating improved the cycles-to-failure results significantly. GEN: H. Graf, E. Sayej, "Reversion Resistance of Engine Mounts," *Rubber World*, February, 2000, p. 55.

HXNBR Coating

Consider coating products of natural rubber with a thin layer of HXNBR to prolong high-temperature service life and provide better protection from ozone attack. GEN: Rani Joseph (Cochin University, India), "HXNBR for Improving Ageing Resistance of Natural Rubber Products," Paper No. 1 presented at the Fall Meeting of the Rubber Division, ACS, November 1–3, 2005, Pittsburgh, PA.

NR/EPDM Blend

Consider covulcanizing an NR/EPDM blend with sulfur/peroxide cure. The EPDM will impart better ozone resistance to the compound. GEN: S. Tobing, "Co-vulcanization in NR/EPDM Blends," *Rubber World*, February, 1988, p. 33.

Increasing the EPDM portion in a compound based on an NR/EPDM blend will improve resistance to ozone. The theoretical optimum ozone resistance might be achieved when the EPDM concentration is increased to 35 to 40% (where the EPDM becomes the continuous phase in an NR/EPDM blend). However, as the concentration of EPDM is increased, more and more cure compatibility problems may result, which can cause a deterioration in some physical properties. GEN: A. Ahmad, "NR/EPDM Blend for Automotive Rubber Component," Rubber Research Institute of Malaysia; E. McDonel, K. Baranwal, J. Andries, *Polymer Blends*, Vol. 2, Chapter 19, "Elastomer Blends in Tires," Academic Press, 1978, p. 287.

E. Andrews reported that the dispersion of "small zones" of EPR in natural rubber helps slow down ozone cracking. Also, using blends of EPR as high as 35 to 45% will greatly improve the compound's resistance to ozone attack. GEN: W. Hess, C. Herd, P. Vegvari, "Characterization of Immiscible Elastomer Blends," *Rubber Chemistry and Technology*, July–August, 1993, Vol. 66, p. 329.

NR/CIIR Blend

Use a higher ratio of chlorobutyl (CIIR) in a blend with natural rubber. The CIIR may display better cure compatibility than EPDM with NR. However, reportedly EPDM serves as a better antiozonant in an NR blend. Sometimes a three-way blend of CIIR and EPDM with NR is used to optimize ozone resistance and other physical properties. GEN: E. McDonel, K. Baranwal, J. Andries, *Polymer Blends*, Vol. 2, Chapter 19, "Elastomer Blends in Tires," Academic Press, 1978, p. 287.

CR

Reportedly, many polychloroprene-based compounds can meet a 50 pphm ozone resistance requirement. The addition of 1 to 2 phr of a mixed diaryl *p*-phenylenediamine antiozonant may improve this ozone resistance to perhaps

100 pphm ozone. However, the use of these phenylenediamines will decrease the CR compound's scorch safety time, and they should be added during mixing with the curatives. RP: L. L. Outzs.

EPDM/CR Alloy

Consider developing a rubber compound based on a blend consisting of 70 phr Neoprene GRT and 30 phr EPDM with a compatibilizer of escor acid terpolymer at 10 phr (ethylene-methacrylate-acrylic acid terpolymer). This compound displays superior DeMattia cut growth resistance, heat resistance, and ozone resistance and is claimed to be favorable for many power transmission belt applications. GEN: P. Arjunan, R. Kusznir, A. Dekmezian, "Compatibilization of CR/EPM Blends for Power Transmission Belt Applications," *Rubber World*, February, 1997, p. 21.

Millable PU/SBR

Adding a given quantity of millable polyurethane incrementally to an SBR-based compound reportedly will improve the compound's ozone resistance. GEN: T. Jablonowski, "Blends of PU with Conventional Rubbers," *Rubber World*, October, 2000, p. 41.

Also, millable polyurethane can be blended with NBR, NBR/PVC, EPDM, SBR, and BR to improve ozone resistance. However, adjustments in the cure package and the additional use of a compatibilizer may be necessary. GEN: T. Jablonowski (Uniroyal Chemical Co.), "Blends of Polyurethane Rubbers with Conventional Rubbers," Paper No. 46 presented at the Spring Meeting of the Rubber Division, ACS, April 13–16, 1999, Chicago, IL.

Polyepichlorohydrin Elastomers

Compounds based on polyepichlorohydrin have very good resistance to ozone attack. RT: Chapter 8, "Specialty Elastomers," C. Cable, p. 216.

Silicone Rubber

Compounds based on silicone rubber possess excellent ozone resistance. RT: Chapter 6, "Elastomer Selection," R. School, p. 136.

EVM

Ethylene vinyl acetate (EVM), sometimes called the "forgotten rubber," possesses good ozone resistance when properly compounded because it possesses a saturated backbone. Because of this fact, EVM is being used more in automotive seal and gasket applications where good ozone resistance is needed.

The EVM grades that contain a vinyl acetate content between 40 and 80% are good elastomers for use in rubber applications. GEN: R. Pazur, L. Ferrari, H. Meisenheimer (Bayer Inc.), "Ethylene Vinyl Acetate Copolymers: The Forgotten Rubber," Paper No. XVI presented at the Spring Meeting of the Rubber Division, ACS, May 17–19, 2004, Grand Rapids, MI.

Polyacrylate Rubber

Polyacrylate rubber imparts very good ozone resistance. RT: Chapter 6, "Elastomer Selection," R. School, p. 138.

CM

Compounds based on chlorinated polyethylene (CM) generally have good ozone resistance. RT: Chapter 8, "Specialty Elastomers," L. Weaver, p. 212.

CSM

Compounds based on chlorosulfonated polyethylene (CSM) have very good resistance to ozone attack. RT: Chapter 8, "Specialty Elastomers," C. Baddorf, p. 214.

Polyurethane Rubber

Polyurethane elastomers have an advantage over conventional diene rubbers in that they can possess superior ozone resistance. RT: Chapter 9, "Polyurethane Elastomers," R. W. Fuest, p. 253.

HNBR

In general, HNBR compounds possess fairly good resistance to ozone attack. GEN: R. Campbell (Greene Tweed Co.), "History of Sealing Products and Future Challenges in the Oil Field," Paper presented at a meeting of the Energy Rubber Group, Educational Symposium, September 13, 2011, Galveston, TX.

Some HNBRs can be sulfur cured because they still possess some unsaturation. However, an HNBR compound will usually possess better ozone resistance if it is cured with peroxide instead of sulfur. RT: Chapter 8, "Specialty Elastomers," M. Wood, p. 202.

NBR/PVC Blend

The ozone resistance of an NBR compound can be greatly improved if the raw NBR is prefluxed with polyvinyl chloride (PVC). The raw nitrile rubber is added with plasticized PVC and fluxed in an internal mixer followed by pelletization. This is then compounded further. J. Dunn et al. indicated that emulsion-polymerized PVC grades (commonly used for plastisols) should be used for fluxing.

Fluxing is reportedly necessary to achieve optimal ozone resistance and stress-strain properties. RT: Chapter 6, "Elastomer Selection," R. School, p. 132. GEN: J. Dunn, D. Coulthard, H. Pfisterer, "Advances in Nitrile Rubber Technology," *Rubber Chemistry and Technology*, July-August, 1978, Vol. 51, p. 389.

Increasing the concentration of PVC in a flux PVC/NBR blend should improve the ozone protection. GEN: M. Chase, "Roll Coverings Past, Present, and Future," Presented at Rubber Roller Group Meeting, New Orleans, May 15-17, 1996, p. 3.

NBR/EPDM Blend

Oil-resistant NBR-based compounds can possess improved resistance against ozone attack if the NBR is blended with perhaps 30 phr of EPDM. The EPDM may improve the compound's ozone resistance; however, this compound will also undergo a loss in its oil resistance. However, EPDM and NBR are very incompatible, and if they are not compounded properly, they will impart poor cured properties to the vulcanizate. Care must be taken in selecting the proper curative package that will be compatible with both elastomers. For example, Woods and Mass reported that some dithiocarbamates possessing long-chain alkyl groups may work. Bergstrom reported that curing with sulfur at 0.2 phr, CBS at 0.5 phr, and 40% active DCP at 2.5 phr also was somewhat successful. Mitchell reported that premixing the EPDM with carbon black and then mixing this masterbatch with the NBR also may improve compatibility and compound performance. GEN: J. Dunn, D. Coulthard, H. Pfisterer, "Advances in Nitrile Rubber Technology," *Rubber Chemistry and Technology*, July-August, 1978, Vol. 51, p. 389.

Fluoroelastomers

Usually, fluoroelastomer-based compounds give very good weatherability resistance as well as ozone resistance. GEN: Jim Denham (3M), "Basic Fluoroelastomer Technology," Presented at the Fall Meeting of the Energy Rubber Group, September 13, 2011, Galveston, TX.

Talc

It is reported that using fine particle size talc as a filler can impart some degree of resistance to ozone degradation. GEN: L. Evans, W. Waddell, "Computerized Optimization of a White Sidewall," *Rubber World*, November, 1993, p. 18.

Silica

Precipitated silica reportedly can be used in a black sidewall compound, which can include a nonstaining EPDM base to significantly improve resistance to ozone aging as well as improve tear strength and cut growth. GEN: W. Waddell, L. Evans, "Use of Nonblack Fillers in Tire Compounds," *Rubber Chemistry and Technology*, July-August, 1996, Vol. 69, p. 377.

3.3 Improving Flex Fatigue Resistance and/or Cutting and Chipping Resistance

The standard test methods for measuring flex fatigue actually give different results. Moreover, flex fatigue resistance is a very different rubber compound property compared to "cutting and chipping" resistance. In some cases these two properties are quite independent. Lastly, it should be mentioned that crack initiation resistance is a very different rubber property when compared to cut growth or propagation resistance.

*The following **experimental** ideas might help improve one or more of these compound properties. Also, all relevant literature sources, including the ones cited below, should be researched and read. **Caution:** these general experimental ideas may not work in all specific situations. Changes to improve flex fatigue resistance and/or cutting and chipping resistance will certainly affect other properties as well, for better or for worse, and this book does not purport to show how these other properties are affected. Also, this book does not purport to address safety and health issues.*

Mixing

Increase dispersion of reinforcing fillers such as carbon black and/or silica through better mixing. RT: Chapter 2, "Compound Processing Characteristics and Testing," J. S. Dick, p. 42. GEN: S. Monthey, T. Reed, "Performance Differences Between Carbon Blacks and CB Blends for Critical IR Applications," *Rubber World*, April, 1999, p. 42

Improving the dispersion of carbon black through longer mix cycles will usually improve DeMattia cut growth resistance. RT: Chapter 12, "Compounding with Carbon Black and Oil," S. Laube, S. Monthey, M-J. Wang, p. 308. GEN: W. Hess, "Characterization of Dispersions," *Rubber Chemistry and Technology*, July-August, 1991, Vol. 64, p. 386.

According to Studebaker and Beatty, cut growth is another form of flex fatigue. They reported that sometimes poor dispersion in a compound might actually result in improved cut growth resistance. They felt that the reason for this was that agglomerates in less well dispersed stock might interfere with the propagation of the cut and might lead to a knotty tear. GEN: F. Eirich, *Science and Technology of Rubber*, Chapter 9, M. Studebaker, J. Beatty, "The Rubber Compound and Its Composition," Academic Press, 1978, p. 367.

Phase Mixing

Reportedly, crack growth is slowed if more carbon black is forced into the SBR phases in an SBR/BR blend by using phase mixing techniques. GEN: E. McDonel, K. Baranwal, J. Andries, *Polymer Blends*, Vol. 2, Chapter 19, "Elastomer Blends in Tires," Academic Press, 1978, p. 287.

It was reported by Ferrandino and Hong that the cut growth resistance of a compound based on an NR/BR/EPDM blend can be improved by preblending the carbon black into the NR phase.

GEN: W. Waddell, "Tire Black Sidewall Surface Discoloration and Non-staining Technology: A Review," *Rubber Chemistry and Technology*, July-August, 1998, Vol. 71, p. 590.

For an NR/BR blend, it is reported that a greater concentration of carbon black in the NR phase through phase mixing can improve cut growth resistance. GEN: J. Massie, R. Hirst, A. Halasa, "Carbon Black Distribution in NR/Polybutadiene Blends," *Rubber Chemistry and Technology*, May-June, 1993, Vol. 66, p. 276.

Hess reported that the optimal fatigue life for an NR/BR blend might be achieved using a larger particle size, lower structure HAF black such as N330 with an approximately even distribution of carbon black between the phases. GEN: W. Hess, C. Herd, P. Vegvari, "Characterization of Immiscible Elastomer Blends," *Rubber Chemistry and Technology*, July-August, 1993, Vol. 66, p. 329.

Also, Hess reported that fatigue life increased when more N299 carbon black was phase distributed into the NR phase of an NR/BR blend. A higher concentration of BR in an NR/BR blend likewise caused an increase in the fatigue life of the compound. GEN: W. Hess, "Characterization of Dispersions," *Rubber Chemistry and Technology*, July-August, 1991, Vol. 64, p. 386.

Using a blend of chlorobutyl and natural rubber with the carbon black dispersed mainly in the NR phase reportedly will impart properties that are beneficial for fatigue crack growth (FCG) resistance. GEN: D. Young, "Application of Fatigue Methods Based on Fracture Mechanics for Tire Compound Development," *Rubber Chemistry and Technology*, September-October, 1990, Vol. 63, p. 567.

High-Temperature Mixes and Gel Formation

Raw synthetic rubber such as SBR usually contains stabilizers. However, mixing SBR compounds at high temperatures such as 325°F (163°C) or above can cause the formation of both redispersible "loose" gel (that can be milled out later) and so-called "tight" gel (that cannot be milled out). Both of these types of gel can dramatically reduce the flex fatigue life of the cured rubber compound as measured by the DeMattia Flexing Test. Therefore, it might be prudent to restrict how high the dump temperature should be. GEN: R. Mazzeo, "Preventing Polymer Degradation During Mixing," *Rubber World*, February, 1995, p. 22.

Crosslink Density

Find the optimum crosslink density from curing that imparts the highest flex fatigue resistance. Often this optimum is lower than the optimum crosslink density for maximum ultimate tensile strength. RT: Chapter 3, "Vulcanizate Physical Properties, Performance Characteristics, and Testing," J. S. Dick, p. 47.

Conventional Cure

Consider using compounds based on conventional cure systems in which the ratio of accelerator to sulfur is lower than with semi-EV and EV cures. These cure systems generate a higher preponderance of polysulfide crosslinks that are thermodynamically less stable than mono- and disulfide crosslinks. However, often unaged compounds with conventional cures will possess somewhat better resistance against flex fatigue than compounds based on EV and semi-EV cures. RT: Chapter 15, "Sulfur Cure Systems," B. H. To, p. 387.

Avoid Sulfurless Cures

Avoid sulfurless cures in which thiuram is used as the sulfur donor because these cures result in mono- and disulfide crosslinks that lower crack resistance from flexing compared to polysulfide crosslinks. GEN: T. Kempermann, "Sulfur-Free Vulcanization Systems for Diene Rubber," *Rubber Chemistry and Technology*, July-August, 1988, Vol. 61, p. 422.

Sulfur vs. Peroxide

An advantage of sulfur vs. peroxide cures is that the sulfur-cured compounds can more easily be made to have better flex fatigue resistance. The reason is that the sulfur bonds break and re-form whereas the carbon-carbon bonds (from peroxide cures) do not. RT: Chapter 17, "Peroxide Cure Systems," L. Palys, p. 434.

Zinc-Based Coagent Peroxide Cures

If peroxide cures must be used to maintain good heat aging resistance, then consider the use of zinc-based coagents with peroxide for a cure that gives better dynamic flex properties as well. For example, a peroxide/ZnDMA system gives better flex fatigue resistance than traditional peroxide cures. The establishment of ionic crosslinks will give good DeMattia flex fatigue properties. The bond dissociation energy for $C-O-Zn^{2+}-O-C$ is 293 kJ/mol compared to a C–C bond (from a normal peroxide cure) with 335 kJ/mol. GEN: S. Henning, R. Costin (Sartomer Co.), "Fundamentals of Curing Elastomers with Peroxides and Coagents," Paper E presented at the Spring Meeting of the Rubber Division, ACS, May 16–18, 2005, San Antonio, TX; L. Palys, P. Callais (Atofina Chemicals), "Understanding Organic Peroxides to Obtain Optimal Crosslinking Performance," *Rubber World*, December, 2003.

Hardness Stability

For a compound to have good flex fatigue resistance, it should also be compounded to avoid marching modulus or the increased hardening of the compound while in service. GEN: R. Ohm, "New Developments in Curing Halogen-Containing Polymers," Presented at ACS Rubber Div. Education Symposium No. 45, "Automotive Applications II," Spring, 1998.

HTS

Reportedly, hexamethylene-1,6-bisthiosulfate (HTS), a post-vulcanization stabilizer, forms special hybrid crosslinks when it is used as a supplement to a sulfenamide cure to impart superior flex fatigue resistance to an NR compound and good heat aging resistance as well. RT: Chapter 15, "Sulfur Cure Systems," B. H. To, p. 391.

BCI-MX Antireversion Agent

1,3-Bis(citraconimidomethyl)benzene (BCI-MX), an antireversion agent, reportedly imparts superior reversion resistance to a natural rubber compound and good compressive flex resistance as measured with a Goodrich Flexometer. RT: Chapter 15, "Sulfur Cure Systems," B. H. To, pp. 391–393.

Optimum Cure

According to Studebaker and Beatty, cut growth is another form of fatigue resistance. They reported that low states of cure can result in poor cut growth resistance. Likewise, extremely high states of cure can also result in poor cut growth, perhaps because the ultimate elongation is reduced too much. Hence there is an optimum state of cure for the best cut growth resistance. GEN: F. Eirich, *Science and Technology of Rubber*, Chapter 9, "The Rubber Compound and Its Composition," M. Studebaker, J. R. Beatty, Academic Press, 1978, p. 367.

Longer Cures

J. R. Beatty and B. J. Miksch developed a special cutting and chipping tester to try to simulate the occurrence of cutting and chipping with OTR tires in the field. Cutting and chipping resistance might have some relation to flex fatigue properties; however, it is generally believed to be a complex property that is quite different from traditional flex fatigue measurements. Beatty reported that his test predicted that cutting and chipping resistance might improve for a natural rubber OTR tread if the tread were exposed to a long overcure at 140°C (284°F), compared to a shorter cure time. However, Beatty also indicated that he knew of no logical explanation for this occurrence. Therefore, a much longer cure at a relatively low cure temperature may not actually help cutting and chipping resistance at all. More work needs to be done in this area to determine the optimal cure time. However, it is quite likely that cure time and cure temperature have an important role in determining the degree of cutting and chipping resistance, whether the optimal cure is a longer or a shorter cure. GEN: J. Beatty, B. Miksch, "A Laboratory Cutting and Chipping Tester for Evaluating Off-the-Road and Heavy-Duty Tire Treads," *Rubber Chemistry and Technology*, November-December, 1982, Vol. 55, p. 1531.

Longer Cures at Lower Temperatures

To maximize cured flex fatigue resistance at *constant energy* for a sulfur cure, consider curing for a longer time at a lower temperature as opposed to curing at a higher temperature for a shorter time because the longer cure may result in a higher ultimate crosslink density and a higher preponderance of polysulfidic crosslinks. However, the longer cure at lower temperature can result in a higher crosslink density, which means that flex fatigue comparisons by *constant strain* might show a disadvantage simply because of the compound's greater hardness. GEN: M. Lemieux, P. Killcoar, "Low Modulus, High Damping, High Fatigue Life Elastomer Compounds for Vibration Isolation," *Rubber Chemistry and Technology*, September-October, 1984, Vol. 57, p. 792.

Blowout from Flexing

Cured compounds that possess C-C or monosulfidic crosslinks reportedly give higher blowout temperatures by the Goodrich Flexometer blowout test than cured compounds that possess mostly polysulfidic crosslinks. GEN: A. Gent, M. Hindi, "Heat Build-up and Blowout of Rubber Blocks," *Rubber Chemistry and Technology*, November-December, 1988, Vol. 61, p. 892.

Strain Crystallization

A strain crystallizing base elastomer such as natural rubber may impart some resistance to cut growth (but not, however, resistance to crack initiation). RT: Chapter 3, "Vulcanizate Physical Properties, Performance Characteristics, and Testing," J. S. Dick, p. 60.

Selecting an IR with high *cis* content will improve the compound's ability to undergo strain-induced crystallization and impart higher DeMattia cut growth resistance. RT: Chapter 7, "General Purpose Elastomers and Blends," G. Day, p. 142.

Higher Molecular Weight

Increase the average molecular weight of the base rubber in a formulation to increase fatigue life. GEN: R. Mastromatteo, E. Morrisey, M. Mastromatteo, H. Day, "Matching Material Properties to Application Requirements," *Rubber World*, February, 1983, p. 25.

Using BR polymers with higher average molecular weights can significantly improve fatigue life. RT: Chapter 7, "General Purpose Elastomers and Blends," G. Day, p. 170.

Increasing the average molecular weight of an SBR used in a compound reportedly will increase the compound's flex fatigue life. GEN: J. Zhao, G. Ghebremeskel, "A Review of Some of the Factors Affecting Fracture and Fatigue in SBR and BR Vulcanizates," *Rubber Chemistry and Technology*, July-August, 2001, Vol. 74, p. 409.

To improve flex fatigue resistance as well as abrasion resistance, consider using an oil-extended general purpose elastomer masterbatch. These oil-extended masterbatches generally have higher average molecular weights but can still be mixed and processed easily because of the oil content that is present. GEN: K. Grosch, "The Rolling Resistance, Wear, and Traction Properties of Tread Compounds," *Rubber Chemistry and Technology*, July-August, 1996, Vol. 69, p. 495.

J. R. Beatty and B. J. Miksch developed a special cutting and chipping tester to try to simulate the occurrence of cutting and chipping with OTR tires in the field. Cutting and chipping resistance might have some relation to flex fatigue properties; however, it is generally believed to be a complex property that is quite different from traditional flex fatigue measurements. Beatty reported that his test predicted that cutting and chipping resistance probably improves when oil-extended SBR and oil-extended BR masterbatches are used. GEN: J. Beatty, B. Miksch, "A Laboratory Cutting and Chipping Tester for Evaluating Off-the-Road and Heavy-Duty Tire Treads," *Rubber Chemistry and Technology*, November-December, 1982, Vol. 55, p. 1531.

Natural Rubber

For NR-based compounds, avoid using a chemical plasticizer (peptizer) such as dibenzamido-diphenyl disulfide (PCTP) or pentachlorothiophenol (PCTP) because they may increase Goodrich Flexometer heat buildup and decrease blowout times. RT: Chapter 14, "Ester Plasticizers and Processing Additives," C. Stone, p. 368.

Flexible Coating Over NR Mount

For an all-natural rubber motor mount, consider applying a curable formulated chloroprene/polyoctenamer undercoat (drying off the solvent), followed by the application of a formulated curable halobutyl/polyoctenamer top coating, followed by drying. When properly applied and cured on the surface of the NR motor mount, this two-layer coating serves as an effective barrier against ozone attack as well as thermooxidative attack of the natural rubber. It significantly slows down the diffusion of ozone and oxygen into the NR and prevents the cleavage of the backbone of the NR molecules. Fatigue cycling testing of NR motor mounts showed that the two-layer coating improved the cycles-to-failure results significantly. GEN: H. Graf, E. Sayej, "Reversion Resistance of Engine Mounts," *Rubber World*, February, 2000, p. 55.

Guayule NR

Guayule natural rubber reportedly imparts better flex fatigue properties to a gum compound than Hevea NR because of the nonpolymeric components in Guayule that have a greater tendency to strain crystallize. On the other hand, when carbon black is added to these formulations, the fatigue life for both types of natural rubber becomes equivalent. GEN: P. Santangelo, C. Roland, "The Fatigue Life of Hevea and Guayule Rubbers," *Rubber Chemistry and Technology*, March-April, 2001, Vol. 74, p. 69.

Deproteinized Natural Rubber

It was reported from a statistical design of experiment that deproteinized natural rubber significantly imparted higher flex life than standard natural rubber, IR, or BR. GEN: R. Del Vecchio, E. Ferro, K. Winkler, "Fatigue Life Comparisons of NR Compounds," Paper No. 106 presented at the Fall Meeting of the Rubber Division, ACS, October 17, 2003, Cleveland, OH.

Polybutadiene Rubber

Consider a neodymium-catalyzed, very high *cis* content BR to achieve very good flex fatigue resistance. These catalysts increase the strain-induced crystallization characteristics of the BR and result in much improved flex fatigue resistance. *Cis* content should be greater than 98%. GEN: E. Lauretti, L. Gargani, "Neodymium Catalysts May Aid BR Products," *Rubber & Plastics News*, March 7, 1988, p. 18; J. Zhao, G. Ghebremeskel, "A Review of Some of the Factors Affecting Fracture and Fatigue in SBR and BR Vulcanizates," *Rubber Chemistry and Technology*, July-August, 2001, Vol. 74, p. 409; J. Zhao, G. Ghebremeskel (Ameripol Synpol Corp.), "A Review of Some of the Factors That Affect Fracture and Fatigue in SBR and BR Vulcanizates," Paper No. 55

presented at the Spring Meeting of the Rubber Division, ACS, April, 2000, Dallas, TX.

Increasing the vinyl content in the BR used in a compound reportedly may increase the compound's flex fatigue life. GEN: J. Zhao, G. Ghebremeskel, "A Review of Some of the Factors Affecting Fracture and Fatigue in SBR and BR Vulcanizates," *Rubber Chemistry and Technology*, July-August, 2001, Vol. 74, p. 409.

NR/BR Blend and SBR/BR Blend

A compound based on a classic 50:50 blend of 1,4-*cis*-polybutadiene and natural rubber usually will possess very good crack growth resistance. This combination appears to provide some protection from both rapid catastrophic conditions as well as slow mechano-chemical conditions that normally would cause crack growth to occur. The 50:50 blend also reportedly has a strong advantage over NR alone at low-temperature conditions. GEN: H. Kim, G. Hamed, "On the Reason That Passenger Tire Sidewalls Are Based on Blends of NR and *cis*-Polybutadiene," Presented at ACS Rubber Div. Meeting, Fall, 1999, Paper No. 184, p. 7.

According to Studebaker and Beatty, fatigue life and tire groove cracking can be improved by adding *cis*-BR to SBR or NR treads. GEN: F. Eirich, *Science and Technology of Rubber*, Chapter 9, "The Rubber Compound and Its Composition," M. Studebaker, J. R. Beatty, Academic Press, 1978, p. 367.

NR/SBR Blend

According to Studebaker and Beatty, fatigue life and tire groove cracking can be improved by adding a small concentration of SBR to a natural rubber tread. GEN: F. Eirich, *Science and Technology of Rubber*, Chapter 9, "The Rubber Compound and Its Composition," M. Studebaker, J. R. Beatty, Academic Press, 1978, p. 367.

NR/BR/SBR Triblend with Homogenizing Agent

Using a C9 hydrocarbon resin (such as Rhenosin TP100®) as a homogenizing agent in a triblend of SMR, BR, and SBR (sidewall compound) reportedly will significantly improve crack growth resistance as measured by the DeMattia crack growth tester. GEN: L. Steger (Rhein Chemie), K. Hillner, S. Schroter, "Resins in Tyre Compounds."

NR/CIIR/EPDM Triblend

Lodocsi and Young reported the use of a triblend of NR, CIIR, and EPDM in a white sidewall compound to achieve a better balance between flex resistance and static/dynamic ozone resistance. GEN: W. Waddell, "Tire Black Sidewall

Surface Discoloration and Non-staining Technology: A Review," *Rubber Chemistry and Technology*, July-August, 1998, Vol. 71, p. 590.

HNBR

HNBR can be compounded to display excellent flex fatigue resistance with good oil and heat aging resistance. It is therefore commonly used in timing belts. RT: Chapter 6, "Elastomer Selection," R. School, p. 131.

SBR

Increasing the bound styrene content in an SBR used in a gum compound may possibly increase the compound's flex fatigue life and tearing resistance. GEN: J. Zhao, G. Ghebremeskel, "A Review of Some of the Factors Affecting Fracture and Fatigue in SBR and BR Vulcanizates," *Rubber Chemistry and Technology*, July-August, 2001, Vol. 74, p. 409. However, this observation may apply only to gum SBR compounds. Others have found that low bound styrene content (or no styrene with EBR) will improve flex fatigue life and tear resistance for commercial compounds containing fillers. RP: J. M. Long.

NBR

When selecting from nitrile elastomer grades with concerns for increased flex life (by the pierced DeMattia method), consider NBR grades that have higher bound ACN content or select NBR grades produced from a cold polymerization rather than a hot polymerization process. GEN: R. Del Vecchio, E. Ferro, "Effects of NBR Polymer Variations on Compound Properties," Presented at Rubber Div. ACS Meeting, Spring, 2001, Paper No. 21.

CR

When using neoprene in compounding, the G-type gives better flex resistance. RT: Chapter 6, "Elastomer Selection," R. School, p. 133; Chapter 8, "Specialty Elastomers," L. L. Outzs, p. 208.

For good flex cracking resistance using polychloroprene, base your compound on G-type CR with MBI or ZMTI antioxidant. GEN: L. Outzs (DuPont), "Neoprene," DuPont Compounding Course, 2006, Akron, OH.

ZMTI and CR

The addition of 2 phr of ZMTI reportedly will significantly improve a CR compound's flex resistance, but the ZMTI additive will also significantly shorten scorch time. Therefore, it may be better to add ZMTI with the curatives later in the mix sequence. RP: L. L. Outzs.

CR/CIIR Compounds

For sulfur-modified polychloroprene/chlorobutyl compounds, it has been reported that the presence of magnesium oxide (as an acid acceptor) actually promotes more C–C bond links than ether and thioether links. By substituting a new heat stabilizer called "DBU/MMBI" in place of magnesium oxide, significantly improved fatigue-to-failure properties (by the Monsanto FTFT) are reported as well as better cut growth resistance for these compounds. The chemical identity of this "DBU/MMBI" heat stabilizer is only partially disclosed as DBU: 1,5-diaza-bicyclo-(4.5.0)-undecen-(11) or $C_9H_{16}N_2$. GEN: R. Musch, R. Schubart, A. Sumner, "Heat Resistant Curing System for Halogen-Containing Polymers," Presented at ACS Rubber Div. Meeting, Spring, 1999.

Reactive EPDM

Substituting a 2% maleic-anhydride modified EPDM in place of unmodified EPDM in a blend with natural rubber reportedly will significantly improve the flex fatigue resistance of the NR/EPDM compound. GEN: A. Coran, "Blends of Dissimilar Rubbers: Cure-Rate Incompatibility," *Rubber Chemistry and Technology*, May–June, 1988, Vol. 61, p. 281.

CR/EPDM Alloy

Consider developing a rubber compound based on a blend of 70 phr Neoprene GRT, 30 phr EPDM, and a compatibilizer of escor acid terpolymer at 10 phr (ethylene–methacrylate–acrylic acid terpolymer), which displays superior DeMattia cut growth resistance, heat resistance, and ozone resistance claimed to be favorable for many power transmission belt applications. GEN: P. Arjunan, R. Kusznir, A. Dekmezian, "Compatibilization of CR/EPM Blends for Power Transmission Belt Applications," *Rubber World*, February, 1997, p. 21.

HIIR

A tire halobutyl innerliner compound with lower modulus can result in higher flex crack resistance. RT: Chapter 8, "Specialty Elastomers," G. Jones, D. Tracey, A. Tesler, p. 187.

Compounds based on BIIR reportedly may give slightly better aged flex fatigue resistance than those based on CIIR. GEN: J. Fusco, "New Isobutylene Polymers for Improved Tire Processing," Presented at Akron Rubber Group Meeting, January 24, 1991.

BIMS

Brominated isobutylene-co-*p*-methylstyrene (BIMS)-based compounds may have better flex fatigue properties than a conventional halobutyl rubber compound when measured by either the Monsanto fatigue-to-failure tester or the DeMattia cut growth test. GEN: G. Jones, "Exxpro Innerliners for Severe Service Tire Applications," Presented at ITEC, 1998, Paper No. 7A.

Flowers and Fusco reported that substituting BIMS in place of halobutyl rubber or EPDM in a triblend with NR in a tire sidewall compound resulted in a significant improvement in resistance to fatigue crack growth and ozone resistance. GEN: W. Waddell, "Tire Black Sidewall Surface Discoloration and Non-staining Technology: A Review," *Rubber Chemistry and Technology*, July–August, 1998, Vol. 71, p. 590.

Fluoroalkoxyphosphazene vs. FKM, FVMQ, and NBR

Reportedly fluoroalkoxyphosphazene (FZ) can give better flexural fatigue resistance than compounds based on FKM, FVMQ, and NBR as measured by the Wallace Fatigue Test. FZ also has a wide service temperature range and unique chemical resistance characteristics. GEN: H. Penton, "P = N for Performance," *European Rubber Journal*, February, 1986, p. 20.

Polyurethanes

DeMattia flex fatigue resistance for a two-part cast polyurethane system usually can be increased by adjusting the curative ratio. The curative ratio refers to the relative amounts of prepolymer and curative. The amount of curative [such as methylene-bis-orthochloroaniline (MBCA)] needed to match exactly the isocyanate groups on the prepolymer is referred to as "100% theory" or "100% stoichiometry." Thus if the curative is reduced by 5%, it is referred to as "95% theory" or "95% stoichiometry." On the other hand, if the curative is increased by 5% from theory, it is referred to as "105% theory" or "105% stoichiometry." Generally, going to a higher stoichiometry, such as 105%, may increase (improve) DeMattia flex resistance. RT: Chapter 9, "Polyurethane Elastomers," R. W. Fuest, p. 251.

TPV

EPDM/PP thermoplastic vulcanizates (TPVs) display superior fatigue resistance (and low set) compared to many conventional thermoset rubbers. In fact, many TPVs can possess both good fatigue resistance and good compression set. (Usually these are trade-offs in conventional rubber compounding.) RT: Chapter 10, "Thermoplastic Elastomers," C. P. Rader, p. 274. RP: C. P. Rader.

Fillers

Find the optimum loading of a filler that will impart the highest fatigue life. GEN: R. Mastromatteo, E. Morrisey, M. Mastromatteo, H. Day, "Matching Material Properties to Application Requirements," *Rubber World*, February, 1983, p. 26.

Consider using a filler with lower surface area (larger particle size) to increase compound fatigue life. This is not always the case. Sometimes the opposite may be true. GEN: R. Mastromatteo, E. Morrisey, M. Mastromatteo, H. Day, "Matching Material Properties to Application Requirements," *Rubber World*, February, 1983, p. 26.

Use a filler with greater surface activity to increase compound fatigue life. GEN: R. Mastromatteo, E. Morrisey, M. Mastromatteo, H. Day, "Matching Material Properties to Application Requirements," *Rubber World*, February, 1983, p. 26.

Carbon Blacks

Increasing the surface area (decreasing the particle size) of the carbon black used in a compound will sometimes increase the cut growth resistance and flex resistance of the compound, at least for a natural rubber based compound. RT: Chapter 12, "Compounding with Carbon Black and Oil," S. Laube, S. Monthey, M-J. Wang, p. 308. However, for an SBR/BR blend, decreasing the surface area of the carbon black might actually be better to achieve improved flex resistance. For example, N660 is generally better than N330 in a NR/BR tire sidewall compound for flex fatigue resistance. RP: J. M. Long.

Decreasing the structure of the carbon black used in a compound will usually increase the cut growth resistance and flex resistance of the compound. RT: Chapter 12, "Compounding with Carbon Black and Oil," S. Laube, S. Monthey, M-J. Wang, p. 308.

Increasing the loading of the carbon black used in a compound through an optimum level will usually increase the cut growth resistance and flex resistance of the compound through an optimum level. This optimum may occur at a lower concentration for a smaller particle size carbon black. RT: Chapter 12, "Compounding with Carbon Black and Oil," S. Laube, S. Monthey, M-J. Wang, p. 308. RP: M-J. Wang.

Consider using carbon black with a low structure and moderately high surface area. The low structure will result in a lower modulus, resulting in a higher ultimate elongation. The moderately high surface area gives high tensile strength and good tear resistance. The resulting "high tensile x elongation" product is thought to relate to improved *cutting and chipping resistance.* Of course, low-structure carbon blacks just described are notoriously difficult to disperse. Therefore, steps to achieve good dispersion must be performed or poor dispersion will result. Poor dispersion will result in poor tear as well as poor cutting and chipping properties. A lower structure N200 type carbon black such as N231 has been used to improve cutting and chipping resistance for medium radial truck treads, for example. RT: Chapter 12, "Compounding with Carbon Black and Oil," S. Laube, S. Monthey, M-J. Wang, pp. 317, 319.

It is known that a compound's fatigue life will increase with a higher loading of carbon black until it reaches a maximum and then its fatigue life will start to decrease with greater loading levels of carbon black. If a carbon black with a higher surface area is used, this maximum for fatigue life will shift to a lower loading level. On the low carbon black loading side of this curve, the improvement in fatigue life is reportedly due to an increase in tearing energy. On the other hand, on the high carbon black loading side of this fatigue life curve, the decrease in fatigue life is reportedly due to an increase in "effective initial flaw size" from carbon black agglomeration. GEN: J. Zhao, G. Ghebremeskel, "A Review of Some of the Factors Affecting Fracture and Fatigue in SBR and BR Vulcanizates," *Rubber Chemistry and Technology*, July-August, 2001, Vol. 74, p. 409.

J. R. Beatty and B. J. Miksch developed a special cutting and chipping tester to try to simulate the occurrence of cutting and chipping with OTR tires in the field. Cutting and chipping resistance might have some relation to flex fatigue properties; however, it is generally believed to be a complex property that is quite different from traditional flex fatigue measurements. Beatty reported that his test predicted that cutting and chipping resistance probably reaches an optimum at a carbon black loading of approximately 55 phr or above for both NR and SBR compounds. GEN: J. Beatty, B. Miksch, "A Laboratory Cutting and Chipping Tester for Evaluating Off-the-Road and Heavy-Duty Tire Treads," *Rubber Chemistry and Technology*, November-December, 1982, Vol. 55, p. 1531.

Special Carbon Blacks

For off-the-road tires (OTR), cutting and chipping properties are very important. It has been hypothesized that the OTR tread should have a dynamic stiffness (E*) great enough to minimize "cutting" or penetration by rocks, road aggregates, and so on. Also, the compound should have a high enough tearing strength to avoid chipping. Tread compounds that contain special carbon blacks with high surface area (small particle size), relatively low structure (oil absorption number), low microporosity, and good surface activity (such as Degussa's Ecorax 1990) and an additional 10 phr of precipitated hydrated silica should possess better cutting and chipping resistance. GEN: W. Niedermier (Degussa AG), "A Tailor-Made Carbon Black to Improve the Balance Between 'Cut and Chip,' Treadwear, and Hysteresis," Paper No. 8A presented at ITEC 2002, September, 2002, Akron, OH.

Silica

Commonly, precipitated silica is used at a 10 to 15 phr level to increase a compound's cutting and chipping resistance. RT: Chapter 12, "Compounding with Carbon Black and Oil," S. Laube, S. Monthey, M-J. Wang, p. 319.

Consider using low levels of precipitated silica in a rubber compound where an improvement in flex fatigue (resistance to cut growth and cracking) is needed. RT: Chapter 13, "Precipitated Silica and Non-black Fillers," W. Waddell, L. Evans, p. 331.

Using a precipitated silica with a higher surface area may lengthen the scorch safety time of a rubber compound as well as slow down the actual cure rate, but it may also decrease cut growth. GEN: W. Waddell, L. Evans, "Use of Nonblack Fillers in Tire Compounds," *Rubber Chemistry and Technology*, July-August, 1996, Vol. 69, p. 377.

Silica/Silane

It is well known that using silane coupling agents with precipitated silica will generally improve the compound's abrasion resistance and lower the hysteresis of the cured compound. However, it is usually better to use an organosilane with polysulfidic sulfane functionality rather than a mercaptosilane to achieve better flex cracking resistance. RP: T. D. Powell.

Talc

Talc has been reported to impart improvements in fatigue and/or cutting and chipping resistance. It is reported that talc has advantages over precipitated silica and clay in improving fatigue performance as measured with the Monsanto Fatigue-to-Failure Test. It is reported that one of the advantages of talc in imparting better fatigue life is that it works well with rubber having razor cuts. It is reported that partial substitution with surface-treated talcs have significantly improved peroxide-cured EPDM hose compound for fatigue life. GEN: O. Noel, Educational Symposium on Fillers, "Talc: A Functional Mineral for Rubber," Presented at ACS Rubber Div. Meeting, May, 1995; O. Noel, *Luzenac America Research Report*, "Automotive Coolant Hose/Fatigue Resistance," March 29, 1999. RP: O. Noel.

Avoid Ground Quartz

The use of ground quartz should be avoided in rubber compounding because its presence in a compound can be quite detrimental to fatigue resistance as well as possibly creating a safety hazard to rubber workers. RT: Chapter 8, "Specialty Elastomers," J. R. Halladay, p. 236.

Starch Filler

It is reported that when 5 phr of a modified starch is used in place of a fully reinforcing carbon black in an SBR-based compound, there was some improvement in tensile fatigue life and crack growth resistance. A starch paste and rubber latex were coagulated and then mixed with other compounding ingredients. GEN: Y. Wu, L. Yang, L. Zhang, X. He, "Role of Starch in Improving the Fatigue Life of Carbon Black Filled SBR Composites," Paper No. 52 presented at the Fall Meeting of the Rubber Division, ACS, October 11–13, 2011, Cleveland, OH.

Fibers and Cutting-Chipping Resistance

Beatty and Hamed suggested that adding about 1% cellulose fibers to an OTR tread compound would improve the tread's resistance to chipping and chunking when in service because the dispersed fibers reportedly reduce crack propagation rates. GEN: L. Goettler, K. Shen, "Short Fiber Reinforced Elastomers," *Rubber Chemistry and Technology*, 1983, Vol. 56, p. 575.

Acetone-Diphenylamine Reaction Product

Antioxidants that are reaction products of acetone and diphenylamine (such as BLE-25 or Superflex) reportedly impart better flex cracking and fatigue resistance than the alkylated diphenylamines. RT: Chapter 19, "Antidegradants," F. Ignatz-Hoover, p. 455.

PPD

p-Phenylenediamines (PPDs) are usually considered useful antifatigue agents. RT: Chapter 19, "Antidegradants," F. Ignatz-Hoover, p. 454.

6PPD

6PPD is excellent for extending flex fatigue life in NR-based compounds. This flex fatigue resistance is imparted even to vulcanizates that have been aged. RT: Chapter 19, "Antidegradants," F. Ignatz-Hoover, p. 448.

6PPD/"Mixed" PPDs

Using 6PPD with "mixed" PPDs (such as Polystay 100) may extend flex fatigue life. RP: R. Dailey.

6PPD/77PPD Combination

A combination of 6PPD (for long-term applications) and 77PPD (for short-term applications) is commonly used for better protection against flex cracking. GEN: L. Walker, J. Luecken, "Antidegradants for Ozone and Fatigue Resistance: Laboratory and Tire Tests," *Elastomerics*, May, 1980, p. 36.

TMQ/BLE Combination

Consider a blend of TMQ and BLE where both good heat aging and flex fatigue properties are needed. GEN: S. Hong, C. Lin, "Improved Flex Fatigue and Dynamic Ozone Crack Resistance Through the Use of Antidegradants or Their Blends in Tire Compounds," Presented at ACS Rubber Div. Meeting, Fall, 1999, Paper No. 27.

ZMTI

The use of zinc 2-mercaptotoluimidazole (ZMTI) reportedly can significantly improve crack growth resistance as measured by DeMattia Flex for NR-based compounds as well as CR-, SBR-, and NBR-based compounds. GEN: R. Ohm, "Accelerators and Antidegradants Influence Fatigue Resistance," *Elastomerics*, January, 1988, p. 19.

Hydrocarbon Resins

For elastomer bases such as SBR, NBR, or CR with low loadings of nonreinforcing fillers such as clay, whiting, lithopone, and so forth, adding 15 to 25 phr of a hydrocarbon resin such as a coal tar resin may improve the cured compound resistance to cut growth. GEN: F. O'Connor, J. Slinger, "Processing Aids: The All-Inclusive Category," *Rubber World*, October, 1982, pp. 21–23.

Reportedly, adding petroleum hydrocarbon resins (DCPD or cyclic diolefin resins) to a general purpose rubber-based compound may improve its DeMattia flex fatigue resistance and/or its chipping and chunking resistance. GEN: "Hydrocarbon Resins for Rubber Compounding," Neville Co., 2011.

Ultimate Elongation

Sometimes cut growth resistance (or flex life) is correlated with ultimate elongation properties. GEN: N. Hewitt, "The Use of Viscoelastic Series for Compound Design," *Rubber World*, 1984; R. Del Vecchio, E. Ferro, "Effects of NBR Polymer Variations on Compound Properties," Presented at ACS Rubber Div. Meeting, Spring, 2001, Paper No. 21.

Avoid Porosity in the Vulcanizate

Porosity in a rubber vulcanizate can significantly shorten the flex fatigue life of a given rubber part because of crack initiation and fatigue (as the failure mode). Bubbles in a vulcanizate function as "large precursor cracks" and increase crack propagation rates. In dynamic applications, fully cured rubber parts that possess porosity may experience shorter service life. Therefore, steps to prevent porosity should be taken (see Section 5.1 on ways to reduce porosity). GEN: A. Kasner, E. Meinecke, "Porosity in Rubber: A Review," *Rubber Chemistry and Technology*, July–August, 1996, Vol. 69, p. 223, 424.

Double Networks

Reportedly, natural rubber cured "double networks" might be used to greatly improve mechanical fatigue life. GEN: P. Santangelo, C. Roland, "Failure Properties of Natural Rubber Double Networks," *Rubber Chemistry and Technology*, March-April, 1995, p. 124.

Avoid Passing Through Zero Strain

Beatty reported on the importance of not allowing rubber parts to pass through zero strain under tension when in service when made of compounds based on natural rubber or synthetic elastomer blends with NR. NR blends can display superior fatigue life compared to some other elastomer systems provided that they are not cycled through zero strain. In addition, Lake and Clapson reported that sometimes BR possesses superior flex fatigue life compared to NR. Hess and Chirico reported that NR/BR blends may possess fatigue life superior to that of NR/SBR blends. GEN: W. Hess, C. Herd, P. Vegvari, "Characterization of Immiscible Elastomer Blends," *Rubber Chemistry and Technology*, July-August, 1993, Vol. 66, p. 329.

Optimal Tan δ and Cutting and Chipping Resistance

J. R. Beatty and B. J. Miksch developed a special cutting and chipping tester to try to simulate the occurrence of cutting and chipping with OTR tires in the field. Cutting and chipping resistance might have some relation to flex fatigue properties; however, it is generally believed to be a complex property that is quite different from traditional flex fatigue measurements. Beatty reported that his test predicted that cutting and chipping resistance probably improves when the tread compound has a higher cured tan δ (more hysteretic). However, if the tan δ value is too high, the tire will run too hot and cutting and chipping resistance might decrease. GEN: J. Beatty, B. Miksch, "A Laboratory Cutting and Chipping Tester for Evaluating Off-the-Road and Heavy-Duty Tire Treads," *Rubber Chemistry and Technology*, November-December, 1982, Vol. 55, p. 1531.

Hardness, Controlled Strain or Controlled Stress, and Flex Fatigue

When comparing rubber compounds with different cured hardness values, the conditions under which the flex fatigue test is performed are important. Flex fatigue properties can be quite different depending on whether the test applies an equal strain condition or an equal strain energy condition. GEN: R. Tabar, P. Killgoar, R. Pett, "A Fatigue Resistant Polychloroprene Compound for High Temperature Dynamic Applications," *Rubber Chemistry and Technology*, September-October, 1979, Vol. 52, p. 781.

3.4 Improving Abrasion and/or Wear Resistance

Abrasion resistance and wear resistance can have different meanings for different applications. The severity of wear or abrasion also can be quite important. Wear is important for applications such as tires, belts, sandblasting hose, rubber rollers, and shoe soles. This phenomenon is very complex and is believed to be related to a variety of other cured compound properties such as cut resistance, thermal stability, tear resistance, hardness, fatigue resistance, and so forth. Several quite different standardized ASTM abrasion tests are used, each providing different test results. There is no universally accepted test that is used by everyone.

The following ***experimental*** *ideas may improve abrasion and/or wear resistance. Also, all relevant literature sources, including the ones cited below, should be researched and read.* ***Caution:*** *these general experimental ideas may not work in all specific situations. Changes to improve abrasion and/or wear resistance will certainly affect other properties as well, for better or for worse, and this book does not purport to show how these other properties are affected. Also, this book does not purport to address safety and health issues.*

Mixing

Increase dispersion of reinforcing fillers such as carbon black and/or silica through better mixing. RT: Chapter 2, "Compound Processing Characteristics and Testing," J. S. Dick, p. 42. GEN: S. Monthey, T. Reed, "Performance Differences Between Carbon Blacks and CB Blends for Critical IR Applications," *Rubber World*, April, 1999.

Improving the dispersion of carbon black through longer mix cycles will usually improve compound abrasion resistance as measured by the Akron Angle Abrader. RT: Chapter 12, "Compounding with Carbon Black and Oil," S. Laube, S. Monthey, M-J. Wang, p. 308.

Interestingly, the smaller particle size carbon blacks and silicas (which can potentially impart better abrasion resistance) are the more difficult fillers to disperse. Therefore, it is important to achieve good dispersion of these small particle size fillers to achieve their full benefits for abrasion resistance. RP: T. D. Powell.

Add the carbon black early in the mixing sequence. Avoid adding the carbon black with oil, stearic acid, or other polar ingredients such as antioxidants because these ingredients may be absorbed onto the surface of the carbon black particles, which will interfere with the polymer absorption onto the carbon black surface. Thus adding oil and other ingredients with the carbon black might interfere with carbon black-rubber interaction or the formation of bound rubber. This would be detrimental to abrasion resistance. Therefore, it might be better to add the carbon black first before some of these other ingredients. GEN: M-J. Wang, T. Wang, K. Mahmud, "Effect of Carbon Black Mixing on Rubber Reinforcement," Proceedings of the 3rd International Conference on Carbon Black, p. 205, Mulhouse, October 25-26, 2000. RP: M-J. Wang.

Heat Treatment from Mixing

For diene-based elastomers such as SBR or BR with relatively high mechanooxidation stability, increase the intensity of mixing and lengthen mixing time in order to apply "heat treatment," which may increase the bound rubber and improve the carbon black dispersion, resulting in better abrasion resistance. GEN: M.-J. Wang, T. Wang, K. Mahmud, "Effect of Carbon Black Mixing on Rubber Reinforcement," Proceedings of the 3rd International Conference on Carbon Black, p. 205, Mulhouse, October 25-26, 2000. RP: M-J. Wang.

Phase Mixing

Through phase mixing techniques of an NR/BR blend, one can achieve an increase in the amount of carbon black dispersed in the BR phase, which may increase the tread wear resistance of the compound. GEN: E. McDonel, K. Baranwal, J. Andries, "Polymer Blends," Vol. 2, Chapter 19, *Elastomer Blends in Tires*, Academic Press, 1978, p. 282; W. Hess, C. Herd, P. Vegvari, "Characterization of Immiscible Elastomer Blends," *Rubber Chemistry and Technology*, July-August, 1993, Vol. 66, p. 329.

Krakowski and Tinker reported that increasing the carbon black concentration in the BR phase in an BR/NR blend caused the DIN abrasion values to improve significantly. GEN: W. Hess, C. Herd, P. Vegvari, "Characterization of Immiscible Elastomer Blends," *Rubber Chemistry and Technology*, July-August, 1993, Vol. 66, p. 329.

Fillers

Optimizing the loading of a filler will increase the compound's abrasion resistance. GEN: R. Mastromatteo, E. Morrisey, M. Mastromatteo, H. Day, "Matching Material Properties to Application Requirements," *Rubber World*, February, 1983, p. 26.

Use a filler with higher surface area to increase compound abrasion resistance. GEN: R. Mastromatteo, E. Morrisey, M. Mastromatteo, H. Day, "Matching Material Properties to Application Requirements," *Rubber World*, February, 1983, p. 26.

Use a filler with higher structure to increase compound abrasion resistance. GEN: R. Mastromatteo, E. Morrisey, M. Mastromatteo, H. Day, “Matching Material Properties to Application Requirements,” *Rubber World*, February, 1983, p. 26.

Use a filler with greater surface activity to increase compound abrasion resistance. GEN: R. Mastromatteo, E. Morrisey, M. Mastromatteo, H. Day, “Matching Material Properties to Application Requirements,” *Rubber World*, February, 1983, p. 26.

Carbon Black

Increasing the loading of the carbon black used in a compound will usually increase the abrasion resistance of the compound through an optimum level after which abrasion resistance will decrease. RT: Chapter 12, “Compounding with Carbon Black and Oil,” S. Laube, S. Monthey, M-J. Wang, pp. 308, 315. GEN: K. Hale, J. West, C. McCormick, “Contribution of Carbon Black Type to Skid and Treadwear Resistance,” Presented at ACS Rubber Div. Meeting, Spring, 1975, Paper No. 6.

Increasing the surface area of the carbon black used in a compound will usually increase both the low- and high-severity abrasion resistance of the compound. In fact, the fineness of the carbon black is generally thought to be one of the most important carbon black characteristics that relate to abrasion and wear resistance. However, if the carbon black has too fine a particle size, it may not disperse well, which may result in a loss in abrasion resistance. Another limiting factor regarding the fineness of the particles is the resulting increase in hysteresis and heat generation of the cured rubber while in service, which results in higher surface temperatures that accelerate surface thermooxidative degradation. M-J. Wang reports that for an SBR/BR blend, the optimal surface area for best abrasion resistance may be between 130 and 150 m^2/g by the CTAB surface area measurement. RT: Chapter 12, “Compounding with Carbon Black and Oil,” S. Laube, S. Monthey, M-J. Wang, p. 308. GEN: C-H. Shieh, M. L. Mace, G. B. Ouyang, J. M. Branan, J. M. Funt, Paper presented at ACS Rubber Div. Meeting, Fall, 1991. RP: M-J. Wang.

Increasing the structure of the carbon black used in a compound will usually increase the abrasion resistance of the compound. RT: Chapter 12, “Compounding with Carbon Black and Oil,” S. Laube, S. Monthey, M-J. Wang, p. 308.

For low-severity tire tread wear conditions, it is important to select carbon blacks with high surface area (small particle size). On the other hand, for the best high severity (or very high severity) wear conditions, it may be important to select a high-structured carbon black according to Studebaker and Beatty.

GEN: F. Eirich, *Science and Technology of Rubber*, Chapter 9, "The Rubber Compound and Its Composition," M. Studebaker, J. Beatty, Academic Press, 1978, p. 367. RP: M-J. Wang.

Determine the optimum loading for a specific carbon black that will impart the best abrasion resistance. It is important to note that this carbon black optimal loading differs greatly from one grade of carbon black to the next based on differences in colloidal properties. GEN: A. Sorcar, "Optimum Loading of Carbon Black in Rubber by Monsanto Oscillating Disc Rheometer," *Rubber World*, November, 1987, p. 30.

To improve low-severity abrasion resistance of a rubber compound, consider using a carbon black with a narrower aggregate distribution. However, differences in the distribution of carbon black aggregates reportedly have little effect on high-severity abrasion resistance. GEN: C-H. Shieh, M. L. Mace, G. B. Ouyang, J. M. Branan, J. M. Funt, Presented at ACS Rubber Div. Meeting, Spring, 1991.

Select a tread grade of carbon black that possesses a fine enough particle size and a high enough structure to impart very good tread wear resistance. For example, the carbon blacks listed here were reported to have the following relative wear resistance:

N351 = 87
N339 = 95
N220 = 100
N234 = 108
N134 = 113

A ±13% range among tread-grade carbon blacks for imparted tread wear has been reported. RT: Chapter 12, "Compounding with Carbon Black and Oil," S. Laube, S. Monthey, M-J. Wang, p. 315.

Nanostructured Carbon Blacks

Nanostructured, low-hysteresis carbon blacks also are reported to impart improved truck tread wear. GEN: W. Niedermeier, B. Freund, "Nanostructure Blacks: A New Carbon Black Family Designed to Meet Truck Tire Performance Demands," Presented at ACS Rubber Div. Meeting, Fall, 1998, Paper No. 28.

LL Carbon Black

The use of a high-structure "LL carbon black" (long linkage carbon blacks that are crush resistant during mixing) in place of regular high-structure carbon black reportedly can be used to impart better wear characteristics to a vulcanizate. According to one study, high-structure carbon blacks are commonly susceptible to "crush" during mixing and convert to ordinary carbon blacks.

These special high-structure LL carbon black grades are crush resistant. GEN: H. Mouri, K. Akutagawa, "Reducing Energy Loss to Improve Tire Rolling Resistance," Presented at ACS Rubber Div. Meeting, Spring, 1997, Paper No. 14.

Chemical Promotors for Carbon Black

Use of a carbon black–rubber coupling agent or "chemical promotor" in a carbon black loaded rubber compound may increase resilience and modulus as well as reduce abrasion loss. In the past, such coupling agents as *N*-(2-methyl-2-nitropropyl)-4-nitrosoaniline, *N*-4-dinitroso-*N*-methylaniline, *p*-nitrosodiphenylamine, and *p*-nitroso-*N*,*N*-dimethylaniline were used. These nitroso compounds are no longer used because they give off nitrosoamines, which are suspected carcinogens. Therefore other coupling agents have been tried. A new coupling agent, *p*-aminobenzenesulfonyl azide (or amine-BSA), has recently been investigated that may impart these performance advantages to a compound. GEN: L. Gonzalez, A. Rodriguez, J. deBenito, A. Marcos, "A new Carbon Black–Rubber Coupling Agent to Improve Wet Grip and Rolling Resistance of Tires," *Rubber Chemistry and Technology*, May–June, 1996, Vol. 69, p. 266.

Consider performing an *in situ* chemical reaction during compound mixing at about 150ºC between an in-chain carboxylated functionalized high-vinyl SSBR and an ozone-treated N234-type carbon black in order to improve rubber-filler interaction and improve compound abrasion resistance. GEN: J. Douglas, S. Crossley, J. Hallett, J. Curtis, D. Hardy, T. Gross, N. Steinhauser, A. Lucassen, H. Kloppenburg (Lanxess and Columbian Chemical), "The Use of Surface-Modified Carbon Black with an In-Chain Functionalized Solution SSBR as an Alternative to Higher Cost Green Tire Technology," Paper No. 38 presented at the Fall Meeting of the Rubber Division, ACS, October 11–13, 2011, Cleveland, OH.

Clay-Carbon Nanocomposites

Consider using a clay-carbon nanocomposite to improve abrasion resistance. GEN: A. Chandra, S. Patel (Apollo Tyres), A. Jineesh, D. Tripathy (Rubber Technology Centre IIT Kharagpur), "Clay-Carbon Nanocomposites for Lower Hysteresis High Abrasion Tread," Paper No. 21B presented at the ITEC 2008 Meeting, September 15–17, 2008, Akron, OH.

Carbon Nanotube Masterbatch

It was reported that a special 10 to 15% modified carbon nanotube in a rubber base was used as a masterbatch for another rubber formulation in a regular mix. This masterbatch technique gave good dispersion of the CNT into an NR/SBR tire tread compound where a measurable improvement in Tabor abrasion was achieved vs. a control. The effective concentration of CNT in the rubber tread compound was between 1 to 3 phr. GEN: C. Bosnyak, K. Swogger (Designed Nanotubes, LLC), "Changing Materials to Change Markets," Paper presented at the India Rubber Expo 2011, January 20, 2011, Chennai, India.

Avoid Carbon Black Blends

Avoid extreme blends of carbon black (carcass and tread grades), which can cause tire treads to have very poor wear characteristics. GEN: W. Hess, W. Klamp, "The Effects of Carbon Black and Other Compounding Variables on the Tire Rolling Resistance and Traction," *Rubber Chemistry and Technology*, Vol. 56, p. 390.

Silica

Consider using precipitated silica in a rubber compound if an improvement in abrasion resistance in shoe soles is needed. RT: Chapter 13, "Precipitated Silica and Non-black Fillers," W. Waddell, L. Evans, p. 331.

The use of silane coupling agents with precipitated silica loading can significantly improve the compound's abrasion resistance. Organosilanes with Sx content can impart very good tire wear resistance. GEN: N. Hewitt, "Compounding with Silica for Tear Strength and Low Heat Build-up," *Rubber World*, June, 1982. RP: T. D. Powell.

S. Wolff reported that using precipitated silica and the coupling agent bis(3-triethoxysilylpropyl)-tetrasulfide resulted in significant improvement in abrasion resistance for earthmover tire treads. GEN: W. Waddell, L. Evans, "Use of Nonblack Fillers in Tire Compounds," *Rubber Chemistry and Technology*, July-August, 1996, Vol. 69, p. 377.

Precipitated hydrated silica is very difficult to disperse. Better abrasion resistance can be achieved by using the new generation of highly dispersible silicas (HDS). GEN: S. Daudey, L. Guy (Rhodia), "High Performance Silica Reinforced Elastomers from Standard Technology to Advanced Solutions," Paper No. 37 presented at the Fall Meeting of the Rubber Division, ACS, October 11–13, 2011, Cleveland, OH.

Fibers

Consider using a low loading of short aramid fiber to improve abrasion resistance in the "cross machine" direction, perpendicular to the fiber ends. In the reference cited, this improvement was achieved with 15 phr of Kevlar pulp in a CR compound. GEN: K. Watson, A. Frances, "Elastomer Reinforcement with Short Kevlar Aramid Fiber for Wear Applications," *Rubber World*, August, 1988, p. 20.

Additives

In polyurethane formulations, molybdenum disulfide, fluorocarbons (Teflon®), or special silicone oils are sometimes used to improve friction properties and wear characteristics. RT: Chapter 9, "Polyurethane Elastomers," R. W. Fuest, p. 252.

PTFE

A PTFE additive such as Alphaflex® may improve the abrasion resistance of a rubber compound. GEN: J. Menough, "A Special Additive," *Rubber World*, May, 1987, p. 12.

Reinforcing Resins

Using reinforcing resins in a rubber compound reportedly can significantly improve abrasion resistance. Use a reinforcing novolak phenol-formaldehyde resin with a methylene donor such as hexamethylenetetramine (HMT) or hexamethoxymethylomelamine (HMMM). These two ingredients react *in situ* with each other during the normal vulcanization of the rubber compound. The use of this reinforcing resin and methylene donor in a rubber compound can significantly increase the hardness of the vulcanizate and its abrasion resistance. RT: Chapter 18, "Tackifying, Curing, and Reinforcing Resins," B. Stuck, p. 440.

Oil

Usually there is an optimum oil loading for a given loading of tread grade carbon black that will impart the best tread wear resistance. GEN: K. Hale, J. West, C. McCormick, "Contribution of Carbon Black Type to Skid and Treadwear Resistance," Presented at ACS Rubber Div. Meeting, Spring, 1975, Paper No. 6.

Excessive levels of processing oil used in the recipe may be detrimental to abrasion resistance. RT: Chapter 12, "Compounding with Carbon Black and Oil," S. Laube, S. Monthey, M-J. Wang, p. 311.

Higher Average Molecular Weight

In order to improve flex fatigue resistance as well as abrasion resistance, consider using an oil-extended general purpose elastomer masterbatch. These oil-extended masterbatches generally have higher average molecular weights but can still be mixed and processed easily because of the oil content that is present. GEN: K. Grosch, "The Rolling Resistance, Wear, and Traction Properties of Tread Compounds," *Rubber Chemistry and Technology*, July-August, 1996, Vol. 69, p. 495.

Polyurethane

Polyurethane elastomers have a great advantage over conventional diene rubbers in that they can impart superior abrasion and wear resistance. RT: Chapter 6, "Elastomer Selection," R. School, p. 137; Chapter 9, "Polyurethane Elastomers," R. W. Fuest, p. 253. GEN: M. Chase, "Roll Coverings Past, Present, and Future," Presented at Rubber Roller Group Meeting, May 15–17, 1996, p. 6.

Abrasion resistance for a two-part cast polyurethane system usually can be increased by adjusting the curative ratio. The curative ratio refers to the relative amounts of prepolymer and curative. The amount of curative [such as methylene-bis-orthochloroaniline (MBCA)] needed to match exactly the isocyanate groups on the prepolymer is referred to as "100% theory" or "100% stoichiometry." Thus if curative is reduced by 5%, it is referred to as "95% theory" or "95% stoichiometry." On the other hand, if the curative is increased by 5% from theory, it is referred to as "105% theory" or "105% stoichiometry." Generally, going to a optimal stoichiometry, perhaps near 100% theory, may improve abrasion resistance. RT: Chapter 9, "Polyurethane Elastomers," R. W. Fuest, p. 251.

Usually, but not always, selecting an ether-type polyurethane based on an MDI prepolymer will give better *impingement* abrasion resistance in demanding applications. Usually, but not always, selecting an ester-type polyurethane will give better *sliding* abrasion resistance in demanding applications. RT: Chapter 9, "Polyurethane Elastomers," R. W. Fuest, p. 257.

Millable PU Blends

It has been reported that adding incrementally a given quantity of millable polyurethane to an NBR/PVC-, SBR-, or EPDM-based compound can improve the compound's abrasion resistance. GEN: T. Jablonowski, "Blends of PU with Conventional Rubbers," *Rubber World*, October, 2000, p. 41.

Cold SBR

Use cold- instead of hot-emulsion SBR for better abrasion resistance. RT: Chapter 6, "Elastomer Selection," R. School, p. 129.

Use ESBR polymerized at 5°C (41°F) (or lower) instead of 50°C (122°F) for a significant increase in tire tread wear resistance. RT: Chapter 7, "General Purpose Elastomers and Blends," G. Day, p. 149.

Polybutadiene

For very good abrasion resistance, use a significant loading of 1,4-*cis*-polybutadiene in a compound, and use especially a *cis*-BR produced using neodymium (Nd), cobalt (Co), or titanium (Ti) catalysts. GEN: A. Niziolek, R. Jones, J. Neilsen,

"Influence of Compounding Materials on Tire Durability," Presented at ACS Rubber Div. Meeting, Spring, 1999, Paper No. 59, p. 7.

In tire treads for better wear resistance, avoid using solution BR with significant 1,2 microstructure. Use another high *cis* BR instead. RT: Chapter 7, "General Purpose Elastomers and Blends," G. Day, p. 157.

Tire tread wear reportedly improved about 1% for each percentage of *cis*-BR that was added to an SBR- or oil-extended SBR-based tread formulation in bias ply tires. It was also reported that tire wear improvements with the addition of *cis*-BR to an NR-based tread formulation were not seen to be significant until the percentage of *cis*-BR exceeded 50% for bias tires. GEN: R. Brown, R. Knill, J. Kerscher, R. Todd, "Compounding *cis*-Polybutadiene," *Rubber World*, November, 1961, pp. 70, 71, 72. RP: T. D. Powell.

Of the three elastomers (*cis*-BR, E-SBR, and NR) that are commonly used in tire tread formulations, the *cis*-BR usually imparts the better wear resistance to the tread. This is thought to be due to the lower glass transition temperature of the *cis*-BR polymers. GEN: E. McDonel, K. Baranwal, J. Andries, *Polymer Blends*, Vol. 2, Chapter 19, "Elastomer Blends in Tires," Academic Press, 1978, p. 283.

SBR/BR vs. NR/BR vs. NR/SBR Blends

On the basis of a comparative study, Hess and Chirico reported that an SBR/BR tread gave the best tread wear resistance, followed by an NR/BR blend, followed by an NR/SBR blend, which had the poorest tread wear of the three. GEN: W. Hess, C. Herd, P. Vegvari, "Characterization of Immiscible Elastomer Blends," *Rubber Chemistry and Technology*, July–August, 1993, Vol. 66, p. 329.

BR/NR and BR/SBR Blends

Sarbach reported that tires with treads based on a 50:50 blend of BR/NR gave 20% better wear resistance than tires based on treads based solely on NR. He also reported 46% better tread wear for 50:50 BR/SBR treads vs. tires with treads based solely on SBR. GEN: W. Hess, C. Herd, P. Vegvari, "Characterization of Immiscible Elastomer Blends," *Rubber Chemistry and Technology*, July–August, 1993, Vol. 66, p. 329.

Functionalized Polymers

Use an alkoxysilane-functionalized SSBR in a tire tread compound for improved tread wear. GEN: T. Hogan, A. Randall, W. Hergenrother, C. Lin (Bridgestone Research), "The Role of Functional Polymers in Improving Tire Performance," Paper No. 113 presented at the Fall Meeting of the Rubber Division, ACS, October 13–15, 2009, Pittsburgh, PA.

Natural Rubber

NR compounds can be formulated to have good abrasion resistance with good dynamic properties as well. GEN: M. Chase, “Roll Coverings Past, Present, and Future,” Presented at Rubber Roller Group Meeting, May 15–17, 1996, p. 6.

Adding ball-milled dispersions of reinforcing carbon black to natural rubber latex before coagulation, and then using the coagulated NR in rubber compounding, will yield higher abrasion and wear resistance than simply adding free carbon black directly into the Banbury mixer. GEN: R. Alex, K. Sasidharan, T. Kurlan, A. Kumarchandra, “Carbon Black/Silica Masterbatch from Fresh Natural Rubber Latex,” Paper No. 27 presented at IRC 11, January 19, 2011, Chennai, India.

NR/SBR Blend

The abrasion resistance of an NR/SBR blend may be improved with the addition of a compatibilizer to the compound. GEN: “Effects of Diblock Copolymer as Compatibilizer on Blends,” Presented at the Spring Meeting of the Rubber Division, ACS, 1999, Chicago, IL.

SSBR

In comparing different solution SBR polymers (SSBR) with the same glass transition temperature value (T_g), it was found that the DIN abrasion values got better when the styrene content was increased with a corresponding reduction in vinyl content. GEN: J. Summer, R. Engehausen, J. Trimbach (Bayer AG), “Polymer Developments to Improve Tire Life and Fuel Economy,” Paper No. 98 presented at the Fall Meeting of the Rubber Division, ACS, October 16–19, 2001, Cleveland, OH.

SIBR

Consider using an SIBR (styrene-isoprene-butadiene terpolymer) in a blend with high *cis*-BR at 30 phr in a tire tread compound to achieve a better combination of wear resistance and wet traction. Consider using SIBRs with 45/45/10 and 40/40/20 I/Bd/S monomer ratios for the best trade-offs between tread wear resistance and wet traction. GEN: A. Halasa, B. Gross, W. Hsu (Goodyear Tire), “Multiple Glass Transition Terpolymers of Isoprene, Butadiene, and Styrene,” Paper No. 91 presented at the Fall Meeting of the Rubber Division, ACS, October 13–15, 2009, Pittsburgh, PA.

NBR

Use NBR grades with higher ACN content for higher compound abrasion resistance. RT: Chapter 8, “Specialty Elastomers,” M. Gozdiff, p. 194.

XNBR

Carboxylated nitrile elastomers (XNBRs) impart better abrasion resistance properties than conventional NBR. RT: Chapter 6, "Elastomer Selection," R. School, p. 131.

Use an XNBR with the appropriate zinc oxide in place of conventional NBR to achieve a higher compound abrasion resistance. RT: Chapter 8, "Specialty Elastomers," M. Gozdiff, p. 199.

HNBR

At high temperatures, HNBR reportedly has better abrasion resistance than polyurethane. GEN: M. Chase, "Roll Coverings Past, Present, and Future," Presented at Rubber Roller Group Meeting, May 15–17, 1996, p. 4.

HXNBR

Hydrogenated carboxylated nitrile rubber (HXNBR) with a mild post-cure reportedly has better abrasion resistance than even XNBR. GEN: R. Pazur, L. Farrari, E. Campomizzi (Lanxess), "HXNBR Compound Property Improvements Through the Use of Post Cure," Paper No. 70 presented at the Spring Meeting of the Rubber Division, ACS, May 16–18, 2005, San Antonio, TX.

Specialty Elastomers: CM, CSM, GECO, and NBR/PVC

A four-way comparison of compounds based on CM, CSM, GECO, and NBR/PVC found that CM gave superior abrasion resistance in relation to the other three elastomers. GEN: C. Hooker, R. Vara, "A Comparison of Chlorinated and Chlorosulfonated Polyethylene Elastomers with Other Materials for Automotive Fuel Hose Covers," Presented at ACS Rubber Div. Meeting, Fall, 2000, Paper No. 128.

TOR Blends with NR, BR, SBR, NBR, CR, or EPDM

Polyoctenamer or *trans*-polyoctenylene rubber (TOR) reportedly can be blended in minor amounts with NR, BR, SBR, NBR, CR, EPDM, and so forth to improve compound abrasion resistance. GEN: A. Draxler, "A New Rubber: *trans*-Polyoctenamer," Chemische Werke Huels AG, Postfach, Germany.

Bound Antioxidants

Some improvement in normal abrasion resistance and abrasion resistance for nitrile-based compounds at elevated temperatures can be achieved if an NBR is substituted that contains a bound antioxidant. RT: Chapter 8, "Specialty Elastomers," M. Gozdiff, p. 199. Similar effects have been reported for NR. GEN: "Science and Technology of NR," Malaysian Rubber Producers Research Association, Brickendonbury, Hertford, UK, 1988. The use of QDI in NR compounds may provide an advantage as well. RP: F. Ignatz-Hoover.

TPE

If considering the use of thermoplastic elastomers (TPEs), keep in mind thermoplastic polyurethanes (TPUs) that possess outstanding abrasion resistance. RT: Chapter 10, "Thermoplastic Elastomers," C. P. Rader, p. 271.

Crosslink Density

In general, increasing the crosslink density may increase the abrasion resistance of an NR vulcanizate. GEN: D. Campbell, A. Chapman, "Relationships Between Vulcanizate Structure and Vulcanizate Performance," Malaysian Rubber Producers Research Association, Brickendonbury, Hertford, UK, p. 5.

Types of Crosslinks

In general the polysulfide crosslinks from sulfur cures impart better abrasion resistance to a compound than do peroxide cures. GEN: P. Dluzneski, "Peroxide Vulcanization of Elastomers," *Rubber Chemistry and Technology*, July-August, 2001, Vol. 74, p. 451. RP: J. R. Halladay.

Conventional cure systems may give better wear and abrasion resistance than EV or semi-EV cures in natural rubber based compounds. GEN: Cheng Shin, "Curative and Carbon Black Effects on NR Truck Tires," Paper No. 131 presented at the Fall Meeting of the Rubber Division, ACS, October, 2001, Cleveland, OH.

New Crosslinking Agent

Reportedly, a new crosslinking agent [1,6-bis(*N,N*-dibenzylthiodicarbamyldithio) hexane] gives stable hybrid crosslinks in truck tread compounds for better tread wear properties as well as better heat aging properies. GEN: T. Kleiner (Bayer AG), "Improvements in Abrasion and Heat Resistance by Using a New Crosslink Agent," Paper No. 12A presented at the ITEC 2002 Meeting, September, 2002, Akron, OH.

Peroxide Cures

Using a coagent with a peroxide cure may improve abrasion resistance. GEN: P. Dluzneski, "Peroxide Vulcanization of Elastomers," *Rubber Chemistry and Technology*, July-August, 2001, Vol. 74, p. 451.

Ultimate Tensile Strength vs. Abrasion Resistance

According to Studebaker and Beatty, directional compounds that have higher ultimate tensile strength values are also more likely to have higher abrasion resistance. However, many exceptions exist. GEN: F. Eirich, *Science and Technology of Rubber*, Chapter 9, "The Rubber Compound and Its Composition," M. Studebaker, J. R. Beatty, Academic Press, 1978, p. 367.

Antidegradants

In general, it is referenced that using some antidegradants such as phenylene diamine AOs or QDI (quinonediimine) may indirectly improve Pico abrasion resistance for natural rubber based compounds. Theoretically, dangling chain ends adsorb onto filler surfaces in the compound. Adsorption enhancement is increased by polar end groups from some of these AOs. Molecular slippage is an important component of viscous modulus. So if more energy is dissipated at high strain conditions, less of that energy is available for tearing or abrading the compound. This may be why Pico abrasion index and loss modulus (G″) are sometimes correlated with each other. GEN: F. Ignatz-Hoover (Flexsys), "Wear Characteristics Predicted Using Loss Modulus at High Strain," Paper No. 46 presented at the Fall Meeting of the Rubber Division, ACS, October 5–8, 2004, Columbus, OH.

3.5 Improving Oil and/or Solvent Resistance

Repeated exposure of a cured rubber compound to an oil or solvent environment can cause swelling and degradation of the compound while it is in service. The degradation effects can vary greatly depending on the type of oil or solvent to which the rubber compound is being exposed. For example, an aromatic oil might swell and degrade compound "A" more than compound "B," whereas a paraffinic oil could possibly have the opposite effect. In addition, different types of solvents can greatly differ in their effects on a rubber compound. Aliphatic hydrocarbon solvents are greatly different from ketones, which are different from aromatic hydrocarbon solvents, and so forth. It is very difficult to generalize about oil resistance and/or solvent resistance because there are so many different types to consider. However, oil and solvent resistance is sometimes of major concern for such applications as "under-the-hood" automotive, for example. Resistance to certain ASTM standard reference oils also is commonly used as one of the important criteria used to classify a rubber compound by SAE J200/ ASTM D2000.

Below are some ***experimental*** *ideas that might be useful to consider trying to improve certain types of oil or solvent resistance characteristics of a given rubber compound. Also, all relevant literature sources, including the ones cited below, should be researched and read.* ***Caution:*** *these general experimental ideas may not work in all specific situations. Changes to improve oil and/or solvent resistance will certainly affect other properties as well, for better or for worse, and this book does not purport to show how these other properties are affected. Also, this book does not purport to address safety and health issues.*

General Ranking

One ranking that can be applied to different base elastomers for "overall" oil resistance (resistance to swelling) is the following:

FKM (Best) > CO/ECO ≈ NBR/PVC ≈ NBR > AEM > CM ≈ CSM ≈ VMQ > CR >> EPDM ≈ IIR ≈ SBR ≈ NR (worst)

GEN: J. Horvath, "Selection of Polymers for Automotive Hose and Tubing Applications," *Rubber World*, December, 1987, p. 21.

Fluoroelastomers

Using FKM as the base elastomer will impart extremely good resistance to oil. RT: Chapter 6, "Elastomer Selection," R. School, p. 135.

Compounds based on fluoroelastomers possess an extraordinary level of resistance to oil even at very high temperatures. RT: Chapter 8, "Specialty Elastomers," R. Stevens, p. 229.

New fluoroelastomers made with vinylidene fluoride, tetrafluoroethylene, and propylene reportedly possess better aging properties for many "aggressive engine oils" at high temperatures compared to conventional FKM elastomers. This is reported to be due to the use of propylene in place of hexafluoropropylene in the polymerization process. GEN: W. Grootaert, R. Kolb, A. Worm, "A Novel Fluorocarbon Elastomer for High-Temperature Sealing Applications in Aggressive Motor-Oil Environments," *Rubber Chemistry and Technology*, September-October, 1990, Vol. 63, p. 516.

FKM and Gasoline/Methanol Blends

Increasing levels of ethanol and particularly methanol combined with gasoline cause greater swelling of old traditional hose tube compounds based on NBR or ECO. However, hose tube compounds based on special, customized grades of FKM may show significantly less swelling from these "gasohol" fuels. GEN: R. Malcolmson, "Elastomers for Cars," *Chemtech*, May, 1983, p. 286.

FKM and "Sour" Fuels

The move toward fuel injection systems for automotive engines means that unburned fuel is mostly recirculated. This recirculation increases the probability that a "sour" fuel is produced. This sour fuel is actually an oxidized form of gasoline that contains hydroperoxides. These hydroperoxides promote the generation of free radicals that can quickly degrade some synthetic rubbers. Experiments show that this "sour" gasoline can cause ECO compounds to revert and soften while causing some NBR compounds to harden and become brittle. On the other hand, FKM compounds (based on special, customized grades of FKM) generally are not as greatly affected by this sour fuel. GEN: R. Malcolmson, "Elastomers for Cars," *Chemtech*, May, 1983, p. 286.

Perfluoroelastomers

Many times perfluoroelastomers (FFKMs) are selected as the base rubber for a compound when a better trade-off is needed between heat resistance and oil resistance. FFKM polymers reportedly can impart useful service life at temperatures as high as 316°C (600°F). GEN: M. Coughlin, R. Schnell, S. Wang, "Perfluoroelastomers in Severe Environments: Properties, Chemistry, and Applications," Presented at ACS Rubber Div. Meeting, Spring, 2001, Paper No. 24.

Fluorine Content for Fluoroelastomers

Reportedly, one should select a fluoroelastomer with a higher bound fluorine content to reduce swelling from methanol or Fuel C or blends of both methanol and Fuel C. In some cases these blends are more likely to increase swelling of the vulcanizates and destroy the cured physical properties than when immersed separately in 100% Fuel C or immersed separately in 100% methanol. Fuel C itself is reported to be a 50:50 blend of isooctane and toluene by volume. GEN: T. Dobel, C. Grant (DuPont), "Fuel and Permeation Resistance of Fluoroelastomers to Methanol Blends," Paper No. 47 presented at the Fall Meeting of the Rubber Division, ACS, October 11–13, 2011, Cleveland, OH.

Hydrin Rubber

Epichlorohydrin rubber imparts good resistance to gasoline. RT: Chapter 6, "Elastomer Selection," R. School, p. 138

Compounds based on CO, ECO, or GECO grades of polyepichlorohydrin have good resistance to oil swelling but are not resistant to polar brake fluids. RT: Chapter 8, "Specialty Elastomers," C. Cable, pp. 216–217.

A reported four-way comparison of compounds based on CM, CSM, GECO, and NBR/PVC found that GECO gave superior oil swell resistance. GEN: C. Hooker, R. Vara, "A Comparison of Chlorinated and Chlorosulfonated Polyethylene Elastomers with Other Materials for Automotive Fuel Hose Covers," Presented at ACS Rubber Div. Meeting, Fall, 2000, Paper No. 128.

The permeation of gasoline components from auto fuel hose may be reduced by using hose tube stocks based on special grades of epichlorohydrin homopolymer (CO), CO-epichlorohydrin copolymer (ECO) blends, or FKM veneers. GEN: R. Malcolmson, "Elastomers for Cars," *Chemtech*, May, 1983, p. 286.

ACM

Polyacrylate rubber gives good oil resistance especially against sulfur-bearing lubricants. RT: Chapter 6, "Elastomer Selection," R. School, p. 138.

CPE

Generally, compounds based on chlorinated polyethylene (CM) possess fair to good resistance to oil. Selecting grades of CM with higher chlorine content will usually increase the compound's resistance to oil attack. RT: Chapter 8, "Specialty Elastomers," L. Weaver, pp. 212, 213.

CSM

Generally, compounds based on chlorosulfonated polyethylene (CSM) possess fair to good resistance to oil. Selecting grades of CSM with higher chlorine content should increase the compound's resistance to oil attack. RT: Chapter 8, "Specialty Elastomers," C. Baddorf, pp. 213, 214.

AEM

For better oil resistance from terpolymer grades of ethylene-acrylic elastomer (AEM), select a higher methyl acrylate containing terpolymer (such as Vamac® GLS), which reportedly significantly improves the compound's oil resistance with only a small loss in low-temperature properties. RT: Chapter 8, "Specialty Elastomers," T. Dobel, p. 223.

FKM/ACM Alloy

A compound based on a special proprietary alloy of FKM and ACM (DAI-EL Alloy AG-1530), using a special peroxide cure, may possibly give equivalent resistance to exposure to some oils compared to either an all-FKM- or all-ACM-based compound. GEN: M. Kishine, T. Noguchi, "New Heat-resistance Elastomers," *Rubber World*, February, 1999, p. 40.

NBR

For good oil swell resistance, base a rubber compound on an NBR with a high percentage of ACN content. The higher the bound ACN content is, the better the oil resistance usually will be. RT: Chapter 6, "Elastomer Selection," R. School, p. 131; Chapter 8, "Specialty Elastomers," M. Gozdiff, p. 194.

One possible way to achieve better oil or solvent resistance for an NBR formulation might be to select a low molecular weight ester plasticizer (such as DOP, DOA, or DOS) that may be extracted out of the NBR compound by the swelling media at a rate similar to that at which the compound is absorbing the swelling media itself. Therefore, if these two effects are "in balance," then the net volume swell could be considerably less. Obviously this approach will require a considerable amount of experimental work and "fine tuning" to make it work. Also, if the compound is used for molding a rubber part for use in only a single specific oil throughout its service life, then the compound could possibly be formulated with as much of that specific oil as is feasible and compatible with the compound. This too might prevent excessive volume swelling in service. RP: J. R. Halladay.

HNBR and Oxidized Fuels

Compounds based on HNBR may have better resistance to oxidized fuels than compounds based on NBR or ECO. GEN: K. Hashimoto, N. Watanabe, M. Oyama, Y. Todani, "Highly Saturated Nitrile Elastomer-HSN," *Automotive Elastomers and Design*, September, 1985, p. 26.

FZ Rubber

Consider using fluoroalkoxyphosphazene (FZ) fluoroelastomer in a rubber compound for very good resistance from oil attack. GEN: Al Worm, Jack Kosmala (3M), "Introduction to Fluorocarbon Elastomers," 1991.

HNBR Coatings

Consider selecting the appropriate HNBR rubber coating that might help protect a given rubber article against oil attack. These special HNBR coatings can be crosslinked at room temperature and may help in protecting rubber articles made of natural rubber and/or polybutadiene against oil attack. GEN: J. Halladay, T. Kohli (Lord Corp.), "Novel Elastomeric Coatings for Use on Rubber Components," Paper No. 29 presented at the Spring Meeting of the Rubber Division, ACS, April 28-30, 2003, San Francisco, CA.

XNBR and XNBR/PVC Blends vs. Fuel C/Methanol Blends

Selection of XNBR or blends of XNBR with PVC may provide some improvement in resistance to permeability of Fuel C and methanol blends compared to normal nitrile rubber or PVC bases. GEN: H. Pfisterer, J. Dunn, "New Factors Affecting the Performance of Automotive Fuel Hose," *Rubber Chemistry and Technology*, May-June, 1980, Vol. 53, p. 357.

Fluorosilicone Rubber

Silicone compounds may not possess very good resistance to some fluids; however, using fluorosilicone elastomers may improve fluid resistance to a certain degree. RT: Chapter 6, "Elastomer Selection," R. School, p. 136.

Poly(trifluoropropyl) methylsiloxane (fluorosilicone) usually provides somewhat better resistance against oil attack to the compound than other silicone elastomers. RT: Chapter 8, "Specialty Elastomers," J. R. Halladay, p. 235.

Polysulfide Rubber

Compounds based on polysulfide rubber can exhibit good resistance to some types of solvents while still having reasonable low-temperature properties. RT: Chapter 6, "Elastomer Selection," R. School, p. 139.

Polyurethane

Polyurethane is reported to have above-average oil resistance. GEN: Tom Jablonowski (TSE Industries), "Millathane Millable Polyurethane for Demanding Applications," Presented at the Indian Rubber Expo, January 19–21, 2011, Chennai, India.

Usually, but not always, selecting an ester-type polyurethane will afford an advantage for oil resistance in many applications compared to ether-based PU. RT: Chapter 9, "Polyurethane Elastomers," R. W. Fuest, p. 257.

Millable PU/SBR

Adding a given quantity of millable polyurethane incrementally to an SBR-based compound reportedly will improve the compound's oil resistance somewhat. GEN: T. Jablonowski, "Blends of PU with Conventional Rubbers," *Rubber World*, October, 2000, p. 41.

Liquid BR/Silicone Compound

Consider the possibility of using low molecular weight (liquid) 1,2-polybutadiene resins (such as special Ricon® grades that contain certain antioxidants for improved heat resistance of the BR) as an additive at low part levels in a peroxide-cured silicone compound to allow this compound to accept more carbon black loading and possibly reduce the compound's swelling in certain types of fluids. GEN: R. Drake, "Using Liquid Polybutadiene Resin to Modify Elastomeric Properties," *Rubber & Plastics News*, February 28 and March 14, 1983.

Liquid NBR as a Plasticizer in NBR Compounds

Using a liquid NBR (such as Hycar® 1312) as the plasticizer in an NBR-based compound instead of other conventional plasticizers may help prevent extraction of this plasticizer from the compound. The reason that the liquid NBR is nonmigrating, nonvolatile, and nonextractible is because it "cures" in the vulcanization process. GEN: "A Comparative Evaluation of Hycar Nitrile Polymers," Manual HM-1 Revised, B. F. Goodrich Chemical Co.

Self-Vulcanizable Blends

Consider "self-vulcanizable" blends (not requiring crosslinking agents) of chlorinated natural rubber and carboxylated NBR to achieve good oil resistance. Other "self-vulcanizable" rubber blends that could also be considered for various degrees of oil resistance include the following:

- Chlorosulfonated polyethylene/epoxidized NR
- Polychloroprene/epoxidized NR

- Chlorosulfonated polyethylene/carboxylated NBR
- Epoxidized NR/carboxylated NBR
- Polychloroprene/carboxylated NBR
- Poly(vinyl chloride)/carboxylated NBR

These self-curable blends may or may not provide the degree of oil resistance desired. GEN: P. Ramesh, S. De, "Self Crosslinkable Polymer Blends Based on Chlorinated Rubber and Carboxylated Nitrile Rubber," *Rubber Chemistry and Technology*, March-April, 1992, Vol. 65, p. 24.

TPV

Select a thermoplastic vulcanizate (TPV) with a higher crosslink density in the rubbery domain from dynamic vulcanization to reduce oil swell. RT: Chapter 10, "Thermoplastic Elastomers," C. P. Rader, p. 274.

When selecting from thermoplastic vulcanizates (TPVs), choose an NBR/PP instead of the traditional EPDM/PP for better resistance to many types of oils and lubricants. RT: Chapter 10, "Thermoplastic Elastomers," C. P. Rader, p. 274.

Also consider using DuPont's thermoplastic elastomers called the ETPV family consisting of a copolyester matrix with highly crosslinked AEM (modified ethylene-acrylate rubber), which reportedly has better oil resistance than traditional TPEs and can handle a temperature of 135 °C continuously and peaks that are as high as 180 °C. GEN: J. Drobny (Drobny Polymer Associates), "High Performance Thermoplastic Elastomers: A Review," Paper No. 69 presented at the Fall Meeting of the Rubber Division, ACS, October 14–16, 2008, Louisville, KY; J. Pike, A. Schantz, K. Cook (DuPont), "Introducing the New Class of TPVs: They Can Do the Work of High-Performance Rubbers, Often for Less," Paper No. 7 presented at the Fall Meeting of the Rubber Division, ACS, October 14–17, 2003, Cleveland, OH.

Plasticizers

Using synthetic plasticizers based on multibasic acid esters having linear alcohol end groups in a neoprene-based compound will usually afford better resistance to weight loss from immersion in ASTM 3 Oil. RT: Chapter 14, "Ester Plasticizers and Processing Additives," W. Whittington, p. 351.

Trimellitate plasticizers reportedly impart good stability from some types of oil immersion of chlorinated polyethylene (CPE)-based compounds. RT: Chapter 14, "Ester Plasticizers and Processing Additives," W. Whittington, pp. 359–361.

Crosslink Density

Increasing the crosslink density of a vulcanizate will reduce the amount of oil or solvent swelling when the vulcanizate is immersed in either the oil or solvent medium. In fact, controlled immersion tests of specific vulcanizates have been used for years as an effective method for quantifying crosslink density. The *larger* the amount of a specific solvent (or oil) that is absorbed by a specific vulcanizate, the *lower* the crosslink density will be (Flory-Rehner equation). GEN: D. Campbell, A. Chapman, "Relationships Between Vulcanizate Structure and Vulcanizate Performance," Malaysian Rubber Producers Research Association, Brickendonbury, Hertford, UK, p. 3.

Peroxide Cure vs. Sulfur Cure

Reportedly, peroxide cures generally impart better oil and chemical resistance to a rubber compound than a sulfur cure. GEN: P. Dluzneski, "Peroxide Vulcanization of Elastomers," *Rubber Chemistry and Technology*, July-August, 2001, Vol. 74, p. 451.

Polymer-Filler Interaction

The use of reinforcing fillers with good polymer-filler interaction should improve resistance to oil swelling because these polymer-filler interactions are similar in effect to chemical crosslinks from curing, provided that the presence of the filler does not reduce the crosslink density in the polymer matrix. If the chemical crosslink density from the cure reaction is independent of the filler loading and its presence (such as with a peroxide cure), then better oil resistance for the compound may result from higher loadings of reinforcing fillers with good polymer-filler interaction. GEN: S. Wolff, *Kautsch. Gummi Kunstst.*, 1970, Vol. 23, p. 7. RP: M-J. Wang.

High-Structure Fillers

The use of higher structure fillers may reduce swelling. GEN: Z. Rigbi, B. B. Boonstra, Presented at ACS Rubber Div. Fall Meeting, September 13-15, 1976; *Carbon Black*, Chapter 9, "Carbon Black Reinforcement of Elastomers," S. Wolf, M-J. Wang, p. 308. RP: M-J. Wang.

Carbon Black

Using a higher loading of one of the new ultra-low-structure, semireinforcing carbon blacks might improve permeability resistance to fluids. GEN: S. Bussolari, S. Laube, "A New Cabot Carbon Black for Improved Performance in Peroxide Cured Injection Molded Compounds," Presented at ACS Rubber Div. Meeting, Fall, 2000, Paper No. 98.

FKM, Low Molecular Weight Liquid Rubber, and Oxidized Thermal Black

Using certain types of fluoroelastomers, with a specially oxidized thermal black and a 1200 molecular weight poly-(1,2-butadiene)diol prepolymer as a coupling agent with a peroxide cure, reportedly can afford further swelling resistance against certain types of solvents and oils. GEN: J. Martin, T. Braswell, H. Green, "Coupling Agents for Certain Types of Fluoroelastomers," *Rubber Chemistry and Technology*, November-December, 1978, Vol. 51, p. 897.

Talc Loading

Partially substituting talc for carbon black N990 and silica in nitrile compounds may somewhat reduce swelling from some types of solvents, provided that the marked reduction in compound modulus and tensile strength is acceptable. GEN: H. Pfisterer, J. Dunn, "New Factors Affecting the Performance of Automotive Fuel Hose," *Rubber Chemistry and Technology*, May-June, 1980, Vol. 53, p. 357.

3.6 Improving Staining and Discoloration Resistance

This property is particularly important for rubber compounds that are nonblack in color or come in contact with other nonblack rubber compounds or "stainable" surfaces.

The following ***experimental*** *ideas may help prevent staining or discoloration problems. Also, all relevant literature sources, including the ones cited below, should be researched and read.* ***Caution:*** *these general experimental ideas may not work in all specific situations. Changes to improve staining or discoloration resistance will certainly affect other properties as well, for better or for worse, and this book does not purport to show how these other properties are affected. Also, this book does not purport to address safety and health issues.*

Oils

Select oils that are not highly aromatic. Generally, aromatic process oils tend to stain more than naphthenic or paraffinic oils. RT: Chapter 12, "Compounding with Carbon Black and Oil," S. Laube, S. Monthey, M-J. Wang, p. 312. GEN: J. S. Dick, Chapter 8, "Oils, Plasticizers, and Other Rubber Chemicals," *Basic Rubber Testing: Selecting Methods for a Rubber Test Program*, ASTM, West Conshohocken, PA, 2003, p. 124.

Cure

An advantage of peroxide vs. sulfur cures is that the peroxide cures can more easily be made nonstaining. RT: Chapter 17, "Peroxide Cure Systems," L. Palys, p. 434.

Antioxidants

Use phenolic antioxidants that are nonstaining instead of staining amine antioxidants. RT: Chapter 19, "Antidegradants," F. Ignatz-Hoover, p. 449.

The "bis" phenol antioxidants are generally considered some of the most effective and persistent nonstaining antioxidants available. RT: Chapter 19, "Antidegradants," F. Ignatz-Hoover, p. 452.

It was reported by Warrach and Tsou that bis(1,2,3,6-tetrahydrobenzaldehyde)-pentaerythrityl acetal imparted good ozone protection to compounds based on CR, IIR, CIIR, or BIIR. The acetal AO is nonstaining. GEN: W. Waddell, "Tire Black Sidewall Surface Discoloration and Non-staining Technology: A Review," *Rubber Chemistry and Technology*, July–August, 1998, Vol. 71, p. 590.

Rollick et al. reported that tetrahydro-1,3,5-tri-(*n*)-butyl(*S*)-triazin-ethione does not significantly discolor compounds based on general purpose rubbers and imparts some protection against ozone attack. GEN: W. Waddell, "Tire Black Sidewall Surface Discoloration and Non-staining Technology: A Review," *Rubber Chemistry and Technology*, July–August, 1998, Vol. 71, p. 590.

Ivan et al. reported the introduction of a nonstaining antiozonant. This antiozonant was 3,5-di-*tert*-butyl-4-hydroxybenzylcyanoacetate and reportedly imparted antiozonant protection to NR and BR compounds similar to that afforded by *N*-isopropyl-*N'*-phenyl-*p*-phenylenediamine. GEN: W. Waddell, "Tire Black Sidewall Surface Discoloration and Non-Staining Technology: A Review," *Rubber Chemistry and Technology*, July–August, 1998, Vol. 71, p. 590.

Wheeler and Hong reported that 2,4,6-tris-(*N*-1,4-dimethylpentyl-*p*-phenylenediamino)-1,3,5-triazine is a "nonstaining" antiozonant that may impart dynamic ozone protection similar to that afforded by the *p*-phenylenediamine antiozonants in NR and BR compounds. GEN: W. Waddell, "Tire Black Sidewall Surface Discoloration and Non-staining Technology: A Review," *Rubber Chemistry and Technology*, July–August, 1998, Vol. 71, p. 590.

Stainless Tire Sidewalls

If BIMSM is used at 35 and 40 phr with other elastomers such as *cis*-BR and NR in a tire black sidewall compound, staining antiozonants may not be needed. GEN: D. Flowers, J. Fusco, D. Tracey (Exxon Chemical), "Advancements in New Tire Sidewalls with a New Isobutylene Based Copolymer," Paper No. 50 presented at the Fall Meeting of the Rubber Division, ACS, October 26–29, 1993, Orlando, FL.

Use Saturated Polymer Backbone

One of the best ways to achieve a "high-gloss" appearance over the rubber product's service life is to use a rubber blend that contains a sufficient quantity of rubber with saturated polymer "backbones." By using enough of this (or these) "saturated" elastomer(s) in the blend, the resulting rubber compound will be inherently resistant to ozone attack without the need to use antiozonants. (When these antiozonants bloom to the compound's surface, they destroy the "high-gloss" appearance.) Ozone-resistant elastomers that could be used in such a blend include EPDM, BIIR, CIIR, and the new brominated isobutylene-co-*p*-methylstyrene elastomer. GEN: W. Waddell, "Tire Black Sidewall Surface Discoloration and Non-staining Technology: A Review," *Rubber Chemistry and Technology*, July–August, 1998, Vol. 71, p. 590.

3.7 Improving Weathering Resistance

Exposure to bright sunlight, rain, dew, and so forth can "weather" a rubber compound. In the case of some rubber products, appearance can be very important. Weathering can also lead to the destruction and failure of a rubber product.

The following ***experimental*** *ideas may slow down or prevent the weathering of a cured rubber compound. Also, all relevant literature sources, including the ones cited below, should be researched and read.* ***Caution:*** *these general experimental ideas may not work in all specific situations. Changes to improve weathering resistance will certainly affect other properties as well, for better or for worse, and this book does not purport to show how these other properties are affected. Also, this book does not purport to address safety and health issues.*

General Principle

Generally, to achieve good weathering resistance, base the compound on an elastomer with a fully or almost fully saturated backbone. Examples of such elastomers are EPDM, HNBR, silicone, IIR, and FKM. RP: J. R. Halladay.

Carbon Black

Carbon black in a rubber compound can provide very good protection against ultraviolet light. RT: Chapter 3, "Vulcanizate Physical Properties, Performance Characteristics, and Testing," J. S. Dick, p. 65.

EPDM

Compounds based on EPDM possess very good resistance to weathering. RT: Chapter 6, "Elastomer Selection," R. School, p. 132.

HIIR and Butyl Rubber

Butyl- and halobutyl-based elastomers will impart very good resistance to weathering. RT: Chapter 6, "Elastomer Selection," R. School, p. 134.

Silicone and Fluorosilicone Rubber

Silicone- and fluorosilicone-based compounds possess excellent weathering resistance. RT: Chapter 6, "Elastomer Selection," R. School, p. 136.

CM

Compounds based on chlorinated polyethylene (CM) generally have good weathering resistance. RT: Chapter 8, "Specialty Elastomers," L. Weaver, p. 212.

CSM

Compounds based on chlorosulfonated polyethylene (CSM) have very good weathering resistance. RT: Chapter 8, "Specialty Elastomers," C. Baddorf, p. 214.

Hypalon® (CSM) is known for its superior weathering resistance; reportedly, Hypalon 40 may give the better protection at a 25X magnification, followed by Hypalon 30 and Hypalon 20. GEN: P. Peffer, R. Radcliff, "Factors Affecting Weather Resistance of 'Hypalon' Chlorosulfonated Polyethylene," *Rubber World*, October, 1960, p. 102.

Hypalon® (CSM) vulcanizates that contain magnesium oxide as the filler reportedly give excellent color retention and surface integrity. GEN: P. Peffer, R. Radcliff, "Factors Affecting Weather Resistance of 'Hypalon' Chlorosulfonated Polyethylene," *Rubber World*, October, 1960, p. 102.

Fluoroelastomers

Usually, fluoroelastomer-based compounds give good weatherability resistance as well as ozone resistance. GEN: Jim Denham (3M), "Basic Fluoroelastomer Technology," Presented at the Fall Meeting of the Energy Rubber Group, September 13, 2011, Galveston, TX.

3.8 Improving Hydrolysis Resistance

Hydrolysis resistance is particularly important for polyurethane-based applications. It is a major concern in particular for elastomers based on vinyl acetate, copolyester TPEs, or any polymers with urethane, amide, or ester linkages.

The following ***experimental*** *ideas may improve hydrolysis resistance. Also, all relevant literature sources, including the ones cited below, should be researched and read.* ***Caution:*** *these general experimental ideas may not work in all specific situations. Changes to improve hydrolysis resistance will certainly affect other properties as well, for better or for worse, and this book does not purport to show how these other properties are affected. Also, this book does not purport to address safety and health issues.*

Usually, but not always, selecting an ether-type polyurethane based on MDI prepolymer will provide an advantage for hydrolysis resistance of polyurethanes in demanding applications. RT: Chapter 9, "Polyurethane Elastomers," R. W. Fuest, p. 257.

Avoid using rubber compounds that possess polymers based on polyurethane, vinyl acetate, copolyester, or any polymer with urethane, amide, or ester linkages. RP: C. P. Rader.

4 Optimizing Measurable Processability Properties

4.1 Reducing Viscosity

Sometimes it is necessary to keep the viscosity of a mixed stock low to ensure good processability downstream in the plant.

*The following **experimental** ideas can be considered and may be needed for good processing downstream in the factory. Also, all relevant literature sources, including the ones cited below, should be researched and read. **Caution:** these general experimental ideas may not work in all specific situations. Changes to reduce viscosity will certainly affect other properties as well, for better or for worse, and this book does not purport to show how these other properties are affected. Also, this book does not purport to address safety and health issues.*

Mixing

Improve state of mix and lower compound viscosity by adding more work history. RT: Chapter 2, "Compound Processing Characteristics and Testing," J. S. Dick, p. 23.

Raw synthetic rubber such as SBR usually contains stabilizers. However, mixing SBR compounds at high temperatures such as 325°F (163°C) or above can cause the formation of both redispersible or "loose" gel that can be milled out and so-called "tight" gel that cannot be milled out. Both of these types of gel can increase the Mooney viscosity of the rubber compound. Therefore it might be prudent to restrict how high the dump temperature can be. GEN: R. Mazzeo, "Preventing Polymer Degradation During Mixing," *Rubber World*, February, 1995, p. 22.

Pin Barrel Extruder

If it is simply not feasible to lower the viscosity of a given compound, one might consider using a pin barrel extruder instead of a conventional extruder because the pins in the extruder barrel continuously divide and split the flow of the compound, thus mixing and homogenizing the compound without exposure to high shear rates. Therefore, high Mooney viscosity compounds can be run on a pin barrel extruder where a conventional extruder might not be able to handle such a "tough" compound. RT: Chapter 23, "Rubber Mixing," W. Hacker, p. 508.

Molecular Weight

Use a lower molecular weight elastomer as the base. RT: Chapter 2, "Compound Processing Characteristics and Testing," J. S. Dick, p. 23; Chapter 7, "General Purpose Elastomers and Blends," G. Day, p. 153. GEN: R. Mastromatteo, E. Morrisey, M. Mastromatteo, H. Day, "Matching Material Properties to Application Requirements," *Rubber World*, February, 1983, p. 26.

SBR

Use ESBR instead of solution SBR because SSBR has the higher average molecular weight and the narrower molecular weight distribution, which usually imparts a higher compound Mooney viscosity. RT: Chapter 7, "General Purpose Elastomers and Blends," G. Day, p. 153.

Low-Viscosity BR

Consider using a specially polymerized Ni-*cis*-BR from Goodyear Chemical Co., which was produced from a special catalyst system. This experimental catalyst system may provide a BR with lower viscosity but acceptable cold flow. GEN: K. Castner (Goodyear Tire), "Improved Processing *cis*-1,4-Polybutadiene," Paper No. 3 presented at the Fall Meeting of the Rubber Division, ACS, April 13–16, 1999, Chicago, IL.

Star-Branched HIIR

Star-branched halobutyl rubbers can be substituted for regular halobutyl rubber at the same carbon black/oil loadings to reduce compound viscosity. RT: Chapter 8, "Specialty Elastomers," G. Jones, D. Tracey, A. Tisler, p. 180.

Liquid Polymeric Plasticizers

Consider using a compatible liquid polymeric plasticizer to reduce the compound's viscosity. For example, one could consider using a liquid EPR plasticizer, such as Trilene®, in an EPDM compound to reduce compound viscosity. GEN: J. Sommer, *Elastomer Molding Technology*, Elastech, Hudson, OH, 2003, p. 178.

Liquid Polychloroprene

When using neoprene in compounding, use a low molecular weight neoprene as a plasticizer (such as Neoprene FB) with other neoprene(s) to reduce the compound's viscosity. RT: Chapter 8, "Specialty Elastomers," L. L. Outzs, p. 210.

Low-Viscosity HNBR

Consider selecting one of the low-viscosity grades of HNBR to lower the viscosity of an HNBR compound. GEN: A. Anderson, M. Jones (Zeon Chemicals), "Usage of Low-Viscosity HNBR Polymers to Increase Productivity in High-Shear Molding Methods," Paper No. 75 presented at the Fall Meeting of the Rubber Division, ACS, October 11–13, 2011, Cleveland, OH.

TOR Additive

Polyoctenamer or *trans*-polyoctenylene rubber (TOR) reportedly can be blended with NR, BR, SBR, NBR, CR, or EPDM, in minor amounts, usually resulting in a reduction in compound viscosity at processing temperatures (but an increase in compound viscosity and hardness at room temperature due to crystallinity). GEN: A. Draxler, "A New Rubber: *trans*-Polyoctenamer," Chemische Werke Huels AG, Postfach, Germany.

Natural Rubber, Peptizers, and Processing Aids

For NR-based compounds, consider using a chemical plasticizer (peptizer) such as dibenzamido-diphenyl disulfide at low levels to reduce the compound viscosity, provided the other negative consequences can be tolerated. RT: Chapter 14, "Ester Plasticizers and Processing Additives," C. Stone, pp. 364–377.

For NR-based compounds, consider using processing additives or aids, for example, fatty acid soaps such as alkyl/aryl zinc soap, alkyl zinc soap, or zinc/potassium soap, to reduce the compound viscosity. RT: Chapter 14, "Ester Plasticizers and Processing Additives," C. Stone, p. 366.

Some new zinc-free processing aids may be useful in reducing the viscosity of an NR-based compound. GEN: K. Menting, J. Bertrand, M. Hensel, H. Umland (Schill + Seilacher), "The Ultimate Way to NR Processing?: High Efficiency and Good Dynamic Properties with a Zinc-Free Novel Additive," Paper No. 70 presented at the Spring Meeting of the Rubber Division, ACS, April 28–30, 2003, San Francisco, CA.

Fillers

Decreasing loading of a large particle size filler will reduce compound viscosity (according to the Einstein, Guth, and Gold equation). Decreasing loading of a

RT: *Rubber Technology, Compounding and Testing for Performance*. 2nd ed. Edited by J. S. Dick, Hanser Publications, 2009 ■ GEN: General References from the Literature ■ RP: Review Panel (Advisors for this book) ■ IMPORTANT: The information contained herein is experimental in nature and is meant for individuals with advanced scientific training and rubber compounding experience. All safety precautions should be taken when replicating or applying this information. The author makes no representation or warranty, either expressed or implied.

very small particle size filler will dramatically reduce compound viscosity. GEN: J. S. Dick, H. Pawlowski, J. Moore, "Viscous Heating and Reinforcement Effects of Fillers Using the Rubber Process Analyzer," *Rubber World*, January, 2000, p. 31.

Use fillers with individual spherical-shaped particles instead of fillers with anisometric aggregates for lower compound viscosity. GEN: *Carbon Black*, Chapter 9, "Carbon Black Reinforcement of Elastomers," S. Wolf, M-J. Wang, p. 301.

"Lubricant" Fillers

For a mineral-filled compound, consider partial replacement of a precipitated silica loading with so-called "lubricant" fillers such as synthetic graphite (such as Dixon® 1176) or magnesium silicate talc (such as Mistron Vapor®). GEN: D. Coulthard, W. Gunter, Presented at ACS Rubber Div. Meeting, Fall, 1975, Paper No. 39.

Surface Treatment of Fillers

Consider using loadings of surface-treated fillers such as stearate-treated calcium carbonate, silane-treated clays, and/or titanate-treated titanium dioxide in place of their untreated counterparts along with an appropriate compatible internal lubricant. Often the surface treatment will reduce the compound viscosity. GEN: R. Grossman, Q & A, *Elastomerics*, January, 1989.

Carbon Black

Use a lower loading of carbon black to lower a compound's viscosity. RT: Chapter 12, "Compounding with Carbon Black and Oil," S. Laube, S. Monthey, M-J. Wang, p. 308.

Use a larger particle size (lower surface area) carbon black to lower a compound's viscosity. RT: Chapter 12, "Compounding with Carbon Black and Oil," S. Laube, S. Monthey, M-J. Wang, p. 308. GEN: S. Monthey, "The Influence of Carbon Blacks on the Extrusion Operation for Hose Production," *Rubber World*, May, 2000, p. 38.

Use a lower structure carbon black to lower a compound's viscosity. RT: Chapter 12, "Compounding with Carbon Black and Oil," S. Laube, S. Monthey, M-J. Wang, p. 308. GEN: S. Monthey, "The Influence of Carbon Blacks on the Extrusion Operation for Hose Production," *Rubber World*, May, 2000, p. 38.

Ultra-Low-Structure Carbon Blacks

Use the new ultra-low-structure semireinforcing carbon blacks to reduce the compound's viscosity. GEN: S. Bussolari, S. Laube, "A New Cabot Carbon Black for Improved Performance in Peroxide Cured Injection Molded Compounds," Presented at ACS Rubber Div. Meeting, Fall, 2000, Paper No. 98.

Lowering "Black Scorch" Viscosity

Certain EPDM masterbatches that do not contain curatives will increase their viscosity when processed. This phenomenon is called "black scorch." It is reversible, however. If these masterbatches were passed through the internal mixer again, they would be restored to their lower viscosity levels. One deterrent to "black scorch" is to use an EPDM base with a lower percentage of ENB. Another way of reducing "black scorch" is to use carbon blacks with lower structure. GEN: C. Daniel, J. Pillow (DuPont Dow Elastomers), "Black Scorch in EPDM Compounds," Paper presented at the IRE & C, June 10, 1999, Manchester, UK.

Silica-Loaded Compounds

Consider using tertiary alkylamine fatty acid salt as an additive for a silica-loaded compound, such as a low rolling resistance tread stock, in order to reduce the compound's viscosity and improve its processability. GEN: K. Yanagisawa (Bridgestone Corp.), "Use of Tertiary Alkylamine-Fatty Acid Salt to Improve Rolling Resistance and Processability on Silica Tread," Paper No. 69 presented at the Fall Meeting of the Rubber Division, ACS, October 8–11, 2002, Pittsburgh, PA.

Silica and Silane Coupling Agents

Use of a silane coupling agent may reduce the viscosity of a silica-filled compound. GEN: L. Evans, J. Dew, L. Hope, T. Krivak, W. Waddell, "Hi-Sil EZ: Easy Dispersing Precipitated Silica," *Rubber and Plastics News*, July 31, 1995, p. 12; J. S. Dick and H. A. Pawlowski, "Application of the Rubber Process Analyzer in Characterizing the Effects of Silica on Uncured and Cured Compound Properties, *ITEC '96 Select* by *Rubber and Plastics News*, September, 1997.

Processing Aids, Compatibilizers, Homogenizing Agents

Sometimes a small loading level of a processing aid, compatibilizer, or homogenizing agent can significantly reduce compound viscosity and improve processability. Many of these agents are proprietary, usually consisting of different combinations of soaps, surfactants, and/or resins. GEN: C. Ryan (ChemSpec, Ltd.), "History of Processing Aids," Presented at the Summer Meeting of the Southern Rubber Group, June 11, 2012, Myrtle Beach, SC.

Zinc soap surfactants are effective at sometimes reducing compound viscosity. GEN: K. Kim, J. Vanderkooi (Struktol), "Effects of Zinc Soaps on TESPT and TESPD: Silica Mixtures in Natural Rubber Compounds," Paper No. 70 presented at the Fall Meeting of the Rubber Division, ACS, October 8–11, 2002, Pittsburgh, PA.

Use Processing Aid for Silica Compounds

To achieve lower viscosity for the remilling of a silica-loaded compound that has already completed the silanization reaction with an organosilane in the earlier first pass, consider using a processing aid based on a zinc/potassium soap (Struktol EF44) during this remilling (second pass). GEN: C. Stone, "Improving the Silica 'Green Tire' Tread Compound by the Use of Special Process Additives," Presented at ACS Rubber Div. Meeting, Fall, 1999, Paper No. 77.

Silicone/Silica

In silicone rubber compounds containing fine particle size silica loading, crepe hardening should be minimized by using either hydrophobic-treated silica or an antistructuring agent that is added *in situ*. In addition, precipitated silica has less tendency to crepe harden than fumed silica has at the same loading. RT: Chapter 8, J. R. Halladay, "Specialty Elastomers," p. 236.

Oil

Reduce compound viscosity by increasing the loading of process oil as a compounding ingredient. GEN: K. Hale, J. West, C. McCormick, "Contribution of Carbon Black Type to Skid and Treadwear Resistance," Presented at ACS Rubber Div. Meeting, Spring, 1975, Paper No. 6, Fig. 35.

Coagents for Peroxide Cures

Certain coagents used in peroxide cures of HNBR and other elastomers are reported to reduce the compound's viscosity. RT: Chapter 8, "Specialty Elastomers," M. Wood, p. 202. GEN: P. Dluzneski, "Peroxide Vulcanization of Elastomers," *Rubber Chemistry and Technology*, July–August, 2001, Vol. 74, p. 451.

To achieve a lower compound viscosity when using peroxides to cure a rubber compound, consider using the new proprietary "high performance" (HP) peroxide formulations. Compounds for which the HP peroxide formulations are used reportedly possess improved flow through narrow mold runners for improved productivity. RT: Chapter 17, "Peroxide Cure Systems," L. Palys, p. 418.

Select an effective low molecular weight liquid coagent or a meltable coagent for a peroxide cure, which may lower the compound's viscosity and improve processability. RT: Chapter 17, "Peroxide Cure Systems," L. Palys, p. 432.

4.2 Increasing Shear Thinning

All rubber compounds are basically non-Newtonian in nature while they are processed in the factory. This means that the viscosity of the compound decreases when it is exposed to a higher shear rate. However, some compounds will shear thin more than other compounds when processed at a higher shear rate. A rubber compound's shear thinning profile greatly determines how well it will behave in many downstream processes such as extrusion, injection molding, and so forth. It is therefore important to control this processing characteristic.

The following ***experimental*** *ideas might be tried if one desires to increase shear thinning behavior of a given compound. Also, all relevant literature sources, including the ones cited below, should be researched and read.* ***Caution:*** *these general experimental ideas may not work in all specific situations. Changes to increase shear thinning will certainly affect other properties as well, for better or for worse, and this book does not purport to show how these other properties are affected. Also, this book does not purport to address safety and health issues.*

Increase Applied Shear Rate

Increasing applied shear rate during mixing or extrusion will reduce the apparent viscosity of a rubber compound through shear thinning. RT: Chapter 2, "Compound Processing Characteristics and Testing," J. S. Dick, p. 28.

Elastomer Base

Different elastomers impart different shear thinning profiles to a rubber compound. For example, polyacrylate elastomers are probably more shear thinning than many other elastomers. Likewise, HNBR is more shear thinning than regular NBR. GEN: J. S. Dick, "Comparison of Shear Thinning Behavior Using Capillary and Rotorless Shear Rheometry," *Rubber World*, 2002, p. 23.

Filler Effects

The filler loading of a rubber compound can have a profound effect on shear thinning behavior. For example, precipitated silica loadings can sometimes shear thin a compound more than carbon black loadings can. GEN: J. S. Dick, H. Pawlowski, "Application of the Rubber Process Analyzer in Characterizing the Effects of Silica on Uncured and Cured Compound Properties," *ITEC '96 Select* by *Rubber & Plastics News*, September, 1997.

Metallocene Catalyst for EPDM

The new metallocene catalyst for commercial EPDM polymerization has made it possible to specifically control independently both molecular weight distribution and long-chain branching to produce commercial grades of EPDM with greater shear thinning behavior and greater extruder output. GEN: D. Parikh, M. Hughes, M. Laughner, L. Meiske, R. Vara, "Next Generation of Ethylene Elastomers," Presented at ACS Rubber Div. Meeting, Fall, 2000.

4.3 Reducing Elasticity (Nerve)

Reducing levels of nerve in an uncured rubber compound can help prevent excessive die swell during extrusions and nonfills during injection molding.

*The following **experimental** ideas may prevent excessive levels of nerve or uncured elasticity in a mixed stock. Also, all relevant literature sources, including the ones cited below, should be researched and read. **Caution:** these general experimental ideas may not work in all specific situations. Changes to reduce uncured elasticity will certainly affect other properties as well, for better or for worse, and this book does not purport to show how these other properties are affected. Also, this book does not purport to address safety and health issues.*

Lower Molecular Weight, Reduce Chain Entanglements

Use raw rubber with lower molecular weight and fewer chain entanglements. RT: Chapter 2, "Compound Processing Characteristics and Testing," J. S. Dick, p. 30.

Carbon Black Loading

Increasing the loading of carbon black in a compound should normally reduce extrusion die swell (although there are some noted exceptions). GEN: K. Hale, J. West, C. McCormick, "Contribution of Carbon Black Type to Skid and Treadwear Resistance," Presented at ACS Rubber Div. Meeting, Spring, 1975, Paper No. 6, Fig. 22.

Oil Loading

Consider selecting a compatible processing oil with the appropriate solubility parameter for use in a rubber compound with filler to lower the nerve of the compound during processing. GEN: J. Dick, "Applications of the Rubber Process Analyzer in Predicting Processability and Cured Dynamic Properties," Paper No. 2 presented at the Spring Meeting of the Rubber Division, ACS, May 18–21, 1993, Denver, CO.

Mixing Work History

Increasing mixing time and total energy at dump (increasing total work history) should reduce the uncured elasticity of the compound. GEN: J. S. Dick, M. Ferraco, K. Immel, T. Mlinar, M. Senskey, J. Sezna, "Utilization of the Rubber Process Analyzer in Six Sigma Programs," *Rubber World*, January, 2003, p. 32.

Talc

The use of talc in a rubber compound can reportedly reduce nerve. GEN: O. Noel, Education Symposium on Fillers, "Talc: A Functional Mineral for Rubber," Presented at ACS Rubber Div. Meeting, Fall, 1995.

Reportedly, adding talc to carbon black compounds may reduce the nerve of the compound for better calendering. GEN: O. Noel, G. Meli (Rio Tinto Minerals/Luzenac), "Synergism of Talc with Carbon Black," Paper No. 13 presented at the Fall Meeting of the Rubber Division, ACS, October 14–16, 2008, Louisville, KY.

Chemical Peptizers

Natural rubber based compounds can be very nervy. Using a very small amount of chemical peptizer during the mastication of natural rubber stock may reduce nerve by breaking down the backbone of the polymer. GEN: C. Ryan (ChemSpec. Ltd.), "History of Additives," Presented at the Summer Meeting of the Southern Rubber Group, June 11, 2012, Myrtle Beach, SC.

4.4 Increasing Tack

Good building tack of an uncured rubber compound is very important for tire building or fabricating a conveyor belt.

The following ***experimental*** *ideas might increase a rubber compound's building tack. Also, all relevant literature sources, including the ones cited below, should be researched and read.* ***Caution:*** *these general experimental ideas may not work in all specific situations. Changes to increase building tack will certainly affect other properties as well, for better or for worse, and this book does not purport to show how these other properties are affected. Also, this book does not purport to address safety and health issues.*

Base Elastomer Molecular Weight

Use a base elastomer with a lower Mooney viscosity. GEN: "Polysar Halobutyl Innerliner Problem Solving Guide," Processing Problem No. 2.

Natural Rubber

Use NR as the base polymer for the compound because NR usually imparts very good building tack properties. RT: Chapter 2, "Compound Processing Characteristics and Testing," J. S. Dick, p. 42; Chapter 6, "Elastomer Selection," R. School, p. 127.

Natural Rubber Blend

Add NR to compounds that are based on SBR, BR, or EPDM to improve building tack. GEN: E. McDonel, K. Baranwal, J. Andries, *Polymer Blends*, Vol. 2, Chapter 19, "Elastomer Blends in Tires," Academic Press, 1978, p. 281.

In blends of NR with synthetic rubbers such as BR and SBR, generally a minimum of 30% NR is needed to impart good tack for tire construction according to Morrisey. In fact he reported a near linear improvement in tack when the concentration of NR is increased. GEN: W. Hess, C. Herd, P. Vegvari, "Characterization of Immiscible Elastomer Blends," *Rubber Chemistry and Technology*, July-August, 1993, Vol. 66, p. 329.

IR vs. NR

Avoid using synthetic IR (polyisoprene) and use natural rubber instead because NR imparts better tack. RT: Chapter 6, "Elastomer Selection," R. School, p. 130.

RT: *Rubber Technology, Compounding and Testing for Performance*. 2nd ed. Edited by J. S. Dick, Hanser Publications, 2009 ■ GEN: General References from the Literature ■ RP: Review Panel (Advisors for this book) ■ IMPORTANT: The information contained herein is experimental in nature and is meant for individuals with advanced scientific training and rubber compounding experience. All safety precautions should be taken when replicating or applying this information. The author makes no representation or warranty, either expressed or implied.

Avoid 3,4-IR Content

Avoid synthetic *cis*-1,4-polyisoprene (IR) with any 3,4 content. Reportedly if 3,4 content is increased to 10%, then almost all tack is lost. GEN: C. Gibbs, S. Horne, J. Macey, H. Tucker, "Effect of Gel and Structure on the Properties of *cis*-1,4-Polyisoprene," *Rubber World*, April, 1961, p. 69.

Avoid EPDM

Avoid EPDM as the base polymer for the compound because EPDM imparts poor building tack properties. RT: Chapter 2, "Compound Processing Characteristics and Testing," J. S. Dick, p. 42.

NR/EPDM Covulcanized Blend

Consider covulcanizing an NR/EPDM blend with a sulfur/peroxide cure. The NR will impart better tack to the compound. GEN: S. Tobing, "Co-vulcanization in NR/EPDM Blends," *Rubber World*, February, 1988, p. 33.

Liquid NBR for NBR-Based Compounds

A liquid NBR (such as Hycar 1312) can be used as an effective tackifier for an NBR-based compound if used at low concentration. GEN: "A Comparative Evaluation of Hycar Nitrile Polymers," Manual HM-1 Revised, B. F. Goodrich Chemical Co.

Polychloroprene

When using neoprene in compounding, selecting G-type gives better building tack. RT: Chapter 8, "Specialty Elastomers," L. L. Outzs, p. 210.

Silicone Rubber

In silicone rubber compounding, avoid extending with diatomaceous earth, which causes a decrease in tack with silicone. RT: Chapter 8, "Specialty Elastomers," J. R. Halladay, p. 236.

Rosin-Based Emulsifiers

Use an emulsion NBR with a rosin acid based emulsification system instead of a fatty acid or synthetic emulsification system. RT: Chapter 8, "Specialty Elastomers," M. Gozdiff, p. 195. Also, use an emulsion SBR with a rosin acid based emulsification system instead of a fatty acid or synthetic system. RP: J. M. Long.

Phenolic Tackifiers

One of the most effective classes of tackifying resins that can be used in general purpose rubber compounding is the phenol-formaldehyde tackifying resin, produced from alkylphenol reacting with formaldehyde. The alkyl group is usually either octyl or *tert*-butyl. These phenolic tackifying resins give a special advantage to the uncured compound by imparting very good tack under aging conditions involving high temperature and humidity. RT: Chapter 18, "Tackifying, Curing, and Reinforcing Resins," B. Stuck, p. 443.

A very important advantage of the *p*-alkylated phenolic resin tackifiers over the older coumarone-indene resins or the C5 terpene hydrocarbon resin tackifiers is that the *p*-alkylated phenolic resins can "promote and maintain tack at elevated temperatures." Therefore, phenolic tackifiers should be selected when good aged tack is required. GEN: J. White, "The Role of Tackifying Resins on the Tackification of Rubber Compounds," *Rubber World*, February, 1984, p. 41.

When using octyl phenol formaldehyde resins as tackifiers, Rhee reported that the optimal molecular weight is 2095 (about 10 alkylphenol units on average). He further reported that generally the optimal loading for these phenolic tackifying resins was about 2 phr. GEN: C. K. Rhee, J. Andries, "Factors Which Influence Autohesion of Elastomers," *Rubber Chemistry and Technology*, March-April, 1981, Vol. 54, p. 101.

Consider using the so-called "super tackifiers" such as phenol acetylene resins. GEN: T. Banach, L. Howard, M. Belill, "Advanced Tackifiers for Rubber Compounding," Paper No. 96 presented at the Fall Meeting of the Rubber Division, ACS, November 1-3, 2005, Pittsburgh, PA.

Under some situations, *para*-tertiary-butyl-phenol-formaldehyde tackifying resins may outperform the standard *para*-tertiary-octyl-phenol-formaldehyde tackifying resins. GEN: T. Banach, L. Howard, M. Belill, "Advanced Tackifiers for Rubber Compounding," Paper No. 96 presented at the Fall Meeting of the Rubber Division, ACS, November 1-3, 2005, Pittsburgh, PA.

Wood Rosin Tackifiers

Tackifiers based on wood rosin can be used as relatively inexpensive tackifying agents in rubber compounding where good original (unaged) tack is needed. RT: Chapter 18, "Tackifying, Curing, and Reinforcing Resins," B. Stuck, p. 443.

C5 Resins

The so-called "C5 resins" or polyterpenes (such as Wingtack 95®) are reportedly growing in popularity as tackifiers because of good aged tack retention and less adverse effects on cured compound dynamic properties. RP: R. Dailey.

Additives That Cause Bloom

Try to avoid too much of certain rubber compounding ingredients, such as certain antidegradants, oils, accelerators, sulfur, waxes, zinc oxide, and so forth at loadings high enough to cause bloom problems (surface exudation), which interferes with building tack. GEN: "Polysar Halobutyl Innerliner Problem Solving Guide," Processing Problem No. 2.

Use Insoluble Sulfur in Place of Regular Rhombic Sulfur

Use insoluble sulfur in place of regular rhombic crystalline sulfur to avoid sulfur bloom and reduction in tack. GEN: S. Tobing, "Covulcanization in NR/EPDM Blends," *Rubber World*, February, 1988, p. 33.

Avoid Higher Oil Loadings

Generally, tack is decreased when oil levels are increased in a rubber compound. Therefore, decreasing the concentration of oil in a compound may actually improve tack. GEN: C. K. Rhee, J. Andries, "Factors Which Influence Autohesion of Elastomers," *Rubber Chemistry and Technology*, March-April, 1981, Vol. 54, p. 101.

Carbon Black Effects

Generally, increasing loadings of carbon black will reduce tack. Small particle size carbon blacks reduce tack more than larger particle size carbon blacks. Therefore, reducing the carbon black loading as well as using a larger particle size carbon black may actually increase the rubber compound's building tack. GEN: C. K. Rhee, J. Andries, "Factors Which Influence Autohesion of Elastomers," *Rubber Chemistry and Technology*, March-April, 1981, Vol. 54, p. 101.

According to Studebaker and Beatty, simply decreasing the total loading of filler and/or oil will usually improve a compound's building tack. GEN: F. Eirich, *Science and Technology of Rubber*, Chapter 9, "The Rubber Compound and Its Composition," M. Studebaker, J. Beatty, Academic Press, 1978, p. 367.

Using carbon blacks with lower structure will generally increase the compound's tack compared to using a high structure carbon black. GEN: C. K. Rhee, J. Andries, "Factors Which influence Autohesion of Elastomers," *Rubber Chemistry and Technology*, March-April, 1981, Vol. 54, p. 101.

Mixing

Avoid premature scorch. Decrease mixing temperature or change from a one-pass to a two-pass mix. GEN: "Polysar Halobutyl Innerliner Problem Solving Guide," Processing Problem No. 2.

Apply more work history to the batch to achieve a lower compound Mooney viscosity after dump. GEN: "Polysar Halobutyl Innerliner Problem Solving Guide," Processing Problem No. 2.

Reclaim Rubber

Consider using limited amounts of reclaim rubber in a rubber compound in order to improve the "green" tack of the compound. GEN: H. Gandhi (Gujaret Reclaim and Rubber Products Ltd.), "Reclaim Rubber: An Alternate Choice in Natural Crisis," Paper No. 10 presented at the Fall Meeting of the Rubber Division, ACS, October 10–12, 2006, Cincinnati, OH.

Cooling Rate

To retain the maximum tack, one should rapidly cool calendered rubber sheet from a calendering process. RT: Chapter 8, "Specialty Elastomers," G. Jones, D. Tracey, A. Tisler, p. 182.

Carefully control the cooling rate of a calendered or extruded stock. Adjust the cooling rate to maximize tack retention. GEN: "Polysar Halobutyl Innerliner Problem Solving Guide," Processing Problem No. 2.

Factory Ambient Humidity and Temperature

Both factory temperature and humidity can affect building tack. Ambient humidity reportedly has a far greater effect on a compound's tack than temperature. If the humidity increases significantly, then tack will decrease. SBR compounds reportedly decrease in aged tack faster than NR-based compounds when the humidity level rises. A likely reason is that the SBR polymers could be more hygroscopic because of the acid salts they contain. Therefore, higher building tack can be achieved in the factory if the humidity can be reduced and controlled. GEN: C. K. Rhee, J. Andries, "Factors Which Influence Autohesion of Elastomers," *Rubber Chemistry and Technology*, March–April, 1981, Vol. 54, p. 101.

Liners

It is recommended that good quality polyethylene or polypropylene liners be used for tire innerliner storage before the tire building process. These liners should be clean in that they contain no foreign material. Also, avoid cotton or other fabric liners that might allow air to propagate through the liner roll. GEN: B. Sharma, B. Rodgers, D. Tracey (ExxonMobil Chemical Co.), "Tire Halobutyl Rubber Innerliner," Paper No. 94 presented at the Fall Meeting of the Rubber Division, ACS, October 10–12, 2006, Cincinnati, OH.

4.5 Reducing Stickiness to Metal Surfaces

A tendency for a rubber compound to stick to a metal surface during processing is not necessarily related to building tack, discussed in the preceding section. In fact these two properties can be quite different. If a mixed rubber batch has an excessive tendency to stick to various metal surfaces, this could pose a possible processing problem. On the other hand, rubber sticking to the barrel wall of an extruder, for example, is a desirable characteristic for good extrusion.

*The following **experimental** ideas may reduce the tendency of a mixed uncured rubber stock to stick excessively to metal surfaces. Also, all relevant literature sources, including the ones cited below, should be researched and read. **Caution:** these general experimental ideas may not work in all specific situations. Changes to reduce stickiness to metal surfaces will certainly affect other properties as well, for better or for worse, and this book does not purport to show how these other properties are affected. Also, this book does not purport to address safety and health issues.*

Mooney Viscosity

Generally, compounds with higher Mooney viscosity values have less of a tendency to stick to metal surfaces. Therefore, increasing a compound's viscosity by increasing fillers, decreasing oils, or using a base elastomer with a higher molecular weight should increase the compound's viscosity and result in fewer tendencies for the compound to stick to metal surfaces. GEN: "Polysar Halobutyl Innerliner Problem Solving Guide," Processing Problem No. 3.

Reduce Tackifiers

Decrease or eliminate the use of tackifiers in a rubber compound. GEN: "Polysar Halobutyl Innerliner Problem Solving Guide," Processing Problem No. 3.

Clays

For halobutyl compounds, avoid the use of untreated clays as compounding ingredients. GEN: "Polysar Halobutyl Innerliner Problem Solving Guide," Processing Problem No. 3.

Roll Temperature for HIIR

In milling or calendering, halobutyl compounds tend to go to the cooler roll. The lowest level of stickiness is usually at about 85°C (185°F). GEN: "Polysar Halobutyl Innerliner Problem Solving Guide," Processing Problem No. 4.

Fatty Acids and Metallic Soaps

Adding fatty acids such as stearic acid or metallic stearates such as zinc stearate to some compounds may help reduce compound stickiness to metal surfaces. Of course, certain compounds, such as those based on CSM or CM, cannot tolerate the presence of *any* zinc at all, as it will cause accelerated degradation. RP: J. R. Halladay.

CR

Adding 5 to 10 phr of *cis*-polybutadiene to a polychloroprene compound should reduce the tendency of the compound to stick to the mill rolls. This will improve the low-temperature properties of the compound as well. RP: L. L. Outzs.

Reducing the mill temperature may also lessen the tendency of a polychloroprene compound to stick to the mill roll surface. RP: L. L. Outzs.

"Platy" Fillers

Replacing calcium carbonate (whiting) or clay fillers with "platy" fillers such as talc or ground oyster shells may also reduce the tendency of the compound to stick to metallic surfaces. RP: L. L. Outzs.

EPDM

Selecting a semicrystalline, high molecular weight EPDM may help reduce the compound's sticking to the mill rolls. RP: L. L. Outzs.

4.6 Increasing Percent Dispersion of Carbon Black and Fillers

Poor carbon black dispersion from the mixing process will result in inferior cured physical properties and poor product performance characteristics such as wear resistance.

*The following **experimental** ideas might improve the dispersion of carbon black and other fillers. Also, all relevant literature sources, including the ones cited below, should be researched and read. **Caution:** these general experimental ideas may not work in all specific situations. Changes to increase percent dispersion of carbon black and fillers will certainly affect other properties as well, for better or for worse, and this book does not purport to show how these other properties are affected. Also, this book does not purport to address safety and health issues.*

Molecular Weight and Molecular Weight Distribution

Selection of solution elastomers with higher average molecular weight and narrower molecular weight distribution may provide greater shear in mixing that may result in better filler dispersion. RT: Chapter 7, "General Purpose Elastomers and Blends," G. Day, p. 165.

Resinous Additives

Consider using resinous homogenizing agents in a rubber compound to improve filler dispersion in compounds based on a polymer blend as well as with those based on a single polymer. RT: Chapter 14, "Ester Plasticizers and Processing Additives," C. Stone, p. 372.

Certain resinous process aids (such as Struktol 40 MS) and mineral rubber are reported to improve processing characteristics and filler dispersion in halobutyl rubber compounds as well as improve the compatibility of halobutyl rubber with highly unsaturated elastomers. RT: Chapter 8, "Specialty Elastomers," G. Jones, D. Tracey, A. Tisler, p. 180.

Other Additives

Fatty amine processing aids can be used to reduce incorporation time for ultrafine particle sized carbon blacks at high loadings, thus increasing the time available for filler disagglomeration (or dispersion). GEN: H. Takino, S. Iwama, Y. Yamada, S. Kohjiya, "Carbon Black Dispersion and Grip Property of High-Performance Tire Tread Compound," Presented at ACS Rubber Div. Meeting, Spring, 1996, Paper No. 2.

Vinyl acetate wax may be used to reduce incorporation time for ultrafine particle sized carbon blacks at high loadings, thus increasing the time available for filler disagglomeration (or dispersion). GEN: H. Takino, S. Iwama, Y. Yamada, S. Kohjiya, "Carbon Black Dispersion and Grip Property of High-Performance Tire Tread Compound," Presented at ACS Rubber Div. Meeting, Spring, 1996, Paper No. 2.

Blends of Carbon Blacks

For the best dispersion, try to avoid using blends of carbon blacks (such as a common blend of N330 and N650) and try to use only a single carbon black. It is reported that poorer dispersion is sometimes achieved by using a blend of carbon blacks relative to using only a single carbon black. GEN: S. Monthey, T. Reed, "Performance Differences Between Carbon Blacks and CB Blends for Critical IR Applications," *Rubber World*, April, 1999.

Carbon Black

Avoid using carbon black with pellets that are too hard because these pellets may not break up during mixing or disperse properly. RT: Chapter 12, "Compounding with Carbon Black and Oil," S. Laube, S. Monthey, M-J. Wang, p. 303.

Use carbon blacks at their specific optimal loading for optimal dispersiblity. RT: Chapter 12, "Compounding with Carbon Black and Oil," S. Laube, S. Monthey, M-J. Wang, p. 308.

Use carbon blacks with lower surface area for better dispersiblity. RT: Chapter 12, "Compounding with Carbon Black and Oil," S. Laube, S. Monthey, M-J. Wang, p. 308.

Use carbon blacks with higher structure for better dispersiblity. RT: Chapter 12, "Compounding with Carbon Black and Oil," S. Laube, S. Monthey, M-J. Wang, p. 308.

Select carbon blacks with high structure and low surface area because as a group these carbon blacks are the easiest to disperse. Conversely, avoid carbon blacks with low structure and high surface area because as a group these carbon blacks are the hardest to disperse. Very low structure carbon blacks may "wet out" easily during mixing; however, they may not disperse properly. RT: Chapter 12, "Compounding with Carbon Black and Oil," S. Laube, S. Monthey, M-J. Wang, p. 303.

Achieving good percent dispersion with an SAF carbon black (N110 type) is very difficult. One way to achieve good dispersion of SAF blacks is to use SAF masterbatches prepared by a polymer supplier. An SAF masterbatch will shorten the mix cycle and allow good dispersion. GEN: F. Eirich, *Science and*

Technology of Rubber, Chapter 9, "The Rubber Compound and Its Composition," M. Studebaker, J. R. Beatty, Academic Press, 1978, p. 367.

Liquid Phase Mixing of Carbon Black in Natural Rubber

For better dispersion of carbon black in natural rubber, one might review Cabot Corporation's patented continuous coagulation/liquid phase mixing technology. This process reportedly produces very good natural rubber/carbon black masterbatches that can be used in rubber compounding to improve carbon black dispersion compared to conventional methods and impart better reinforcement properties to the rubber compounds. GEN: T. Wang, M. Wang, J. Shell, Y. Wong, V. Vejins (Cabot Corp.), "Liquid Phase Mixing: The Future of Natural Rubber Compounding for Productivity and Performance," Paper No. 24 presented at the Fall Meeting of the Rubber Division, ACS, October 14–17, 2003, Cleveland, OH.

Upside-Down Mixes

Normally a high-structure, large particle size carbon black disperses best. However, anomalies have been observed with *upside-down* mixes of highly loaded EPDM compounds with a 120:105 black-to-oil phr ratio in which a lower structured N326 gave better dispersion than a higher structured N351 black. Hess explains that if the black's absorption capacity is either well above or well below the loading of oil present, then high structure and large particle size black will disperse best. However, if the black's absorption capacity is close to the quantity of oil used in the compound, then reportedly the black will absorb most of the oil, which reduces the black's incorporation rate. Topcik suggests that this problem can be corrected by modifying the upside-down mixing procedure so that some of the oil is held back until the carbon black is incorporated. GEN: W. Hess, "Characterization of Dispersions," *Rubber Chemistry and Technology*, July–August, 1991, Vol. 64, p. 386.

Phase Mixing

For rubber blends such as natural rubber/polybutadiene, consider using phase mixing techniques to effectively force fine particle size carbon black into the NR domains for better dispersion and homogeneity. RT: Chapter 23, "Rubber Mixing," W. Hacker, pp. 515–516.

Affinities of Carbon Black for Different Rubber Domains in Mixing

In trying to achieve ultimate carbon black dispersion in rubber blends, one should consider that different elastomers are not perfectly miscible with each other and usually establish both continuous and discontinuous domains throughout the compound. It is also important to realize that these domains of different elastomers have different affinities for an HAF-type carbon black. It has been reported that BR and SBR have a very high affinity for fully reinforcing carbon blacks, CR and NBR reportedly with a somewhat lower affinity, NR with a still lower affinity, with butyl rubber usually showing the lowest affinity for carbon black. Therefore, to try to disperse carbon black more evenly among different rubber phases, phase mixing techniques should be tried. GEN: J. Callan, W. Hess, C. Scott, "Elastomer Blends, Compatibility and Relative Response to Fillers," *Rubber Chemistry and Technology*, June, 1971, Vol. 44, p. 814.

In a BR/BIIR blend, the carbon black is attracted mainly to the BR domains. By using a compatibilizer (a mixture of aliphatic-naphthenic-aromatic resins), one can improve the tendency for the bromobutyl to wet out the carbon black in a BR matrix. GEN: T. Kohli, J. Halladay, "BR/BIIR Blends for Low-Temperature Damping," *Rubber & Plastics News*, July 19, 1993, p. 14.

Oil

Sometimes selecting a highly aromatic oil as the process oil in a given rubber compound may result in better carbon black dispersion from mixing for an SBR- and BR-based compound. RT: Chapter 12, "Compounding with Carbon Black and Oil," S. Laube, S. Monthey, M-J. Wang, p. 312.

Order of Additions of Carbon Black and Oil to a Mix

Normally, better carbon black dispersion is achieved if the oil is not added with the carbon black. It is usually better if one adds the oil to a mix after the carbon black is added. GEN: W. Hess, "Characterization of Dispersions," *Rubber Chemistry and Technology*, July-August, 1991, Vol. 64, p. 386.

Cooling Batches Overnight

Generally, the percent dispersion of fillers can be improved if the batch from the first-pass mix is cooled overnight and then remilled. GEN: W. Hess, "Characterization of Dispersions," *Rubber Chemistry and Technology*, July-August, 1991, Vol. 64, p. 386.

Silica

Consider using highly dispersible grades of precipitated silicas (such as Hi-Sil EZ®) for better dispersion of silica during mixing. GEN: L. Evans, J. Dew, L. Hope, T. Krivak, W. Waddell, "Hi-Sil EZ: Easy Dispersing Precipitated Silica," *Rubber & Plastics News*, July 31, 1995, p. 12.

Silica fillers are more "rubber phobic" than carbon black. Therefore, silica fillers are harder to disperse than carbon blacks in general purpose elastomers.

However, using functionalized solution SBRs can greatly improve the dispersibility of silica. It is reported that S-SBR, which is functionalized with primary amine and alkoxysilane during anionic polymerization, is better at achieving well-dispersed silica than using S-SBR functionalized with only alkoxysilane during the polymerization process. However, just using S-SBR with only alkoxysilane functionalization is still better than S-SBR without any functionalization at all. GEN: M. Iwano, T. Sone, T, Tominaga (JSR Corp.), "Improved Silica Dispersibility with Functionalized S-SBR for Lower Rolling Resistance," Paper No. 19A presented at the ITEC 2008, September 15–17, 2008, Akron, OH.

Also, for better silica dispersion, consider using epoxidized natural rubber (ENR). GEN: A. Chapman, S. Cook, R. Davies, J. Patel, J. Clark (Malaysian Rubber Board), "Microdispersion of Silica in Tire Tread Compounds Based on Epoxidized Natural Rubber," Paper No. 73 presented at the Fall Meeting of the Rubber Division, ACS, October 12–15, 2012, Pittsburgh, PA.

Processing Aid for Silica Dispersion

To achieve better silica dispersion in the first pass with the silanization reaction with an organosilane, consider using a processing aid based either on a resin blend (Struktol XP1343) or on a fatty acid ester (Struktol XP1335). These processing aids were reportedly specially designed to have a minimal effect on the organosilane coupling chemistry. GEN: C. Stone, "Improving the Silica 'Green Tire' Tread Compound by the Use of Special Process Additives," Presented at ACS Rubber Div. Meeting, Fall, 1999, Paper No. 77.

Silica Dispersion in Polar Elastomers

Conventional mixing of silica will usually disperse better in a polar rubber than in a nonpolar rubber and may produce better reinforcement. GEN: H. Tanahashi, S. Osanai, M. Shigekuni, K. Murakami, Y. Ikaeda, S. Kohjiya, "Reinforcement of Acrylonitrile-butadiene Rubber by Silica Generated In Situ," *Rubber Chemistry and Technology*, March–April, 1998, Vol. 71, p. 38

In Situ Silica Formation in Rubber Matrix

Achieving good dispersion of precipitated silica in rubber is sometimes difficult. One experimental method of possibly achieving very good silica dispersion in rubber is to have an *in situ* formation of silica from silica tetraethoxysilane (TEOS). This involves both a hydrolysis and a condensation reaction to achieve the *in situ* formation of silica that is well dispersed. GEN: S. Kohjiya, K. Murakami, S. Iio, T. Tanahashi, Y. Ikeda, "In Situ Filling of Silica onto 'Green' Natural Rubber by the Sol-Gel Process," *Rubber Chemistry and Technology*, March–April, 2001, Vol. 74, p. 16.

Reactive Polymer and Fiber Dispersion

The use of low molecular weight maleated polybutadiene (PBDMA) with 5 phr of fiber pulps such as cotton, nylon-6, and polyester will significantly improve incorporation time for these fibers in an EPR compound. The PBDMA may shorten the fiber incorporation time for other compounds based on other elastomer systems as well. The PBDMA reportedly functions as a compatibilizer for the rubber and these polymer fibers and reportedly aids in the wetting of these fibers during mixing. GEN: A. Estrin, "Application of PBDMA for Enhancement of EPR Loaded with Chopped Fibers," *Rubber World*, April 2000, p. 39.

Carbon Black Masterbatches

Consider using a carbon black masterbatch of the base elastomer where the polymer manufacturer has already incorporated a specified carbon black into the rubber. Examples of such carbon black masterbatches are seen with SBR 1606 and so forth. The use of these carbon black masterbatches will shorten the required mixing time and may result in better ultimate dispersion. RP: J. M. Long.

Fiber Masterbatch

When trying to achieve good dispersion of aramid fibers into a rubber compound, consider using a "masterbatch concentrate" that contains the Kevlar pulp. These are commercially available from the fiber manufacturers. GEN: K. Watson, A. Frances, "Elastomer Reinforcement with Short Kevlar Aramid Fiber for Wear Applications," *Rubber World*, August, 1988, p. 20.

Commonly, aramid fiber dispersion can be improved in a rubber compound by using aramid fiber predispersions. GEN: S. Monthey (Rhein Chemie Corp.), "Improving Aramid Fiber Dispersion in Elastomeric Compounds Using the Next Generation of Polymer Bound Predispersion," Paper No. 27 presented at the Fall Meeting of the Rubber Division, ACS, October 10–12, 2006, Cincinnati, OH.

Gelled Polymer and Breakdown

A small amount of "loose" gel in synthetic polyisoprene will usually increase the bulk viscosity of the IR, which will break down during mixing and may actually help in achieving better filler dispersion (somewhat analogous to higher gel content in natural rubber). On the other hand, if the gel is a "tight" gel, it will not break down during mixing and will remain intact, resulting in poor quality. GEN: C. Gibbs, S. Horne, J. Macey, H. Tucker, "Effect of Gel and Structure on the Properties of *cis*-1,4-Polyisoprene," *Rubber World*, April, 1961, p. 69.

Nanofillers

Nanofillers represent a particular challenge in achieving good dispersion. It has been reported that by using a special surface coating by a plasma polymer film based on pyrrole or thiophene (monomers) to naturally occurring halloysite nanotubes (HNT)–a naturally occurring aluminosilicate nanotube–greatly reduces the filler-filler interaction and increases rubber-filler interactions between HNT and the rubber medium, with better dispersion. GEN: M. Poikelispaa, A. Fas, W. Dierkes, and J. Vuorinen (Tampere University of Technology, Finland), Paper No. 53 presented at the Fall Meeting of the Rubber Division, ACS, October 10–13, 2011, Cleveland, OH.

Mill Mixing

In open mill mixing, avoid directly mixing hard and soft polymers together. Usually, polymers of the same softness tend to mix and blend more readily than do hard and soft polymers. Also, one does not want to achieve too rapid a breakdown of the elastomer(s) until good dispersion of compounding ingredients is achieved. For highly loaded compounds, one does not want to add dry fillers on a mill too rapidly. One should allow time for the earlier filler additions to work their way into the compound. The addition of the oils and plasticizers is usually held off until some dispersion of fillers is achieved. However, this is not always the case if the recipe calls for extremely high levels of oil to be added. RT: Chapter 23, "Rubber Mixing," W. Hacker, p. 510.

Banbury Mixing

In Banbury mixing, the challenge is to maintain stiffness so that too much breakdown of the rubber does not take place until good dispersion occurs. Therefore oils and plasticizers are usually added later in the mix cycle. A common mixing sequence for a Banbury mix is (1) load rubbers and part of fillers, (2) add remaining filler, and (3) add oils and plasticizers. However, if the batch softens too rapidly, it may be necessary to follow an "upside-down" mixing sequence, which is quite different. RT: Chapter 23, "Rubber Mixing," W. Hacker, p. 513.

Mixing, Two-Pass

To achieve better dispersion, one may consider establishing a two-pass mix in which a so-called "nonproductive" or masterbatch is prepared (no curatives) so that more work history can be applied in a "second-pass" mix to achieve better filler dispersion, especially for very fine particle size carbon blacks or silicas. RT: Chapter 23, "Rubber Mixing," W. Hacker, p. 515.

Cooling System for Mixer

Use an internal mixer that has an effective cooling system so that the batch temperature does not rise too quickly. Rapidly rising batch temperature adds to the batch's heat history, contributing to shorter scorch times (for final batch curatives) and causing the compound viscosity to also rapidly decrease, which results in lower shear mixing forces and less effective dispersion at the higher temperature range. Therefore keeping the batch at a lower temperature range for a longer time period should improve the ultimate dispersion and state-of-mix. Besides assuming that the optimum batch size has been selected and that one is working with an effective cooling system, one can also gain some additional control of mixing temperature by adjusting either ram pressure or rotor speed. RT: Chapter 23, "Rubber Mixing," W. Hacker, p. 514.

Continuous Mixing Systems

If one is using a continuous compounding or mixing system, the challenge is to create the proper conditions to achieve adequate distribution and dispersion of compounding ingredients. Often this is very difficult to do. RT: Chapter 23, "Rubber Mixing," W. Hacker, p. 508.

4.7 Increasing Green Strength

Green strength (or uncured compound strength) is very important to prevent "blowouts" in the second-stage tire building machine or to keep a complicated extruded profile from collapsing from the force of gravity.

The following ***experimental*** *ideas might help increase the green strength of a rubber compound. Also, all relevant literature sources, including the ones cited below, should be researched and read.* **Caution:** *these general experimental ideas may not work in all specific situations. Changes to increase green strength will certainly affect other properties as well, for better or for worse, and this book does not purport to show how these other properties are affected. Also, this book does not purport to address safety and health issues.*

Molecular Weight Effects

Use a higher molecular weight base elastomer. RT: Chapter 2, "Compound Processing Characteristics and Testing," J. S. Dick, p. 41.

The green strength of a compound based on SBR can be improved somewhat by selecting an SBR with a higher average molecular weight; however, very high molecular weight may hurt other processing characteristics. GEN: G. Hamed, "Tack and Green Strength of NR, SBR, and NR/SBR Blends," *Rubber Chemistry and Technology*, May–June, 1981, Vol. 54, p. 403.

Long-Chain Branching

Synthetic elastomers with more long-chain branching (LCB, six or more carbon atoms branched off the backbone) generally have higher green strength. GEN: D. Hofkens, G. Zandyoort, S. Baird, M. Boggelen, P. Knape, G. Choonoo, M. Koch, S. Bhattacharjee (DSM Elastomers), "A New EPDM to Meet Compounding and Processing Challenges," Paper No. 24 presented at the Fall Meeting of the Rubber Division, ACS, October 13–15, 2009, Pittsburgh, PA.

Post-polymerization Treatment to Form "Labile Crosslinks"

To improve green strength of a synthetic rubber, post-polymerization treatments can be considered that will impart a very low crosslink density of labile crosslinks (perhaps only 1 per 3000 monomer units), which are destroyed from mechanical shear but may re-form at rest. Some of these polymer treatments involve maleic anhydride, thioglycolic acid derivatives, or carboxylic acids for nonemulsion polymers. Also, emulsion SBR having a third monomer with an amino group was polymerized. A quaternary ammonium halide salt was added to the latex to allow labile crosslinks to form later for improved green strength. GEN: E. Buckler, G. Briggs, J. Dunn, E. Lasis, Y. Wei, "Green Strength in Emulsion SBR," *Rubber Chemistry and Technology*, November-December, 1978, Vol. 51, p. 872.

Strain Crystallization

Use strain-crystallizing elastomers to impart better green strength to a compound. RT: Chapter 2, "Compound Processing Characteristics and Testing," J. S. Dick, p. 41.

Natural Rubber

Use NR as the base elastomer. RT: Chapter 2, "Compound Processing Characteristics and Testing," J. S. Dick, p. 41; Chapter 6, "Elastomer Selection," R. School, p. 127; Chapter 7, "General Purpose Elastomers and Blends," G. Day, p. 142.

Normally, natural rubber will impart good green strength to a rubber compound because NR tends to crystallize on stretching. However, NR grades that contain higher levels of linked fatty acid ester groups tend to promote crystallization more rapidly and thus have greater green strength. Generally speaking, the minimum level of linked fatty acid ester groups to promote this crystallization is approximately 2.8 mmol/kg. GEN: S. Kawahara, Y. Isono, T. Kakubo, Y. Tanaka, E. Aik-Hwee, "Crystallization Behavior and Strength of Natural Rubber Isolated from Different Hevea Clone," *Rubber Chemistry and Technology*, March-April, 2000, Vol. 73, p. 39.

Avoid IR

Avoid using synthetic IR; use NR instead because it imparts better green strength. RT: Chapter 6, "Elastomer Selection," R. School, p. 130.

Block Polymers

Small amounts of block styrene in what is normally random SBR may improve green strength. RT: Chapter 7, "General Purpose Elastomer and Blends," G. Day, p. 148.

Star-Branched Polymers

Consider using butyl rubber with a "star" molecular structure in place of regular butyl rubber to improve green strength. RT: Chapter 8, "Specialty Elastomers," G. Jones, D. Tracey, A. Tesler, pp. 174, 180, 182.

Generally, higher molecular weight increases green strength, but, in addition, stresses relaxation rates become slower, affecting processing characteristics. To achieve a better balance between green strength and stress relaxation, one might consider trying a star-branched elastomer such as star-branched butyl rubber. GEN: W-C. Wang, K. Powers, J. Fusco, "Star Branched Butyl-Novel Butyl Rubber for Improved Processability. I. Concepts, Structure and Synthesis," Presented at ACS Rubber Div. Meeting, Spring, 1989, Paper No. 21.

Semicrystalline EPDM

Use semicrystalline EPDM grades (with higher ethylene content) to impart higher room-temperature green strength. RT: Chapter 8, "Specialty Elastomers," R. Vara, J. Laird, p. 191.

Metallocene-Catalyzed EPDM Grades

Single-site constrained-geometry metallocene catalyst technology for commercial polymerization has made it possible to efficiently produce commercial grades of EPDM with very high ethylene content and high crystallinity. These special grades with higher ethylene content impart higher green strength. Using this unique technology, it is also possible to manipulate the ethylene composition distribution to enhance green strength further. GEN: D. Parikh, M. Hughes, M. Laughner, L. Meiske, R. Vara, "Next Generation of Ethylene Elastomers," Presented at ACS Rubber Div. Meeting, Fall, 2000, Cincinnati, OH.

New Proprietary Ziegler–Natta Catalyzed EPDM

Through new proprietary Ziegler–Natta catalyst technologies, unique high-temperature crystallinity through ethylene sequencing in EPDM reportedly gives better "high-temperature" green strength. With this ethylene sequencing, some crystallinity undergoes multiple crystallinity transitions up to 75°C (167°F). GEN: S. Brignac, H. Young, "EPDM with Better Low-Temperature Performance," *Rubber & Plastics News*, August 11, 1997, p. 14.

Gas-Phase EPDM

Try compounding with a ultra-low Mooney viscosity gas-phase polymerized EPDM with very high ethylene content and high filler loading (a higher ethylene content increases the green strength). GEN: A. Paeglis, "Very Low Mooney Granular Gas-Phase EPDM," Presented at ACS Rubber Div. Meeting, Fall, 2000, Paper No. 12.

Molecular Weight Distribution

Consider NBR grades with more narrow molecular weight distribution for higher green strength. RT: Chapter 8, "Specialty Elastomers," M. Gozdiff, p. 197.

CR

Selecting fast-crystallizing grades of polychloroprene may improve the green strength of the compound. Adding a high-styrene SBR resin to a CR compound may also improve the compound's green strength. RP: L. L. Outzs.

BIIR

Adding a specially prepared BIMSM ionomer to a bromobutyl rubber (BIIR) compound reportedly significantly increased the compound's measured green strength. GEN: A. Tsou, I. Duvdevani, P. Agarwal (ExxonMobil Chemical Co.), "Quaternary Ammonium BIMS Ionomers," Paper No. 23 presented at the Spring Meeting of the Rubber Division, ACS, April 28–30, 2003, San Francisco, CA.

"T-Type" Neoprene

When compounding with neoprenes, T-types resist collapse and distortion better than extrudates from W-types, with G-types imparting the least resistance for neoprenes. RT: Chapter 8, "Specialty Elastomers," L. L. Outzs, p. 211.

Polynorbornene as an Additive

The green strength of compounds based on NR, SBR, BR, CR, and/or NBR can be significantly increased by blending them with small amounts of polynorbornene. RT: Chapter 8, "Specialty Elastomers," C. Cable, p. 226.

Polyoctenamer as an Additive

Polyoctenamer or *trans*-polyoctenylene rubber (TOR) reportedly can be blended with NR, BR, SBR, NBR, CR, EPDM, and so forth in minor amounts to improve green strength at room temperature (as a result of crystallinity). GEN: A. Draxler, "A New Rubber: *trans*-Polyoctenamer," Chemische Werke Huels AG, Postfach, Germany; J. Sommer, *Elastomer Molding Technology*, Elastech, Hudson, OH, 2003, p. 180.

K. Nedden reported that polyoctanomer (TOR) significantly improved the green strength of a natural rubber tire formulation. Reportedly, a 10% addition of TOR gave a 50% improvement in green strength. GEN: W. Hess, C. Herd, P. Vegvari, "Characterization of Immiscible Elastomer Blends," *Rubber Chemistry and Technology*, July–August, 1993, Vol. 66, p. 329.

PTFE

Try Alphaflex®, a PTFE additive, to increase green strength of a rubber compound. GEN: J. Menough, "A Special Additive," *Rubber World*, May, 1987, p. 12.

Carbon Black

Selecting a carbon black with higher surface area reportedly will increase the compound's green strength. RT: Chapter 12, "Compounding with Carbon Black and Oil," S. Laube, S. Monthey, M-J. Wang, p. 321.

Selecting a carbon black with a higher structure reportedly will increase the compound's green strength. RT: Chapter 12, "Compounding with Carbon Black and Oil," S. Laube, S. Monthey, M-J. Wang, p. 321.

Increasing the carbon black loadings of epichlorohydrin compounds with blacks such as N330, N550, N762, and so forth should improve the compound's green strength in extrusion operations. RT: Chapter 8, "Specialty Elastomers," C. Cable, p. 219.

N326 carbon black reportedly is used in many wire coat or skim compounds for tires because it imparts relatively high green strength while maintaining a relatively low compound viscosity level (which is needed for penetration into the steel tire cord). RT: Chapter 12, "Compounding with Carbon Black and Oil," S. Laube, S. Monthey, M-J. Wang, p. 319.

Carbon Black Loadability

For good green strength or collapse resistance, use a carbon black with a higher structure and a lower surface area. The lower surface area will impart to this carbon black higher loadability. Higher loadings of carbon black will also improve the compound's green strength. GEN: S. Monthey, "The Influence of Carbon Blacks on the Extrusion Operation for Hose Production," *Rubber World*, May, 2000, p. 38.

Nanofillers

It is reported that using a partially ion-exchanged montmorillonite clay (nanoclay) in an NBR masterbatch significantly increased the measured green strength of the compound. GEN: R. Faulkner, C. McAfee, "Partially Exfoliated Nanoclay/NBR Composites," Paper No. 30 presented at the Fall Meeting of the Rubber Division, ACS, October 10–12, 2006, Cincinnati, OH.

Chemical Promoters

Adding chemical promoters to rubber compounds based on IIR, SBR, or IR reportedly may improve green strength. (Caution: Care must be taken to avoid exposure of workers to nitrosamines.) GEN: L. Ramos de Valle, M. Montelongo, "Cohesive Strength in Guayule Rubber and Its Improvement Through Chemical Promotion," *Rubber Chemistry and Technology*, November-December, 1978, Vol. 51, p. 863.

Mixing

One common cause of too-low compound green strength is too much mastication of the base polymer during the mixing process. GEN: S. Monthey, "The Influence of Carbon Blacks on the Extrusion Operation for Hose Production," *Rubber World*, May, 2000, p. 38

Phase Mixing

In a patent, B. Lee reported an improvement in green strength when carbon black was forced into the minor NR phase of a blend with synthetic elastomers. GEN: W. Hess, C. Herd, P. Vegvari, "Characterization of Immiscible Elastomer Blends," *Rubber Chemistry and Technology*, July-August, 1993, Vol. 66, p. 329; US Patent 4,455,399, June 19, 1984.

Electron Beam

An electron beam (accelerated electrons) can be used with an EVM compound to increase the Mooney viscosity and green strength. GEN: D. Keller, L. Bryant, J. Dewar, "Enhanced Viscosity EVM Elastomers for GP Molded and Extruded Applications," *Rubber World*, May, 1999, p. 34.

An electron beam reportedly can be used to irradiate bromobutyl compounds to improve green strength. GEN: S. Mohammed, J. Timar, J. Walker, "Green Strength Development by Electron Beam Irradiation of Halobutyl Rubbers," *Rubber Chemistry and Technology*, March-April, 1983, Vol. 56, p. 276.

Commonly, an electron beam is also used to improve the green strength of NR/SBR blends (tire carcass compounds). RP: J. M. Long.

4.8 Increasing Scorch Safety Time

Compound scorch safety time is the time required at a specified temperature before incipient crosslinks form (the onset of vulcanization). If the measured scorch safety time is too short, then the mixed stock may "burn up" (scorch) somewhere downstream in the rubber plant before reaching the curing process, thus producing scrap. Therefore, it may be necessary to investigate ways of providing rubber compounds with additional scorch safety time.

*The following **experimental** ideas may increase a rubber compound's scorch safety time. Also, all relevant literature sources, including the ones cited below, should be researched and read. **Caution**: these general experimental ideas may not work in all specific situations. Changes to increase scorch safety time will certainly affect other properties as well, for better or for worse, and this book does not purport to show how these other properties are affected. Also, this book does not purport to address safety and health issues.*

SBR

Extruding a rubber compound based on SBR generally is less "scorchy" than extruding a similar compound based on natural rubber. RT: Chapter 6, "Elastomer Selection," R. School, p. 129.

NBR

When selecting nitriles with concerns for longer scorch safety times, consider NBR grades that have lower ACN content or lower Mooney viscosity. GEN: R. Del Vecchio, E. Ferro, "Effects of NBR Polymer Variations on Compound Properties," Presented at ACS Rubber Division Meeting, Spring, 2001, Paper No. 21.

Natural Rubber

Adding polybutadiene rubber (BR) to an NR-based compound will normally increase the scorch safety time for the compound. GEN: Lim Yew Swee (Lanxess), "Benefits of Butadiene Rubber in Natural Rubber-Based Truck Tread and Sidewall," Presented at the India Rubber Exposition and Conference, January 2011, Chennai, India.

Carbon Black

Compounds loaded with lower structure carbon black will tend to have longer scorch safety times. RT: Chapter 12, "Compounding with Carbon Black and Oil," S. Laube, S. Monthey, M-J. Wang, p. 308.

Compounds loaded with lower surface area (larger particle size) carbon black will tend to have longer scorch safety times. RT: Chapter 12, "Compounding with Carbon Black and Oil," S. Laube, S. Monthey, M-J. Wang, p. 308.

Compounds with lower loadings of carbon black will tend to have longer scorch safety times. RT: Chapter 12, "Compounding with Carbon Black and Oil," S. Laube, S. Monthey, M-J. Wang, p. 308.

Silica

Using a precipitated silica with a higher surface area may lengthen the scorch safety time of a nonhalogenated rubber compound as well as slow down the actual cure rate, but it may also decrease cut growth. GEN: W. Waddell, L. Evans, "Use of Nonblack Fillers in Tire Compounds," *Rubber Chemistry and Technology*, July-August, 1996, Vol. 69, p. 377.

Use Higher Concentration of Zinc Oxide

Using too low a concentration of zinc oxide in a sulfenamide sulfur cure can sometimes cause scorch problems with compounds based on diene rubber. GEN: W. Hall, H. Jones, "The Effect of Zinc Oxide and Other Curatives on the Physical Properties of a Bus and Truck Tread Compound," Presented at ACS Rubber Div. Meeting, Fall, 1970.

Selection of Accelerators

Consider cure systems based on the following accelerators:

- Sulfenamides: Good scorch safety
- Sulfenimides: Good scorch safety
- Thiazoles: Some scorch safety
- Thiurams: Poor scorch safety
- Dithiocarbamates: Poor scorch safety

RT: Chapter 15, "Sulfur Cure Systems," B. H. To, p. 383.

Ordinal Ranking of Scorch Safety Time

When measured by total scorch safety time in seconds, sulfenamide accelerators at equal concentration can be ranked as follows:

(Longest) DCBS > MBS > TBBS ≅ CBS > MBTS (Shortest)

GEN: F. Ignatz-Hoover, R. Genetti, B. To, "Vulcanization of General Purpose Elastomers," Paper No. D presented at the Spring Meeting of the Rubber Division, ACS, May 16-18, 2005, San Antonio, TX.

Sulfenamide Cures

An MBS accelerator usually imparts more scorch safety to a rubber compound than does a TBBS accelerator. However, MBS is a nitrosamine generator, which poses safety and health concerns in the workplace. RT: Chapter 15, "Sulfur Cure Systems," B. H. To, p. 383.

Selection of "Kickers" for Sulfenamide Cures

With sulfenamide-based cures, use tetra-alkyl thiuram disulfide (TATD) as the "kicker" (secondary accelerator) instead of other lower molecular weight thiurams to prevent excess shortening of scorch safety time while still achieving a faster cure rate than when only a sulfenamide accelerator is used. GEN: S. Hong, M. Hannon, J. Kounavis, P. Greene, Presented at ACS Rubber Div. Meeting, Spring, 2001, Paper No. 37.

Ethylac as a "Kicker"

Often, using a secondary accelerator ("kicker") with a primary accelerator will increase both the cure rate and the propensity to scorch. Studebaker and Beatty reported that Ethylac, when used as a secondary accelerator, has less tendency to make the compound scorchy compared to other secondary accelerators. GEN: F. Eirich, *Science and Technology of Rubber*, Chapter 9, "The Rubber Compound and Its Composition," M. Studebaker, J. R. Beatty, Academic Press, 1978, p. 367.

Retarders

For some conventional thiazole-based cure systems for diene-based elastomers, salicylic acid, benzoic acid, or phthalic anhydride may be used as "retarders" to delay the scorch (and retard the cure rate and state of cure). RT: Chapter 15, "Sulfur Cure Systems," B. H. To, p. 387.

Inhibitors

Careful use of the prevulcanization inhibitor cyclohexylthiophthalimide (CTP) will enable a compounder to specifically adjust the scorch safety time needed for a given compound using a sulfenamide cure system. RT: Chapter 15, "Sulfur Cure Systems," B. H. To, p. 387.

DTDM as a Sulfur Donor for EV Cures

For low or no free sulfur cures, consider using DTDM instead of TMTD as the sulfur donor to provide longer scorch safety times. RT: Chapter 16, "Cures for Specialty Elastomers," B. H. To, pp. 400–401, 404, 405.

ISB Accelerator

A new experimental accelerator, 2-(isopropylsulfinyl)-benzothiazole (ISB), when used with certain secondary accelerators such as tetrabenzylthiuram disulfide or the less environmentally acceptable tetramethylthiuram monosulfide, can provide very good scorch safety with a very fast cure rate. GEN: R. Hopper, "2-(Isopropylsulfinyl)-benzothiazole as a Delayed Action Thiazole Accelerator," *Rubber Chemistry and Technology*, September-October, 1993, p. 623.

Cure Kinetics and Downstream Processing Heat History

Reducing the total downstream heat history (usually from the work history), after the curatives are added during the mixing of the rubber batch, will better preserve the rubber stock's scorch safety time before going into the curing process. GEN: J. Dick, M. Ferraco, K. Immel, T. Mlinar, M. Senskey, J. Sezna, "Utilization of the Rubber Process Analyzer in Six Sigma Programs to Improve Quality and Reduce Production Costs," Paper No. 15 presented at the Fall Meeting of the Rubber Division, ACS, October 16-19, 2001, Cleveland, OH.

Concentrations of Accelerator vs. Sulfur

At low cure temperatures, sometimes a higher concentration of a sulfenamide accelerator in relation to a constant sulfur level gives shorter incubation time (scorch time). However, at much higher cure temperatures, the opposite may result. GEN: T. Kleiner, R. Schuster, "The Influence of Cure System and Polymer Structure on Network Properties," Paper No. C presented at the Spring Meeting of the Rubber Division, ACS, April 29-May 1, 2002, Savannah, GA.

Avoid Moisture Contamination of Sulfenamide Accelerators

To increase scorch safety time, it is important that moisture contamination in compounding ingredients be minimized because the presence of water in a sulfenamide cure such as MBS can hydrolyze the sulfenamide, causing a shorter scorch time. All compounding ingredients should be stored in a dry location. GEN: J. Butler, P. Freakley, "Effect of Humidity and Water Content on the Cure Behavior of a Natural Rubber Accelerated Sulfur Compound," *Rubber Chemistry and Technology*, May-June, 1992, Vol. 65, p. 374.

Avoid Aged Sulfenamides

Using aged sulfenamides in rubber compounding can cause scorch problems. GEN: W. Cole (Flexsys), "Controlling Raw Material Quality Stability," Paper No. 16 presented at the Fall Meeting of the Rubber Division, ACS, October 16-19, 2001, Cleveland, OH.

Peroxide Cures

To achieve more scorch safety when using peroxides to cure a rubber compound, consider using the new proprietary "high performance" (HP) peroxide formulations. RT: Chapter 17, "Peroxide Cure Systems," L. Palys, p. 418.

For peroxide cures, consider selecting BBPIB at a *lower* concentration instead of DCP for a longer scorch safety time. RT: Chapter 17, "Peroxide Cure Systems," L. Palys, p. 418.

When curing with peroxides, one should select a suitable rubber curing peroxide that is more thermally stable and has a higher half-life. RT: Chapter 17, "Peroxide Cure Systems," L. Palys, p. 418.

Peroxides such as DMBPHy, with higher thermal stability than DCP, may provide better scorch safety, especially at higher compounding temperatures. RT: Chapter 17, "Peroxide Cure Systems," L. Palys, p. 428.

To increase the scorch safety time of a peroxide cure, it is sometimes useful to consider the use of a "cure suppressor" such as butylated hydroxytoluene (BHT). This can experimentally be added to the compound until the curemeter maximum torque is reduced by 25 to 50%. This reportedly might require between 0.1 and 1 phr BHT. Then an effective selected coagent is added to return the curemeter maximum torque to its original value, with hopefully a retained improvement in scorch safety time (over the original) and still ensuring a good cure rate as well. GEN: P. Dluzneski, "Peroxide Vulcanization of Elastomers," Rubber Chemistry and Technology, July–August, 2001, Vol. 74, p. 451.

Rubber compounds cured with peroxides such as dicumyl peroxide (DCP) can be somewhat scorchy. By adding 2,2,6,6-tetramethylpiperidinyloxyl (under the trade name of "Tempo") may significantly increase the scorch safety time of the peroxide-cured compound. This chemical functions as a nitroxyl free radical that reacts with other carbon free radicals. GEN: Benny George, Rosamma Alex, "Scorch Control in Peroxide Vulcanization Using a Stable Free Radical," Paper presented at India Rubber Expo 2011, January 19, 2011, Chennai, India.

Selection of Coagents for Peroxide Cures

Be careful when selecting coagents for a peroxide cure because many coagents will significantly reduce scorch safety time for a given compound. RT: Chapter 17, "Peroxide Cure Systems," L. Palys, p. 432.

Use low molecular weight (liquid) high-vinyl 1,2-polybutadiene resins (such as Ricon® grades) as a coagent in a peroxide cure of EPDM compounds to obtain significantly better scorch safety than some other traditional coagents might impart. GEN: R. Drake, "Using Liquid Polybutadiene Resin to Modify Elastomeric Properties," *Rubber & Plastics News*, February 28 and March 14, 1983.

For a peroxide cure, avoid selecting a coagent such as trimethylolpropane triacrylate (TMPTA), trimethylolpropane trimethacrylate (TMPTMA), or *N,N′*-*m*-phenylene dimaleimide (HVA-2) because these coagents may significantly shorten scorch safety time. Coagents such as 1,2-polybutadiene (PB), diallyl phthalate (DAP), triallyl isocyanurate (TAIC), or triallyl cyanurate (TAC) will usually have less effect on scorch safety. GEN: P. Dluzneski, "Peroxide Vulcanization of Elastomers," *Rubber Chemistry and Technology*, July–August, 2001, Vol. 74, p. 451.

Work has been reported where the crosslink density of a peroxide cure system can be increased with more delayed scorch by using a cure system based on dicumyl peroxide, a coagent such as *N,N′*-*m*-phenylenedimaleimide (BMI-MP), and a sulfur "spender" such as dipentamethylenethiuram tetrasulfide (DPTT) for curing either saturated or low unsaturated rubber. GEN: M. A. Grima, J. G. Eriksson, A. G. Talma, R. N. Datta, and J. W. M. Noordermeer, "Mechanistic Studies into the New Concept of Co-agents for Scorch Delay and Property Improvement in Peroxide Vulcanization," Paper No. 86 presented at the Fall Meeting of the Rubber Division, ACS, October 10, 2006.

Peroxide/Sulfur Cures

Using about 4 phr *N,N′*-*p*-phenylenedimaleimide (BMI-PP) as a coagent with 0.7 to 0.96 phr of a sulfur-containing compound, dipentamethylenethiuram tetrasulfide (DPTT), with 2.4 phr dicumyl peroxide (DCP), for curing an EPM-based compound may greatly improve the compound's scorch safety. The sulfur-containing compound helps improve the scorch safety of this peroxide cure. GEN: M. Grima, A. Talma, R. Datta, J. Noordermeer, "New Concept of Co-agents for Scorch Delay and Property Improvement in Peroxide Vulcanization," Paper No. 33 presented at the Fall Meeting of the Rubber Division, ACS, November 1–3, 2005, Pittsburgh, PA.

Avoid Diurethane Cures

Avoid using diurethane cures because they reportedly make an NR rubber compound very scorchy. GEN: T. Kempermann, "Sulfur-Free Vulcanization Systems for Diene Rubber," *Rubber Chemistry and Technology*, July–August, 1988, Vol. 61, p. 422.

AOs

When selecting PPD antiozonant(s), if possible, avoid dialkyl PPDs because they are the scorchiest of the three major PPD classes. RT: Chapter 19, "Antidegradants," F. Ignatz-Hoover, p. 457.

In tire sidewall compounds, 6PPD AO used alone reportedly may reduce scorch safety; however, blends of 6PPD and TAPDT (higher molecular weight AO) may preserve some scorch protection and still impart good flex fatigue protection and dynamic ozone crack resistance. GEN: S. Hong, C. Lin, "Improved Flex Fatigue and Dynamic Ozone Crack Resistance Through the Use of Antidegradants or Their Blends in Tire Compounds," Presented at ACS Rubber Div. Meeting, Fall, 1999, Paper No. 27.

Mixing

Avoid high dump temperatures from the Banbury for final mixes because this will contribute to the total heat history for the compound and shorten the scorch safety time. RT: Chapter 8, "Specialty Elastomers," J. Jones, D. Tracey, A. Tisler, p. 181.

Use an internal mixer that has an effective cooling system so that the batch temperature does not rise too quickly. A rising batch temperature adds to the batch's heat history, contributing to a shortening of the scorch safety time for a final mix. In addition, a rising batch temperature causes the compound viscosity to decrease, which causes lower shear mixing forces and less effective dispersion at the higher temperature range. Therefore, keeping the batch at a lower temperature range longer should improve the ultimate dispersion and state of mix. Besides assuming that the optimum batch size has been selected and working with an effective cooling system, one can also gain some additional control of mixing temperature by adjusting either ram pressure or rotor speed. Minimizing the total heat history of a final or one-pass batch should improve the scorch safety time of the batch. RT: Chapter 23, "Rubber Mixing," W. Hacker, p. 514.

Improve Cooling After Mixing

Improve the mixed stock cooling after dump for longer scorch safety time. GEN: "Polysar Halobutyl Innerliner Problem Solving Guide," Mixing Problem No. 3.

Compound for Lower Viscosity and Lower Viscous Heating

Compounds based on lower structure carbon blacks may have less tendency to have scorch problems downstream in the factory because these compounds have a lower viscosity and are usually exposed to less viscous heating during processing. Because these compounds run cooler than those stocks containing higher structure carbon blacks, they are less likely to scorch prematurely. The same may apply for low surface area carbon blacks vs. high surface area carbon blacks. The low surface area carbon blacks run cooler. GEN: F. Eirich, *Science and Technology of Rubber*, Chapter 9, "The Rubber Compound and

Its Composition," M. Studebaker, J. R. Beatty, Academic Press, 1978, p. 367; J. S. Dick, H. Pawlowski, "A Short Rheological Test to Overcome the Inadequacies of Traditional Scorch Measurements," Presented at ACS Rubber Div. Meeting, Fall, 2002; J. Dick, H. Pawlowski, J. Moore, "Viscous Heating and Reinforcement Effects of Different Fillers using the Rubber Process Analyzer," *Rubber World*, January, 2000, p. 22.

HIIR

For halobutyl compounds, acidic ingredients usually reduce scorch time whereas basic ingredients increase it–exactly the opposite of what is experienced for many conventional nonhalogenated elastomer cures. Therefore, review all compounding ingredients for the halobutyl cure and consider possible adjustments that might provide longer scorch safety time without the loss of other needed properties. GEN: "Polysar Halobutyl Innerliner Problem Solving Guide," Mixing Problem No. 3.

Use chlorobutyl instead of bromobutyl when a longer scorch safety time is desired. RT: Chapter 6, "Elastomer Selection," R. School, p. 134; Chapter 8, "Specialty Elastomers," J. Jones, D. Tracey, A. Tisler, p. 183.

The use of alkaline fillers such as calcium stearate may retard the cure of a halobutyl rubber compound. RT: Chapter 8, "Specialty Elastomers," J. Jones, D. Tracey, A. Tisler, p. 180.

The use of acidic clays in a halobutyl rubber compound can promote a very fast cure that may require the use of halobutyl rubber scorch retarders such as magnesium oxide. RT: Chapter 8, "Specialty Elastomers," J. Jones, D. Tracey, A. Tisler, p. 180.

For halobutyl rubber compounds, exercise caution when using phenolic and some other types of tackifying resins that react with the halobutyl and can shorten the scorch safety time. RT: Chapter 8, "Specialty Elastomers," J. Jones, D. Tracey, A. Tisler, p. 180.

Be careful of phenolic resins, aromatic oils, and wood rosins in a halobutyl formulation because each of these ingredients can potentially shorten the compound's scorch safety time. GEN: "Polysar Halobutyl Innerliner Problem Solving Guide," Mixing Problem No. 3.

Magnesium oxide is a scorch retarder in halobutyl rubber cures except where an amine cure is applied. RT: Chapter 8, "Specialty Elastomers," J. Jones, D. Tracey, A. Tisler, p. 185.

Be careful of amine antioxidants and antiozonants in halobutyl formulations because they may shorten scorch safety time. GEN: "Polysar Halobutyl Innerliner Problem Solving Guide," Mixing Problem No. 3.

Be very careful in using specific curatives in halobutyl compounds because some may make the compound too scorchy. Vultac is probably too scorchy for bromobutyl compounds. GEN: "Polysar Halobutyl Innerliner Problem Solving Guide," Processing Problem No. 1.

Do not use zinc stearates in a slab dip for halobutyl compounds because zinc is a curing agent and can cause scorch problems. GEN: "Polysar Halobutyl Innerliner Problem Solving Guide," Processing Problem No. 1.

BIMS

Brominated isobutylene-co-*p*-methylstyrene (BIMS)-based compounds may have better scorch safety properties than a conventional halobutyl rubber compound possesses. GEN: G. Jones, "Exxpro Innerliners for Severe Service Tire Applications," Presented at ITEC, 1998, Paper No. 7A.

Curing Polychloroprene

When using neoprene in compounding, a W-type can be more effectively compounded to give better scorch safety. RT: Chapter 8, "Specialty Elastomers," L. L. Outzs, p. 211.

Consider using a TMTM/DOTG/sulfur cure instead of a traditional ETU cure for a possible improvement in scorch safety when curing a polychloroprene-based compound. RT: Chapter 16, "Cures for Specialty Elastomers," B. H. To, pp. 400–401, 406.

Increase the loading of magnesium oxide or the surface area of the magnesium oxide used for a polychloroprene cure to lengthen the scorch safety time. Typically high-activity MgO (surface area >100 m^2/g) is used to cure polychloroprene. The typical loading for MgO for a polychloroprene cure is 4 phr. GEN: R. Ohm, "New Developments in Curing Halogen-Containing Polymers," ACS Rubber Div. Education Symposium No. 45, "Automotive Applications II," Spring, 1998, p. 3.

Consider using an "encapsulated" form of high-activity magnesium oxide in a CR compound. This is because the regular free powder form of high-activity magnesium oxide when exposed to a hot, humid atmosphere will react with the humidity in the air quickly and become quite inactive in only 30 minutes (or less) of exposure. RP: L. L. Outzs.

In many polychloroprene cures, stearic acid may function as a retarder, resulting in longer scorch and cure time. GEN: R. Ohm, "New Developments in Curing Halogen-Containing Polymers," Presented at ACS Rubber Div. Education Symposium No. 45, "Automotive Applications II," Spring, 1998, p. 3.

In curing polychloroprene compounds, select a higher concentration of MBTS to be used with ETU to achieve a longer scorch safety time as well as possibly a faster cure rate. GEN: R. Grossman, Q & A, *Elastomerics*, January, 1989, p. 37; R. M. Murray, D. C. Thompson, *The Neoprenes*, DuPont, 1963, p. 28.

Polychloroprene compounds that contain ZMTI can be scorchy. One possible way to lengthen scorch safety of a CR compound is to add a small quantity of CBS accelerator. GEN: R. Ohm, "Accelerators and Antidegradants Influence Fatigue Resistance," *Elastomerics*, January, 1988, p. 19.

In polychloroprene compounding with an ETU cure, *N*-cyclohexylthiophthalimide can function as a good scorch inhibitor to increase scorch safety. GEN: R. Tabar, P. Killgoar, R. Pett, "A Fatigue Resistant Polychloroprene Compound for High Temperature Dynamic Applications," *Rubber Chemistry and Technology*, September-October, 1979, Vol. 52, p. 781.

N-(Cyclohexylthio)phthalimide (CTP) is effective as a prevulcanization inhibitor for curing a W-type neoprene with thiocarbanilide (A-1) acceleration in order to achieve an increase in scorch safety. RT: Chapter 16, "Cures for Specialty Elastomers," B. H. To, p. 401.

Proprietary Thiadiazole Accelerator for CR and HIIR for Better Scorch Safety

Consider using a proprietary accelerator called Vanax® 189 from R. T. Vanderbilt for curing either halobutyl or polychloroprene. Vanax 189 is a derivative of 1,2-dimercapto-1,3,4-thiadiazole. This accelerator reportedly imparts better scorch safety to CR and halobutyl compounds than many other accelerators used. GEN: R. Ohm, "New Developments in Curing Halogen-Containing Polymers," Presented at ACS Rubber Div. Education Symposium No. 45, "Automotive Applications II," Spring, 1998, pp. 2-4.

Fluoroelastomers

For the best scorch protection in curing an FKM-based compound, use a bisphenol instead of a diamine cure. RT: Chapter 6, "Elastomer Selection," R. School, p. 134; Chapter 8, "Specialty Elastomers," R. Stevens, p. 230.

EPDM

Selecting grades of EPDM with broader molecular weight distribution in some cases may improve the scorch safety imparted to the compound. GEN: R. Grossman, Q & A, *Elastomerics*, January, 1989, p. 37.

"Black Scorch" with EPDM Compounds

"Black scorch" with EPDM compounds is not related to curatives; it can occur with EPDM masterbatches that contain no curatives. This so-called "black scorch" that occasionally occurs in black-loaded EPDM compounds can be lengthened by using an EPDM with a lower 5-ethylidene-2-norbornene (ENB) content. GEN: C. Daniel, J. Pillow, "Black Scorch in EPDM Compounds," Presented at the IRC meeting at Manchester, UK, June 10, 1999.

"Black scorch" in EPDM compounds can be lengthened by using carbon blacks with lower structure. GEN: C. Daniel, J. Pillow, "Black Scorch in EPDM Compounds," Presented at the IRC meeting at Manchester, UK, June 10, 1999.

Ironically, "black scorch" with EPDM compounds may be improved by adding early a very small amount of sulfur. Also, it has been reported that sometimes a small amount of certain sulfur-based accelerators may also help inhibit black scorch. However, some other sulfur-based accelerators have no effect or can even make black scorch worse. This problem will require experimentation. GEN: C. Daniel, J. Pillow, "Black Scorch in EPDM Compounds," Presented at the IRC meeting at Manchester, UK, June 10, 1999.

"Black scorch" with EPDM compounds can be lengthened by remixing the batch. GEN: C. Daniel, J. Pillow, "Black Scorch in EPDM Compounds," Presented at the IRC meeting at Manchester, UK, June 10, 1999.

"Black scorch" with EPDM compounds can be lengthened by using sources of carbon black that have less polar surfaces on the carbon black particles. Carbon blacks that have too high a surface polarity have a greater propensity to promote "black scorch" because of the basic incompatibility with the relatively nonpolar EPDM polymer. This surface polarity can differ from one source of carbon black to another. GEN: X. Zhang, R. Whitehouse, C. Liauw, "Compatibility of Carbon Blacks with Typical EPDM Extrusion Compounds: Phenomenon and Root Causes," Presented at the 1997 IRC Meeting in Malaysia.

4.9 Increasing Cure Rate

Cure rate of a rubber compound can be a very important characteristic. The rate at which a rubber compound cures greatly determines how quickly that rubber product can be removed from the mold. The shorter the residence time in the mold is, the lower the molding costs will be.

The following ***experimental*** *ideas may increase a compound's cure rate. Also, all relevant literature sources, including the ones cited below, should be researched and read.* ***Caution:*** *these general experimental ideas may not work in all specific situations. Changes to increase cure rate will certainly affect other properties as well, for better or for worse, and this book does not purport to show how these other properties are affected. Also, this book does not purport to address safety and health issues.*

Accelerator Selection

Consider cure systems based on the following types of accelerators:

- Aldehyde amine: Slow cure rate
- Guanidines: Medium cure rate
- Thiazoles: Semifast cure rate
- Sulfenamides: Fast-delayed action cure rate
- Sulfenimides: Fast-delayed action cure rate
- Dithiophosphates: Fast cure rate
- Thiurams: Very fast cure rate
- Dithiocarbamates: Very fast cure rate

RT: Chapter 15, "Sulfur Cure Systems," B. H. To, p. 383.

Ordinal Ranking for Cure Rate

When measured by cure rate time in seconds, sulfenamide accelerators at equal concentration can be ranked as follows:

$$\text{(Fastest) TBBS} \cong \text{CBS} > \text{MBS} > \text{DCBS (Slowest)}$$

GEN: F. Ignatz-Hoover, R. Genetti, B. To, "Vulcanization of General Purpose Elastomers," Paper No. D presented at the Spring Meeting of the Rubber Division, ACS, May 16-18, 2005, San Antonio, TX.

TBBS Accelerator

TBBS accelerator usually imparts a faster cure rate as a sulfenamide cure to a rubber compound than do other sulfenamide accelerators. RT: Chapter 15, "Sulfur Cure Systems," B. H. To, p. 384. GEN: J. S. Dick, "Oils, Plasticizers, and Other Rubber Chemicals," *Basic Rubber Testing*, ASTM International, 2003, p. 146.

"Kickers"

The use of DPG, TATM, TBzTD, TMTD, TMTM, TETD, ZDMC, ZDEC, ZDBC, or DOTG as a secondary accelerator (or "kicker") with a primary sulfenamide accelerator such as TBBS, MBS, or CBS can significantly increase the cure rate over what it would have been if only the primary sulfenamide accelerator was used alone. RT: Chapter 15, "Sulfur Cure Systems," B. H. To, p. 383. GEN: S. Hong, M. Hannon, J. Kounavis, P. Greene, Presented at ACS Rubber Div. Meeting, Spring, 2001, Paper No. 37.

The cure rate of a compound based on a cure system with TBBS as the only accelerator reportedly can be increased significantly by adding TMTM as a "kicker." For example, one might achieve a 25% faster cure rate by cutting the TBBS concentration from 1.2 phr to 0.6 phr and adding 35% of the remaining total as 0.2 phr of TMTM ("kicker"). This may yield approximately a 25% improvement in cure rate with maybe a small change in modulus. However, results can differ from one compound to another. RP: T. D. Powell.

ISB Accelerator

A new experimental accelerator, 2-(isopropylsulfinyl)-benzothiazole (ISB), when used with certain secondary accelerators such as tetrabenzylthiuram disulfide or the less environmentally acceptable tetramethylthiuram monosulfide, can provide very good scorch safety with a very fast cure rate. GEN: R. Hopper, "2-(Isopropylsulfinyl)-benzothiazole as a Delayed Action Thiazole Accelerator," *Rubber Chemistry and Technology*, September–October, 1993, p. 623.

TATD Accelerator

Consider using TATD in place of other thiuram disulfide accerators to achieve a faster cure cycle. GEN: M. Boisseau (Crompton Corp.), "New Ultra Accelerator Good for Both NR, SR," *Rubber & Plastics News*, June 17, 2002.

Cure Kinetics Effects from Cure Temperature and Concentration of Sulfur and Accelerator

Higher cure temperatures increase cure rate. In addition, increasing the concentration of sulfur and/or accelerator will usually result in an increase in compound cure rate. GEN: T. Kleiner, R. Schuster, "The Influence of Cure System and Polymer Structure on Network Properties," Paper No. C presented at the Spring Meeting of the Rubber Division, ACS, April 29–May 1, 2002, Savannah, GA.

Higher Zinc Oxide Loading

For compounds with high accelerator concentration and sufficient concentration of stearic acid, a higher concentration of zinc oxide can sometimes shorten cure times. GEN: W. Hall, H. Jones, "The Effect of Zinc Oxide and Other Curatives on the Physical Properties of a Bus and Truck Tread Compound," Presented at ACS Rubber Div. Meeting, Fall, 1970.

Avoid Retarders

For some conventional thiazole-based cure systems for diene rubber(s), avoid salicylic acid, benzoic acid, phthalic anhydride, or other organic acids that may retard a cure and slow down the cure rate. RT: Chapter 15, "Sulfur Cure Systems," B. H. To, p. 387.

Peroxide Cures

To achieve faster cure rate when using peroxides to cure a rubber compound, consider using the new proprietary "high performance" (HP) peroxide formulations. RT: Chapter 17, "Peroxide Cure Systems," L. Palys, p. 418.

For a shorter cure time for peroxide cures, consider selecting DCP at a higher concentration instead of BBPIB. RT: Chapter 17, "Peroxide Cure Systems," L. Palys, p. 418.

Peroxyketal peroxides may impart a faster cure rate to a compound than dialkyl peroxides. RT: Chapter 17, "Peroxide Cure Systems," L. Palys, p. 425.

Peroxide Cure Coagents

For a peroxide cure, select a coagent such as trimethylolpropane triacrylate (TMPTA), trimethylolpropane trimethacrylate (TMPTMA), or *N,N'-m*-phenylene dimaleimide (HVA-2) for an improvement in cure rate. Coagents such as 1,2-polybutadiene (PB), diallyl phthalate (DAP), triallyl isocyanurate (TAIC), or triallyl cyanurate (TAC) may not be effective in increasing the cure rate. GEN: P. Dluzneski, "Peroxide Vulcanization of Elastomers," *Rubber Chemistry and Technology*, July–August, 2001, Vol. 74, p. 451.

Using low molecular weight (liquid) high-vinyl 1,2-polybutadiene resin (such as special Ricon grades that contain certain antioxidants for improved heat resistance of the BR) as an additive at low part levels in a peroxide-cured silicone compound reportedly can increase cure rate. GEN: R. Drake, "Using Liquid Polybutadiene Resin to Modify Elastomeric Properties," *Rubber & Plastics News*, February 28 and March 14, 1983.

Platinum Cures for Silicone Rubber

Curing silicone rubber with a platinum catalyst instead of with a peroxide can reduce cure times by as much as 70%. GEN: J. Sommer, *Elastomer Molding Technology*, Elastech, Hudson, OH, 2003, p. 178.

IR

With IR polymers, higher *cis* or *trans* content will result in a faster sulfur cure rate than higher 3,4-isoprene content. RT: Chapter 7, "General Purpose Elastomers and Blends," G. Day, p. 153.

BR

In BR polymers, higher *cis* or *trans* content will result in a faster sulfur cure rate than higher 1,2-polybutadiene content. Likewise, when BR *cis* or *trans* content is diluted with the addition of styrene to SBR, the cure rate is decreased. RT: Chapter 7, "General Purpose Elastomers and Blends," G. Day, p. 153.

Solution *cis*-BR and solution *trans*-BR based compounds will sulfur cure faster than emulsion BR-based compounds. The fatty acids that are used in emulsion polymerization cause retardation of the sulfur cure for these polymers. RT: Chapter 7, "General Purpose Elastomers and Blends," G. Day, p. 156.

In a pure peroxide cure, 3,4-IR and 1,2-BR microstructures will increase the peroxide cure rate. RT: Chapter 7, "General Purpose Elastomers and Blends," G. Day, p. 156.

Increasing the 1,2-BR content can increase the cure rate in a microwave cure. RT: Chapter 7, "General Purpose Elastomers and Blends," G. Day, p. 156.

SBR

SBR with higher bound styrene content will usually have a slower sulfur cure rate. For example, E-SBR 1013 with over 40% styrene will cure slower than E-SBR 1006 with 23.5% styrene. RT: Chapter 7, "General Purpose Elastomers and Blends," G. Day, p. 153.

S-SBR (such as 711) generally cures faster with sulfur than hot emulsion SBR (such as E-SBR 1006), which cures faster than cold emulsion SBR (such as

E-SBR 1500). The fatty acids and rosin acids that are used in emulsion polymerization cause retardation of the sulfur cure for these polymers. RT: Chapter 7, "General Purpose Elastomers and Blends," G. Day, p. 153.

NR vs. BR vs. SBR Cure Rates

Generally NR cures faster than *cis*-polybutadiene, which is slightly faster than SBR. GEN: M. Studebaker, J. R. Beatty, "Vulcanization," *Elastomerics*, February, 1977, p. 41.

ENR

Generally, the epoxidized natural rubber will sulfur cure faster than natural rubber itself because the epoxy units will also crosslink via nonsulfur chemistry. RT: Chapter 7, "General Purpose Elastomers and Blends," G. Day, p. 156.

HIIR

Consider using bromobutyl in place of chlorobutyl rubber to achieve a faster cure. RT: Chapter 8, "Specialty Elastomers," J. Jones, D. Tracey, A. Tisler, p. 183.

In contrast to "normal" unsaturated rubber/sulfur cures, halobutyl cures are usually accelerated by acidic additives (and retarded by bases). RT: Chapter 8, "Specialty Elastomers," J. Jones, D. Tracey, A. Tisler, p. 183.

If the concentration of magnesium oxide starts to exceed 0.5 phr in a halobutyl rubber cure, then the cure rate might be depressed. RT: Chapter 8, "Specialty Elastomers," J. Jones, D. Tracey, A. Tisler, p. 185.

EPDM

Selecting grades of EPDM with a particular nonrandom distribution of the third monomer ("judicious third monomer introduction") reportedly may increase the cure rate imparted to the compound. GEN: R. Grossman, Q & A, *Elastomerics*, January, 1989, p. 37.

Select EPDM elastomer grades with ethylidene norbornene (ENB) for the fastest cure rate, followed by 1,4-hexadiene (HD) and then dicyclopentadiene (DCPD), which is the slowest of the commonly used third monomers. RT: Chapter 8, "Specialty Elastomers," R. Vara, J. Laird, p. 191.

Select a grade of EPDM with a higher amount of diene third monomer. RT: Chapter 8, "Specialty Elastomers," R. Vara, J. Laird, p. 192.

NBR

Increasing the bound acrylonitrile content (ACN) of the NBR grade may result in a faster cure rate for a sulfur cure system. Decreasing ACN content of the

NBR grade may result in a faster cure rate for a peroxide cure system. RT: Chapter 8, "Specialty Elastomers," M. Gozdiff, p. 194.

Use an emulsion NBR with fatty acid based emulsification system for the fastest cure rate instead of a rosin acid (slowest) or mixed system or synthetic. RT: Chapter 8, "Specialty Elastomers," M. Gozdiff, p. 195.

Use an emulsion NBR that is produced from a coagulating system based on calcium chloride for a faster cure rate. RT: Chapter 8, "Specialty Elastomers," M. Gozdiff, p. 195.

ACM

In the compounding of polyacrylate polymers, usually basic nonblack fillers are selected because acidic fillers retard most cure systems used for ACM. RT: Chapter 8, "Specialty Elastomers," P. Manley, C. Smith, p. 206.

Polychloroprene

Consider curing a W-type neoprene with thiocarbanilide (A-1) acceleration to achieve a very fast cure rate. RT: Chapter 16, "Cures for Specialty Elastomers," B. H. To, p. 401.

Select the optimum loading of magnesium oxide and the optimum surface area of the magnesium oxide used for a polychloroprene cure to maximize the cure rate. Sometimes lowering the MgO loading from the traditional 4 phr can increase the compound's cure rate while sacrificing some scorch safety time. GEN: R. Ohm, "New Developments in Curing Halogen-Containing Polymers," ACS Rubber Div. Education Symposium No. 45, "Automotive Applications II," Spring, 1998, p. 3.

In curing polychloroprene compounds, select more MBTS with ETU to achieve faster cure rate as well as longer scorch safety time. GEN: R. Grossman, Q & A, *Elastomerics*, January, 1989, p. 37.

AEM

To significantly increase the cure rate for terpolymers of AEM (under the trade name of Vamac® from DuPont) that use a traditional diamine cure system based on hexamethylenediamine carbamate (HMDC) and DPG, add dicumyl peroxide and a 1,2-polybutadiene (Ricon® 152), which reportedly can increase the cure rate by as much as 160%. GEN: H. Barager, K. Kammerer, E. McBride, "Increased Cure Rates of Vamac® Dipolymers and Terpolymers Using Peroxides," Presented at ACS Rubber Div. Meeting, Fall, 2000, Paper No. 115.

FKM

In selecting a cure system for fluoroelastomer-based compounds, generally a cure system based on peroxide will provide a faster cure rate than a bisphenol package. GEN: P. Surette (Daikin America), "New DAI-EL Fluoroelastomers for Extreme Environments," Presented at the Fall Meeting of the Energy Rubber Group, September 17–18, 2008, San Antonio, TX.

Silicone Rubber

In selecting a cure system for silicone rubber compounds, consider the use of the faster platinum cure system instead of the slower peroxide-based system in order to achieve a faster cure rate. GEN: S. Richardson (Wacker Silicones), "Silicone Rubber," Presented at a meeting of the Southern Rubber Group, March 8, 2011, Greenville, SC.

Proprietary Thiadiazole Accelerator for HIIR and CR

Consider using a proprietary accelerator called Vanax® 189 from R. T. Vanderbilt for curing either halobutyl or polychloroprene. Vanax 189 is a derivative of 1,2-dimercapto-1,3,4-thiadiazole. This accelerator reportedly imparts a faster cure rate with better scorch safety time to CR and halobutyl compounds than many other accelerators used. GEN: R. Ohm, "New Developments in Curing Halogen-Containing Polymers," Presented at ACS Rubber Div. Education Symposium No. 45, "Automotive Applications II," Spring, 1998, pp. 2–4.

Carbon Black

Increasing the carbon black loading will increase the cure rate of the compound; however, the level of increase is somewhat dependent on the base elastomer. GEN: W. Wampler, M. Gerspacher, H. Yang, "CB's Role in Compound Curing Behavior," *Rubber World*, April, 1994, p. 39.

Cure Time for Thick Articles with an Increase in Thermal Conductivity

Use powdered metals as a compounding ingredient to improve a compound's thermal conductivity used in producing a cured thick rubber article. Among the metal powders to select from, silver, copper, and aluminum are usually the best thermal conductors. Silver is usually too expensive. Copper is a very strong pro-oxidant and degradant of elastomers such as natural rubber. Aluminum powder is used to improve both thermal and electrical conductivity; however, the use of this material can pose special safety problems such as fire hazards and so forth. (In fact, the use of aluminum powder or some other metallic powders in rubber can make the rubber compound highly flammable.) The total cure time for thick article cures is significantly shortened if the thermal conductivity of the compound is increased. Normally, rubber compounds are poor conductors of heat, and there is less heat history in the center of a thick article. Sometimes an additional 5 minutes is required for every additional 6 mm thickness for a thick article being cured at 150°C (302°F) in order to complete the cure. By improving the thermal conductivity of the compound, the cure time for a thick article can be significantly shortened. GEN: V. Vinod, S. Varghese, R. Alex, B. Kuriakose, "Effect of Aluminum Powder on Filled Natural Rubber Composites," *Rubber Chemistry and Technology*, May–June, 2001, Vol. 74, p. 236.

Induction Heating

Consider curing a given rubber compound with an alternating electromagnetic field (induction heating) where the subject compound is loaded with particles of a special silica-coated iron oxide (trade name "Magsilica"), which will generate heat internally and speed up the cure rate throughout the thickness of the compound. GEN: O. Taikum, A. Korch, R. Friehmeit, F. Minister, M. Schoiz, H. Herzog, S. Katusic (Evonik Degussa Corp.), "Novel Silica Coated Iron Oxide 'Magsilica' to Speed Up Crosslinking in Rubber," Paper No. 16 presented at the Fall Meeting of the Rubber Division, ACS, October 11–13, 2011, Cleveland, OH.

4.10 Reducing Reversion

Reversion that can occur during the curing process for some compounds can cause a deterioration in cured physical properties. Compounds based on natural rubber are particularly susceptible to this effect.

*The following **experimental** ideas can be considered to reduce reversion. Also, all relevant literature sources, including the ones cited below, should be researched and read. **Caution:** these general experimental ideas may not work in all specific situations. Changes to reduce reversion will certainly affect other properties as well, for better or for worse, and this book does not purport to show how these other properties are affected. Also, this book does not purport to address safety and health issues.*

Natural Rubber

Avoid natural rubber because NR is very susceptible to reversion. RT: Chapter 2, "Compound Processing Characteristics and Testing," J. S. Dick, p. 41.

For NR compounds, cure at a lower temperature for a longer period of time to increase the preponderance of monosulfidic crosslinks for better resistance to reversion. GEN: M. Studebaker, J. R. Beatty, "Vulcanization," *Elastomerics*, February, 1977, p. 41.

Guayule rubber might give somewhat better resistance to reversion than traditional Hevea natural rubber under certain circumstances in specific recipes. GEN: C. McMahan, K. Cornish, H. Pawlowski, J. Williams, "Dynamic Mechanical Properties of Latex Films," Paper No. 62 presented at the Spring Meeting of the Rubber Division, ACS, May 16–18, 2005, San Antonio, TX.

CR in Place of NR

Polychloroprene compounds can possess significantly better resistance to reversion than natural rubber based compounds. GEN: R. Tabar, P. Killgoar, R. Pett, "A Fatigue Resistant Polychloroprene Compound for High Temperature Dynamic Applications," *Rubber Chemistry and Technology*, September-October, 1979, Vol. 52, p. 781.

Addition of BR or SBR to an NR Compound

Consider adding some SBR or BR to a natural rubber compound to reduce the tendency of NR to revert. The tendency of NR to revert can be countered by the tendency of SBR or BR to increase in crosslink density. GEN: M. Studebaker, J. R. Beatty, "Vulcanization," *Elastomerics*, February, 1977, p. 41.

Addition of Vinyl BR to an NR Compound

One might consider blending vinyl BR with natural rubber to attempt to reduce reversion. GEN: K. Nordsiek, "Rubber Microstructure and Reversion," *Rubber World*, December, 1987, p. 30.

Addition of TOR to an NR Compound

One might consider blending a minor amount of polyoctenamer [or *trans*-polyoctenylene rubber (TOR)] with natural rubber to attempt to reduce reversion. GEN: K. Nordsiek, "Rubber Microstructure and Reversion," *Rubber World*, December, 1987, p. 30.

Peroxide vs. Sulfur Cures for Millable Polyurethane

Using peroxide instead of sulfur cures reportedly will improve the reversion resistance of some compounds based on millable polyurethanes. GEN: C. Gibbs, S. Horne, J. Macey, H. Tucker, "Effect of Gel and Structure on the Properties of *cis*-1,4-Polyisoprene," *Rubber World*, April, 1961, p. 69.

Addition of NBR to a Millable Polyurethane Compound

Adding a given quantity of NBR incrementally to a millable polyurethane compound that is sulfur cured reportedly will improve the compound's reversion resistance. GEN: T. Jablonowski, "Blends of PU with Conventional Rubbers," *Rubber World*, October, 2000, p. 41.

Lower Cure Temperature

If a compound is based on NR, cure at a lower temperature such as below 140°C (284°F). RT: Chapter 2, "Compound Processing Characteristics and Testing," J. S. Dick, p. 41.

Resin Cure

The use of a methylol phenol-formaldehyde resin cure in place of a sulfur cure for a butyl compound should significantly improve the compound's reversion resistance. RT: Chapter 8, "Specialty Elastomer," G. Jones, D. Tracey, A. Tisler, p. 178.

Diurethane Cure for an NR Compound

Consider using a diurethane cure (Novor® crosslinking agents from Hughson Chemical Co.), which reportedly imparts a degree of reversion resistance to an NR vulcanizate. GEN: T. Kempermann, "Sulfur-Free Vulcanization Systems for Diene Rubber," *Rubber Chemistry and Technology*, July–August, 1988, Vol. 61, p. 422.

EV Cures

Use natural rubber compounds based on an efficient vulcanization (EV) or semiefficient (semi-EV) cure system where the ratio of accelerator to sulfur is high and sulfur donors are sometimes used in place of free sulfur. These cure systems generate a higher preponderance of monosulfide and disulfide crosslinks, which are thermodynamically more stable than polysulfidic crosslinks. Therefore, semi-EV and EV cures impart better reversion resistance to an NR compound. RT: Chapter 15, "Sulfur Cure Systems," B. H. To, p. 387.

TBSI Cure

N-tert-butyl-2-benzothiazole sulfenimide (TBSI) accelerator reportedly will provide better resistance to reversion than other sulfenamide accelerators such as DIBS, DCBS, or MBS. RT: Chapter 15, "Sulfur Cure Systems," B. H. To, p. 389.

In regards to comparing delayed-action accelerators for reversion resistance, TBSI is reported to be one of the best, followed by DCBS, followed by MBS, which is closely followed by TBBS. GEN: F. Ignatz-Hoover, "Vulcanization of General Purpose Elastomers," Paper No. D presented at the Spring Meeting of the Rubber Division, ACS, May 16–18, 2005, San Antonio, TX.

Increase ZnO Loading

Sulfur/sulfenamide-cured compounds that contain higher than usual levels of zinc oxide may possess better heat aging characteristics and better resistance to reversion. GEN: W. Hall, H. Jones, "The Effect of Zinc Oxide and Other Curatives on the Physical Properties of a Bus and Truck Tread Compound," Presented at ACS Rubber Div. Meeting, Fall 1970.

ZBPD

Zinc-*O,O*-di-*n*-butylphosphorodithioate (ZBPD), as a supplement to a sulfenamide cure, reportedly will impart very good reversion resistance to an NR compound. RT: Chapter 15, "Sulfur Cure Systems," B. H. To, p. 391.

Alkylphenol Disulfide

Using alkylphenol disulfide accelerators in the cure system for a halobutyl rubber compound reportedly gives better reversion resistance. GEN: B. Rodgers, N. Tambe, S. Solis, B. Sharma (ExxonMobil Chemical Co.), "Alkylphenol Disulfide Polymer Accelerators and the Vulcanization of Isobutylene Based Elastomers," Paper No. 160 presented at the Fall Meeting of the Rubber Division, ACS, October 15–18, 2007, Cleveland, OH.

HTS

Hexamethylene-1,6-bis-thiosulfate (HTS), a post-vulcanization stabilizer, reportedly forms special hybrid crosslinks when it is used as a supplement to a sulfenamide cure to impart superior reversion resistance to an NR compound and good flex fatigue resistance. RT: Chapter 15, "Sulfur Cure Systems," B. H. To, p. 391.

BCI-MX

1,3-bis(Citraconimidomethyl)benzene (BCI-MX), an antireversion agent, reportedly imparts superior reversion resistance to an NR compound and good compressive flex resistance as measured with a Goodrich Flexometer. RT: Chapter 15, "Sulfur Cure Systems," B. H. To, pp. 391–393.

New Crosslinking Agent

Lanxess introduced a new crosslinking agent, 1,6-bis(*N,N*-dibenzylthiodicarbamyldithio) hexane, under the trade name of "Vulcuren," which also can be used to form more stable crosslinks and resist thermal reversion. GEN: T. Kleiner (Bayer), "Improvements in Abrasion and Heat Resistance by Using a New Crosslink Agent," Paper No. 12A presented at the ITEC Meeting, September, 2002, Akron, OH.

New Triacrylate Additives for NR Compounds

As in the case of BCI-MX, it was also discovered that pentaerythritol triacrylate (SR444 from Sartomer) and trimethylolpropane triacrylate (SR351 from Sartomer) both may reduce reversion by forming new crosslinks in an NR compound. GEN: E. Blok, M. Kralevich, J. Varner, "Preliminary Studies on New Anti-reversion Agents for the Sulfur Vulcanization of Diene Rubbers," *Rubber Chemistry and Technology*, March–April, 2000, Vol. 73, p. 114.

It has also been reported that zinc methacrylate salts can function as reversion inhibitors in sulfur-cured rubber compounds. GEN: S. Henning (Sartomer Co.), "The Use of Coagents in Sulfur Vulcanization: Functional Zinc Salts," Paper No. 10 presented at the Spring Meeting of the Rubber Division, ACS, May 16–18, 2005, San Antonio, TX.

"Equilibrium Cure"

Consider using the so-called "equilibrium cure" or a bis(3-triethoxysilylpropyl)-tetrasulfane (TESPT) cure to prevent reversion with an NR-based compound. GEN: S. Wolff, "Chemical Aspects of Rubber Reinforcement," *Rubber Chemistry and Technology*, July–August, 1996, Vol. 69, p. 325; S. Wolff, *Kautsch. Gummi Kunstst.*, 1979, Vol. 32, p. 760.

DNPD and ZnO Cure for a BIIR Compound

Consider using a substituted *p*-phenylenediamine AO such as *N,N'*-di-beta-naphthyl-*p*-phenylene-diamine (DNPD or Agerite White®) and zinc oxide as crosslinking agents for a bromobutyl-based compound, because this combination may improve vulcanizate thermal stability. GEN: D. Edwards, "A High-Pressure Curing System for Halobutyl Elastomers," *Rubber Chemistry and Technology*, March-April, 1987, Vol. 60, p. 62.

Thiuram Cure and Peroxide Cure

Use a cure system for an NR compound that is "consumed" during the curing process (such as DCP or TMTD). GEN: M. Studebaker, J. R. Beatty, "Vulcanization," *Elastomerics*, February, 1977, p. 41.

Consider using a "thiuram cure" for improved reversion resistance provided that problems with nitrosoamines can be avoided through good ventilation and so forth. GEN: T. Kempermann, "Sulfur-Free Vulcanization Systems for Diene Rubber," *Rubber Chemistry and Technology*, July-August, 1988, Vol. 61, p. 422.

Post-cure

Silicone compounds may experience reversion over time and use if they are not post-cured (a post-vulcanization "baking") to drive off low molecular weight silicone species and the peroxide decomposition products. This is particularly true for acidic residues (decomposition products based on carboxylic acids), which must be driven off after cure because they may cause reversion of the silicone compound with time and use. RT: Chapter 8, "Specialty Elastomers," J. R. Halladay, p. 237.

Carbon Black Loading

Use a higher loading of carbon black because carbon black helps inhibit reversion in a natural rubber compound. GEN: M. Studebaker, J. R. Beatty, "Vulcanization," *Elastomerics*, February, 1977, p. 41. However, Boonkerd, Phasook, and Deeprasertkul report that using lower loadings of carbon black in a natural rubber compound may actually improve the compound's reversion resistance. GEN: K. Boonkerd, W. Phasook, C. Deepraseertkul (Thailand), "Effects of Carbon Black on the Reversion Behavior of a Cured Natural Rubber Compound," Paper No. 64 presented at the Fall Meeting of the Rubber Division, ACS, October 11-13, 2011, Cleveland, OH.

Using a thermal black instead of a furnace black in a natural rubber compound may improve resistance against reversion. GEN: K. Boonkerd, W. Phasook, C. Deepraseertkul (Thailand), "Effects of Carbon Black on the Reversion

Behavior of a Cured Natural Rubber Compound," Paper No. 64 presented at the Fall Meeting of the Rubber Division, ACS, October 11–13, 2011, Cleveland, OH.

Dithiophosphate vs. Dithiocarbamate Cures

Functionally, dithiophosphate rubber accelerators are similar to dithiocarbamate accelerators except the phosphorus atoms replace the nitrogen atoms for these traditional dithiocarbamate cure packages. The dithiophosphate cures reportedly give potentially better heat aging resistance and reversion resistance as well compared to dithiocarbamate cures. GEN: S. Monthey, M. Saewe, V. Meenenga (Rhein Chemie), "Using Dithiophosphate Accelerators to Improve Dynamic Properties in Vibration Isolation Applications," Presented to the Spring Meeting of the Southern Rubber Group, June 11–14, 2012, Myrtle Beach, SC.

Also, ZBPD (zinc dibutylphosphorodithiate) should be considered for curing in a BIIR-based compound. GEN: N. Tambe, B. Rodgers, S. Solis, B. Sharma, W. Waddell (ExxonMobil), "Phosphate Accelerators in the Vulcanization of Isobutylene Based Elastomers," Paper No. 88 presented at the Fall Meeting of the Rubber Division, ACS, October 10–12, 2006, Cincinnati, OH.

4.11 Reducing Marching Modulus

Marching modulus occurs when the modulus or stiffness of the rubber compound continues to increase during cure and seems to never reach a plateau.

*The following **experimental** ideas may prevent marching modulus from occurring. **Caution:** these general experimental ideas may not work in all specific situations. Changes to reduce marching modulus will certainly affect other properties as well, for better or for worse, and this book does not purport to show how these other properties are affected. Also, this book does not purport to address safety and health issues.*

Consider using a proprietary accelerator called Vanax® 189 from R. T. Vanderbilt for curing polychloroprene. Vanax 189 is a derivative of 1,2-dimercapto-1,3,4-thiadiazole. This accelerator reportedly imparts a faster cure rate with better scorch safety with less "marching modulus" for CR-based compounds compared to some other cure systems. GEN: R. Ohm, "New Developments in Curing Halogen-Containing Polymers," Presented at ACS Rubber Div. Education Symposium No. 45, "Automotive Applications II," Spring, 1998, pp. 2–4.

4.12 Reducing Cold Flow

Many think of stored bales of rubber in a warehouse as being "solid." However, they discover that these bales will flow if a long enough period of time passes. This flow that can occur over a long storage time is called "cold flow." Some raw rubber will manifest cold flow properties more than other types of raw rubber.

*The following **experimental** ideas might work to reduce cold flow characteristics. Also, all relevant literature sources, including the ones cited below, should be researched and read. **Caution:** these general experimental ideas may not work in all specific situations. Changes to reduce cold flow will certainly affect other properties as well, for better or for worse, and this book does not purport to show how these other properties are affected. Also, this book does not purport to address safety and health issues.*

BR

Use a *cis*-BR that contains a branching agent such as divinyl benzene to improve cold flow resistance. RT: Chapter 6, "Elastomer Selection," R. School, p. 130.

Use anionic polymerized BR that has been copolymerized with small amounts of divinylbenzene to add more long-chain branching and reduce cold flow. RT: Chapter 7, "General Purpose Elastomers and Blends," G. Day, p. 146.

Avoid highly linear *cis*-BR with very little long-chain branching. RT: Chapter 7, "General Purpose Elastomers and Blends," G. Day, p. 145.

Consider using Budene 1280® for improved cold flow and improved processability over other BRs with higher linearity. RP: R. Dailey.

NBR

Consider NBR grades with more narrow molecular weight distribution for higher cold flow resistance. RT: Chapter 8, "Specialty Elastomers," M. Gozdiff, p. 197.

5 Minimizing Adverse Processing Attributes

5.1 Reducing or Eliminating Porosity, Bubbles, or Voids in Both the Uncured and the Cured State

The factory problem of porosity is very common with many different rubber fabricators and costs corporations millions of dollars each year.

The following ***experimental*** *ideas can be considered to try to reduce this menacing problem. Also, all relevant literature sources, including the ones cited below, should be researched and read.* ***Caution:*** *these general experimental ideas may not work in all specific situations. Changes to reduce or eliminate porosity, bubbles, or voids will certainly affect other properties as well, for better or for worse, and this book does not purport to show how these other properties are affected. Also, this book does not purport to address safety and health issues.*

Curing Pressure

Make sure that the rubber compound is always cured under sufficient pressure to avoid porosity. RT: Chapter 8, "Specialty Elastomers," J. R. Halladay, p. 237.

Blow Point

Increasing outside pressure on a rubber part during cure will shrink bubbles. Enough pressure will almost completely shrink these bubbles. Maintaining sufficiently high external pressure on a rubber part during the curing process will minimize porosity. If "complete" shrinkage of bubbles occurs before say 30% of the optimum cure for the center of a rubber part, perhaps no visible bubbles (to the human eye) may normally be observed. This 30% is just a very rough approximation and is very compound-dependent. For example, with some compounds, this point has been reported as low as 25% of the optimum cure for the center of a rubber part. On the other hand, for compounds with very low cured modulus, this point might be as high as 98% of the optimum state of cure for the center of a rubber part. As modulus increases during cure, it becomes more and more difficult for bubbles to grow. Sometimes microvoids can exist after cure (0.005–5 microns in diameter) that are not visible to the unaided eye. Blow point refers to the minimum time in the cure cycle where no voids appear at the center of the rubber part upon release of presssure. GEN: A. Kasner, E. Meinecke, "Porosity in Rubber: A Review," *Rubber Chemistry and Technology*, July–August, 1996, Vol. 69, p. 424.

Step-Down Cure in Autoclave

For autoclave-cured products, avoid porosity and lower tensile strength by slowly reducing pressure toward the end of the cure cycle. This is called a "step-down cure." By reducing the pressure toward the end, it reportedly allows any trapped gases to slowly permeate out while the compound cools. RP: L. L. Outzs.

Peroxide Cures

Especially with peroxide-cured compounds, higher processing temperatures may lead to bubble formation at the rubber surface because the rubber cures too rapidly, thus not allowing sufficient time for the gas products from the peroxide cure to escape. This problem might be partially solved by lowering processing temperatures. GEN: P. Dluzneski, "Peroxide Vulcanization of Elastomers," *Rubber Chemistry and Technology*, July–August, 2001, Vol. 74, p. 451.

The peroxide used in a rubber cure can generate low molecular weight decomposition byproducts during the vulcanization process, which can produce off-gassing and porosity. One method to help minimize this type of porosity is to reduce the peroxide level and increase the coagent level to try to maintain the cured physical properties. RT: Chapter 17, "Peroxide Cure Systems," L. Palys, p. 434. RP: L. L. Outzs.

Volatile Vulcanization Byproducts

Other contributors to porosity problems are volatile byproducts from certain vulcanization reactions. Some reactions that generate volatiles are from the vulcanization of FKM. Also, decomposition products from peroxide cures (such as alcohols, acetophenone, or other ketones) generate volatiles. In addition, *in situ* formed water from diurethane (Novor) crosslinking agents used in curing and carbon dioxide formed from reaction of diisocyanates with water during the curing of PU also create volatility problems. Even some water formed from the reaction of stearic acid with zinc oxide during curing may cause a problem. This *in situ* formation of volatiles during curing must be addressed using some of the other suggestions mentioned in this section. For example, one might consider using zinc stearate instead of zinc oxide and stearic acid in some special situations. GEN: A. Kasner, E. Meinecke, "Porosity in Rubber: A Review," *Rubber Chemistry and Technology*, July-August, 1996, Vol. 69, p. 424.

Mixing

Check the fill factor and loading levels used in Banbury mixing procedures. Make sure they are not too low. GEN: B. Shama, B. Rodgers, D. Tracey, N. Tambe, D. Rouckhout (ExxonMobil), "Tire Halobutyl Rubber Innerliner," Paper No. 94 presented at the Fall Meeting of the Rubber Division, ACS, October 10-12, 2006, Cincinnati, OH.

Viscosity

Keeping the compound viscosity on the high side may help reduce blistering. GEN: B. Shama, B. Rodgers, D. Tracey, N. Tambe, D. Rouckhout (ExxonMobil), "Tire Halobutyl Rubber Innerliner," Paper No. 94 presented at the Fall Meeting of the Rubber Division, ACS, October 10-12, 2006, Cincinnati, OH.

Effects of Dispersion on Porosity

Ensure that good dispersion of compounding ingredients and especially fillers is achieved during mixing. Poor dispersion of fillers can lead to bubbles and results in porosity problems. Undispersed agglomerates can act as "nucleation sites" for the formation of porosity. GEN: A. Kasner, E. Meinecke, "Porosity in Rubber: A Review," *Rubber Chemistry and Technology*, July-August, 1996, Vol. 69, p. 424.

Poor Dispersion vs. Overmastication

Overmastication of a rubber stock can lead to porosity problems. On the other hand, poor dispersion can also lead to porosity problems. GEN: A. Kasner, E. Meinecke, "Porosity in Rubber: A Review," *Rubber Chemistry and Technology*, July-August, 1996, Vol. 69, p. 424.

Avoid Moisture in Raw Materials

Keep moisture content very low in all raw materials.

- This is especially true for the following:
  - Raw rubber
  - Carbon black
  - Silica
  - Clay
  - Nylon fibers

  - Cellulosic fibers
  - Stearic acid
  - Antioxidants
- Store raw rubber and compounding ingredients in dry areas.
- If necessary, consider drying or vacuum drying ingredients.

GEN: A. Kasner, E. Meinecke, "Porosity in Rubber: A Review," *Rubber Chemistry and Technology*, July-August, 1996, Vol. 69, p. 424; J. S. Dick, Chapters 7 and 8, *Basic Rubber Testing*, ASTM, 2003. RP: J. M. Long.

Avoid Volatiles in Raw Materials

Consider the following:

- Avoid process oils with high volatiles.
- Avoid extender oils with high volatiles.
- Avoid plasticizers with high volatiles.

GEN: A. Kasner, E. Meinecke, "Porosity in Rubber: A Review," *Rubber Chemistry and Technology*, July-August, 1996, Vol. 69, p. 424; J. S. Dick, Chapter 8, *Basic Rubber Testing*, ASTM, 2003.

Eliminate Moisture in Mixed Stock

Take steps to keep out moisture in a mixed stock downstream. The following are some suggestions:

- Allow sufficient drying time for slab-dipped stock to remove water from the slab dip itself.
- Store mixed batches in a dry, low-humidity location.
- Do not store mixed stock for lengthy time periods such as for two or more days.
- If necessary, consider further milling on an open mill to drive off moisture.
- Consider using calcium oxide as a compounding ingredient to serve as a moisture scavenger, while understanding that even a small concentration of calcium oxide may greatly degrade some of the compound's cured physical properties.

GEN: A. Kasner, E. Meinecke, "Porosity in Rubber: A Review," *Rubber Chemistry and Technology*, July-August, 1996, Vol. 69, p. 424.

Downstream Process Changes to Avoid Porosity

Consider the following to help reduce porosity problems:

- To avoid trapped air, reduce the gage thickness of the calendered sheets. To achieve a given thickness, try using more sheets with thinner gage to build up to a given thickness.
- Try making thicker sheets with a roller head die instead of building up thickness with a calender.
- Avoid underweight preforms in a compression molding operation (lowers the pressure).
- Avoid oversized molds (in relation to press size) in compression molding.
- Avoid too long a cure time for compression molding (too much reduction in pressure).
- Avoid having too low a compound viscosity for extrusion or injection molding (air becomes trapped).
- Avoid compounds with too high a shear thinning profile for extrusion or injection molding.
- Avoid too small a shot size for injection molding.
- Avoid too short a mold residence time in injection molding (a major cause of porosity problems).
- Use an extruder with vented barrels connected to a vacuum.
- Use a slower screw speed on a vented extruder.
- The external pressure in a steam autoclave cure of an extrudate can sometimes reduce or eliminate porosity.
- Using hot air/steam cures can provide pressure to reduce porosity while still maintaining a lower cure temperature.
- Wrap rubber articles for autoclave curing in fabric liners or cover with sheaths to help maintain higher external pressure.
- Avoid very high cure temperatures in continuous vulcanization processes at ambient pressure.
- Consider using "pressurized liquid medium" in a continuous vulcanization process.
- Select a continuous vulcanization process that will form a "crosslinked skin." (The crosslinked skin can help reduce porosity; fast cure systems form crosslinked skins better.)

GEN: A. Kasner, E. Meinecke, "Porosity in Rubber: A Review," *Rubber Chemistry and Technology*, July–August, 1996, Vol. 69, p. 424.

Careful Control of Extruder Temperature

When using a vented extruder with a vacuum, one might consider a higher "safe" temperature to help promote the faster removal of volatiles. For nonvented extruders, however, it may be better to extrude at a lower temperature to keep the compound viscosity higher and avoid porosity problems. GEN: A. Kasner, E. Meinecke, "Porosity in Rubber: A Review," *Rubber Chemistry and Technology*, July-August, 1996, Vol. 69, p. 424.

Talc

The use of talc in a rubber compound reportedly can reduce the volume of trapped air. However, if talc is poorly dispersed with talc agglomeration, then greater porosity can result. GEN: O. Noel, Education Symposium on Fillers, "Talc: Functional Mineral for Rubber," ACS Rubber Div. Meeting, Spring, 1995. RP: O. Noel.

Use Vulcanized Vegetable Oil

V. Lue reported that the use of factice (vulcanized vegetable oil) as a compounding ingredient can reduce porosity downstream. GEN: A. Kasner, E. Meinecke, "Porosity in Rubber: A Review," *Rubber Chemistry and Technology*, July-August, 1996, Vol. 69, p. 424.

Press Bumping

Irregular shaped preforms may sometimes cause trapped air problems. Therefore "bumping" the press can help somewhat. Bumping is when the press is briefly opened and then closed. GEN: A. Kasner, E. Meinecke, J. Sommer, *Elastomer Molding Technology*, Elastech, Hudson, OH, 2003, p. 107; "Porosity in Rubber: A Review," *Rubber Chemistry and Technology*, July-August, 1996, Vol. 69, p. 424.

Applying Vacuum During Molding

Trapped air can sometimes be eliminated through the application of a vacuum to the molding operation. GEN: J. Sommer, *Elastomer Molding Technology*, Elastech, Hudson, OH, 2003, pp. 107-08; A. Kasner, E. Meinecke, "Porosity in Rubber: A Review," *Rubber Chemistry and Technology*, July-August, 1996, Vol. 69, p. 424.

Vent Holes

When molding articles with complex shapes such as tires, consider using vent holes. GEN: J. Sommer, *Elastomer Molding Technology*, Elastech, Hudson, OH, 2003, p. 107.

Tire Innerliner

To prevent blisters from forming in the tire innerliner production process, consider minimizing the compound waiting time between the calender or extruder and the tire building operation and the curing stage. Also, blisters in the innerliner can be prevented by using a squeeze roll with uniform contact pressure accross the liner while it is plied up with a barrier or squeegee. In addition, using a higher viscosity halobutyl rubber or increasing the ratio of carbon black to oil used in the compound may reduce the number of blisters. Also, using higher molecular weight processing oils or oils that contain only a very small fraction of low molecular weight components may also help minimize blister occurrence. GEN: B. Shama, B. Rodgers, D. Tracey, N. Tambe, D. Rouckhout (ExxonMobil), "Tire Halobutyl Rubber Innerliner," Paper No. 94 presented at the Fall Meeting of the Rubber Division, ACS, October 10-12, 2006, Cincinnati, OH.

5.2 Mixing: Reducing Filler or Black Incorporation Times (BITs)

The black incorporation time in a mixing operation is a very important factor for determining how fast a given mixing procedure can be completed. However, fast incorporation time does not necessarily guarantee good filler or carbon black dispersion.

*The following **experimental** ideas may shorten the BIT. Also, all relevant literature sources, including the ones cited below, should be researched and read. **Caution:** these general experimental ideas may not work in all specific situations. Changes to reduce filler incorporation time will certainly affect other properties as well, for better or for worse, and this book does not purport to show how these other properties are affected. Also, this book does not purport to address safety and health issues.*

Carbon Black

Use carbon blacks with lower surface area for faster BIT. RT: Chapter 12, "Compounding with Carbon Black and Oil," S. Laube, S. Monthey, M-J. Wang, p. 308.

Use carbon blacks with lower structure for faster BIT (but poorer dispersibility). RT: Chapter 12, "Compounding with Carbon Black and Oil," S. Laube, S. Monthey, M-J. Wang, p. 308.

Use lower loadings for faster BIT. RT: Chapter 12, "Compounding with Carbon Black and Oil," S. Laube, S. Monthey, M-J. Wang, p. 308.

Oil

Sometimes selecting a highly aromatic oil as the process oil in a given rubber compound may result in faster oil take-up times during mixing for SBR- and BR-based compounds. RT: Chapter 12, "Compounding with Carbon Black and Oil," S. Laube, S. Monthey, M-J. Wang, p. 312.

NBR

Using cold emulsion NBRs with greater linearity enables compounds to have faster filler incorporation. RT: Chapter 8, "Specialty Elastomers," M. Gozdiff, p. 198.

Certain Metallocene-Catalyzed EPDM Grades

Single-site constrained-geometry metallocene catalyst technology for commercial polymerization makes it possible to produce commercial grades of EPDM with high ethylene content. Through this technology, it is possible to manipulate the ethylene distribution in such a way as to affect the distribution of the melting endotherm. Thus this technology can provide commercial EPDM grades with improved black incorporation time. GEN: D. Parikh, M. Hughes, M. Laughner, L. Meiske, R. Vara, "Next Generation of Ethylene Elastomers," Presented at ACS Rubber Div. Meeting, Fall, 2000, Cincinnati, OH.

TOR as an Additive

Polyoctenamer or *trans*-polyoctenylene rubber (TOR) reportedly can be blended with NR, BR, SBR, NBR, CR, EPDM, and so forth in minor amounts to improve incorporation of some fillers. GEN: A. Draxler, "A New Rubber: *trans*-Polyoctenamer," Chemische Werke Huels AG, Postfach, Germany.

Fatty Amine Processing Aids

Fatty amine processing aids can be used to shorten incorporation time for ultrafine particle sized carbon blacks at high loadings, thus increasing the time available for filler disagglomeration (or dispersion). GEN: H. Takino, S. Iwama, Y. Yamada, S. Kohjiya, "Carbon Black Dispersion and Grip Property of High-Performance Tire Tread Compound," Presented at ACS Rubber Div. Meeting, Spring, 1996, Paper No. 2.

Vinyl Acetate Wax as an Additive

Vinyl acetate wax may be used to reduce incorporation time for ultrafine particle sized carbon blacks at high loadings, thus increasing the time available for filler disagglomeration or dispersion. GEN: H. Takino, S. Iwama, Y. Yamada, S. Kohjiya, "Carbon Black Dispersion and Grip Property of High-Performance Tire Tread Compound," Presented at ACS Rubber Div. Meeting, Spring, 1996, Paper No. 2.

Lower MW Elastomers

Consider selecting a low Mooney viscosity elastomer such as a low viscosity HNBR. GEN: E. Campomizzi, L. Ferrari, R. Pazur (Lanxess), "Enhancing Compound Properties and Aging Resistance by Using Low Viscosity HNBR," Paper No. 69 presented at the Spring Meeting of the Rubber Division, ACS, May, 2005, San Antonio, TX.

5.3 Mixing: Decrease Total Mixing Time

Achieving a faster mix cycle with good quality of the dumped stock will result in significant savings in manufacturing costs.

The following ***experimental*** *ideas may reduce the total mixing time. Also, all relevant literature sources, including the ones cited below, should be researched and read.* ***Caution:*** *these general experimental ideas may not work in all specific situations. Changes to decrease total mixing time will certainly affect other properties as well, for better or for worse, and this book does not purport to show how these other properties are affected. Also, this book does not purport to address safety and health issues.*

Use Higher Molecular Weight Polymers

Use NR and IR polymers with higher molecular weights instead of lower molecular weights. RT: Chapter 7, "General Purpose Elastomers and Blends," G. Day, p. 145.

Use S-SBR instead of E-SBR because S-SBR has the highest number average molecular weight and the narrowest molecular weight distribution, which usually imparts a faster Banbury mix cycle. RT: Chapter 7, "General Purpose Elastomers and Blends," G. Day, p. 153.

EPDM

For EPDM-based compounds, select an EPDM with a narrower molecular weight distribution and less long-chain branching because usually EPDMs with broad molecular weight distribution and high branching will shear thin more when oil is added to the mix. Thus the carbon black is wetted by the polymer at a slower rate. In addition, the greater shear thinning causes less shear stresses in the mix cycle, resulting in a slower mix. GEN: K. Beardsley, R. Tomlinson, "Processing of EPDM Polymers as Related to Structure and Rheology," *Rubber Chemistry and Technology*, September-October, 1990, Vol. 63, p. 540.

Metallocene-Catalyzed EPDM

Through the new metallocene catalyst for commercial EPDM polymerization, it is possible to specifically control independently both the molecular weight distribution and long-chain branching to produce commercial grades of EPDM with faster mixing times. GEN: D. Parikh, M. Hughes, M. Laughner, L. Meiske, R. Vara, "Next Generation of Ethylene Elastomers," Presented at ACS Rubber Div. Meeting, Fall, 2000.

RT: *Rubber Technology, Compounding and Testing for Performance*. 2nd ed. Edited by J. S. Dick, Hanser Publications, 2009 ■ GEN: General References from the Literature ■ RP: Review Panel (Advisors for this book) ■ IMPORTANT: The information contained herein is experimental in nature and is meant for individuals with advanced scientific training and rubber compounding experience. All safety precautions should be taken when replicating or applying this information. The author makes no representation or warranty, either expressed or implied.

Star-Branched Polymers

Star-branched halobutyl polymers reportedly mix faster than regular halobutyl polymers. RT: Chapter 8, "Specialty Elastomers," G. Jones, D. Tracey, A. Tisler, p. 181.

Powdered Rubber

Blend powdered rubber with regular solid rubber during conventional Banbury mixing to achieve a shorter mix cycle. RT: Chapter 23, "Rubber Mixing," W. Hacker, p. 516.

Processing Aids

Consider using resinous homogenizing agents in compounds based on a polymer blend as well as with compounds based on a single polymer. RT: Chapter 14, "Ester Plasticizers and Processing Additives," C. Stone, p. 372.

Treated Sulfur

When mixing compounds based on NBR polymers, sometimes it is difficult to achieve good dispersion of conventional "rubber maker's sulfur." Therefore, commonly magnesium carbonate treated sulfur is used as the curative for nitrile-based compounds because it disperses better in this polar elastomer. RT: Chapter 16, "Cures for Specialty Elastomers," B. H. To, p. 398.

Oil or Wax Treatment of Rubber Chemicals

It is normally difficult to achieve good dispersion of sodium bicarbonate as a blowing agent. For better dispersion, one should probably use sodium bicarbonate that has been coated with an oil or wax to improve dispersion. Organic blowing agents generally disperse better than inorganic blowing agents. RT: Chapter 21, "Chemical Blowing Agents," R. Annicelli, pp. 478–479.

Banbury Batch Size

To achieve the shortest Banbury mixing time, it is important to use the optimum batch size for a specific mixer. It is usually assumed that the optimum batch size is reached when an increase or decrease from that size results in a longer mix time. Too small a batch size can take longer to mix, as can too large a batch size. An internal mixer has a specific batch size for a specific compound at which it is most efficient. RT: Chapter 23, "Rubber Mixing," W. Hacker, p. 511.

Banbury Rotor Speed

Increasing the shear rate during mixing by increasing the Banbury rotor speed will usually decrease the mixing time. However, the effectiveness of the Banbury cooling system is also important. GEN: W. Hess, "Characterization of Dispersions," *Rubber Chemistry and Technology*, July-August, 1991, Vol. 64, p. 386.

Banbury Rotors

Consider selecting the patented six-wing VCMT rotor, which will very likely give a faster mix cycle time. GEN: R. Jorkasky II (Kobelco Stewart Bolling, Inc.), "Effect of Rotor Type on Cycle Times of Various Rubber Systems," Paper No. 23A presented at ITEC 2002, September, 2002, Akron, OH.

Black/Oil Masterbatch

Use carbon black/oil masterbatches (such as SBR 1606) in place of free black mixing with clear SBR polymer (such as SBR 1500). RT: Chapter 4, "Rubber Compound Economics," J. Long, p. 80. Chapter 23, "Rubber Mixing," W. Hacker, p. 516.

Carbon Black

Avoid using carbon black with pellets that are too soft because they will generate too high a "fines" level (dust), which will take too long a time to wet out during the mixing process. RT: Chapter 12, "Compounding with Carbon Black and Oil," S. Laube, S. Monthey, M-J. Wang, p. 303.

Use carbon blacks with lower surface area for better dispersibility. RT: Chapter 12, "Compounding with Carbon Black and Oil," S. Laube, S. Monthey, M-J. Wang, p. 308.

For the fastest carbon black dispersion, select a carbon black with large particle size for faster wettability and high structure (because the high structure will increase the compound's viscosity, thus generating more shear stresses during mixing). GEN: W. Hess, "Characterization of Dispersions," *Rubber Chemistry and Technology*, July-August, 1991, Vol. 64, p. 386.

Use carbon blacks at their specific optimal loading for optimal dispersibility. RT: Chapter 12, "Compounding with Carbon Black and Oil," S. Laube, S. Monthey, M-J. Wang, p. 308.

Talc

Talc reportedly has an advantage over fillers in that it is described as "organophilic" and will incorporate significantly faster than many other white fillers. Therefore its use in a rubber compound may result in a shorter mix time.

GEN: O. Noel, Education Symposium on Fillers, "Talc: A Functional Mineral for Rubber," Presented at ACS Rubber Div. Meeting, Spring, 1995. RP: O. Noel.

It is also reported that the presence of talc in a compound can improve the dispersion of carbon black that is also present in the compound, thus reducing the mixing time by as much as 20%. GEN: O. Noel, G. Meli (Rio Tinto Minerals/ Luzenac), "Synergism of Talc with Carbon Black," Paper No. 13 presented at the Fall Meeting of the Rubber Division, ACS, October 14–16, 2008, Louisville, KY.

Nanomicron Filler

It is claimed that a fibrous lath-like crystal structure nanomicron filler under the trade name of "Vaporlink" can be used in place of other traditional fillers in order to reduce mixing time. GEN: "Specialty Fillers for Rubber Application," Paper No. 20 presented at the India Rubber Expo 2011, January 19, 2011, Chennai, India.

Oil

Sometimes, selecting a highly aromatic oil as the process oil for a given rubber compound may result in a faster mix time for SBR- and BR-based compounds. RT: Chapter 12, "Compounding with Carbon Black and Oil," S. Laube, S. Monthey, M-J. Wang, p. 312.

Processing Aids

The careful use of processing aids, peptizers, or dispersing agents can shorten the mixing cycle. GEN: C. Ryan (ChemSpec), "Definition of a Processing Additive," Presented at the Summer Meeting of the Southern Rubber Group, June 11–13, 2012, Myrtle Beach, SC.

5.4 Mixing: Reducing or Eliminating Lumps in Dumped Batches

Lumps contained in dumped batches can be a major quality problem.

*The following **experimental** ideas might help eliminate this problem. Also, all relevant literature sources, including the ones cited below, should be researched and read. **Caution:** these general experimental ideas may not work in all specific situations.Changes to reduce or eliminate lumps will certainly affect other properties as well, for better or for worse, and this book does not purport to show how these other properties are affected. Also, this book does not purport to address safety and health issues.*

By using star-branched halobutyl rubber in place of regular halobutyl rubber in mixing a rubber compound, lumps are reportedly reduced. RT: Chapter 8, "Specialty Elastomers," G. Jones, D. Tracey, A. Tisler, p. 181.

5.5 Mixing: Reducing Bagging on the Mill

When a stock "bags" on the mill, it can be a major nuisance.

The following ***experimental*** *ideas can help eliminate this problem. Also, all relevant literature sources, including the ones cited below, should be researched and read.* ***Caution:*** *these general experimental ideas may not work in all specific situations. Changes to reduce bagging will certainly affect other properties as well, for better or for worse, and this book does not purport to show how these other properties are affected. Also, this book does not purport to address safety and health issues.*

Avoid using very high levels of *cis*-BR. RT: Chapter 7, "General Purpose Elastomers and Blends," G. Day, p. 145.

Use E-SBR instead of S-SBR because S-SBR usually has the higher number average molecular weight and the narrower molecular weight distribution, which imparts bagging on the mill. RT: Chapter 7, "General Purpose Elastomers and Blends," G. Day, p. 153.

Consider applying milk to the hot mill rolls to temporarily create a tacky mill surface and eliminate bagging. Because this is a temporary condition, milk may have to be reapplied every three or four batches. It is reported that powdered milk works also. RP: L. L. Outzs.

5.6 Mixing: Reducing Back Rolling on the Mill

Another common nuisance that can occur in the factory is a stock going to the back roll when placed on a mill.

*The following **experimental** ideas may help reduce this tendency. Also, all relevant literature sources, including the ones cited below, should be researched and read. **Caution:** these general experimental ideas may not work in all specific situations. Changes to reduce back rolling on the mill will certainly affect other properties as well, for better or for worse, and this book does not purport to show how these other properties are affected. Also, this book does not purport to address safety and health issues.*

The nip distance for a two-roll mill can influence whether the rubber goes to the front or back roll while mixing. Adjusting this nip setting can cause the rubber compound to travel to the front or the back roll. The nip position where this change occurs is called the "front-back transition point." The mill temperature, friction ratio, and radius can also have an effect. GEN: N. Tokita, "Analysis of Band Formation in Mill Operation," *Rubber Chemistry and Technology*, May-June, 1979, Vol. 52, p. 387.

5.7 Extrusion: Reducing Die Swell (Improving Dimensional Stability)

Usually high elasticity in an uncured compound can contribute to die swell and dimensional stability problems during an extrusion operation.

*The following **experimental** ideas may contribute to reducing die swell. Also, all relevant literature sources, including the ones cited below, should be researched and read. **Caution:** these general experimental ideas may not work in all specific situations. Changes to reduce die swell will certainly affect other properties as well, for better or for worse, and this book does not purport to show how these other properties are affected. Also, this book does not purport to address safety and health issues.*

Fillers

Generally, extending a rubber compound with a filler will reduce the die swell. RT: Chapter 8, "Specialty Elastomers," G. Jones, D. Tracey, A. Tisler, p. 182.

Carbon Black

Compounds with higher loadings of carbon black tend to have less extrusion shrinkage than compounds with lower loadings of the same carbon black. RT: Chapter 12, "Compounding with Carbon Black and Oil," S. Laube, S. Monthey, M-J. Wang, p. 308. GEN: R. Kannabrian, "Correlation Between End Correction and Extrudate Swell for Some Raw Elastomers and Black-Filled Rubber Compounds," *Rubber Chemistry and Technology*, November-December, 1984, Vol. 57, p. 1001.

Compounds loaded with carbon blacks of higher structure tend to have less die swell and extrusion shrinkage than compounds loaded with carbon blacks of lower structure. This is because the rubber that becomes occluded within the aggregates is somewhat "dead" in terms of elastic memory. Compounds with higher structure carbon blacks possess more of this occluded rubber. Lower die swell will allow for better wall-gage control or dimensional stability because if there is less die swell there will be less variation. RT: Chapter 12, "Compounding with Carbon Black and Oil," S. Laube, S. Monthey, M-J. Wang, pp. 308, 321. GEN: S. Monthey, "The Influence of Carbon Blacks on the Extrusion Operation for Hose Production," *Rubber World*, May, 2000, p. 38. RP: M-J. Wang.

Compounds loaded with carbon blacks of higher surface area (smaller particle size) tend to have less extrusion shrinkage than compounds loaded with carbon blacks of lower surface area. However, carbon black surface area variations have less effect on shrinkage than carbon black structure variations. RT:

Chapter 12, "Compounding with Carbon Black and Oil," S. Laube, S. Monthey, M-J. Wang, p. 308.

It is not yet completely understood why, but sometimes substituting an N110 fully reinforcing carbon black in place of a semireinforcing black such as N762 at a 60-phr loading in NR can actually cause an increase in the 60-second relaxed extrudate swell. On the other hand, substituting N110 for N762 at 40 phr can result in the opposite effect of "lower swell," which would be the expected result. GEN: J. Leblanc, "Factors Affecting the Extrudate Swell and Melt Fracture Phenomena of Rubber Compounds," *Rubber Chemistry and Technology*, November–December, 1981, Vol. 54, p. 905.

Molecular Weight

Selecting NBR grades with lower Mooney plasticity usually results in compounds with less shrinkage. GEN: "A Comparative Evaluation of Hycar Nitrile Polymers," Manual HM-1 Revised, B. F. Goodrich Chemical Co.

Natural Rubber Grades

Reportedly, different grades of NR will impart different degrees of die swell to a compound based on an NR/BR blend. The 60-second relaxed extrudate swell is lower when 5L grade is used. Also, the 5CV grade imparts a low die swell, but not quite as low as 5L. However, SMR 10 and SMR 20 grades of NR imparted more die swell to the compound. GEN: J. Leblanc, "Factors Affecting the Extrudate Swell and Melt Fracture Phenomena of Rubber Compounds," *Rubber Chemistry and Technology*, November–December, 1981, Vol. 54, p. 905.

IR vs. NR

Generally, an extrudate based on synthetic IR possesses less die swell and lower nerve level than a similar compound based on natural rubber. RT: Chapter 6, "Elastomer Selection," R. School, p. 131.

BR/NR Blend

The 60-second relaxed extrudate swell reportedly decreases when the concentration of BR is reduced in a compound based on a natural rubber/polybutadiene blend. GEN: J. Leblanc, "Factors Affecting the Extrudate Swell and Melt Fracture Phenomena of Rubber Compounds," *Rubber Chemistry and Technology*, November–December, 1981, Vol. 54, p. 905; J. Douglas, D. Hardy, A. Lucassen, T. Gross, J. Kroll (Lanxess), "Processing Behaviour of High-*Cis* Polybutadienes in Rubber Compounds: Highlighting Neodymium Polybutadienes," Paper No. 143 presented at the Fall Meeting of the Rubber Division, ACS, October 13–15, 2008, Pittsburgh, PA.

NBR

Usually, but not always, high-acrylonitrile NBRs have less shrinkage than low-acrylonitrile NBR grades. GEN: "A Comparative Evaluation of Hycar Nitrile Polymers," Manual HM-1 Revised, B. F. Goodrich Chemical Co.

CR

The addition of 10 to 20 phr of Neoprene WB® to a polychloroprene compound is reported to reduce die swell in an extrusion operation. Neoprene T grades reportedly work as well. RP: L. L. Outzs.

NBR with Gel

Consider using special NBR grades with high gel content for better dimensional stability. RT: Chapter 8, "Specialty Elastomers," M. Gozdiff, p. 197.

Consider using hot NBRs that are crosslinked with a difunctional monomer (gelled) and use at about 10 to 25 phr with other polymers, such as XNBR, SBR, or cold NBR to control die swell or stabilize extruded profiles. RT: Chapter 8, "Specialty Elastomers," M. Gozdiff, p. 198.

Long-Chain Branching

The level of LCB contained in a raw elastomer can affect the imparted die swell found in the mixed compound. GEN: D. Reynolds (Bytewise), "Mistake-Proofing the Tire Manufacturing Process," Presented at a Rubber Seminar in Qingdao, China, August 23, 2011.

TOR as an Additive

Polyoctenamer or *trans*-polyoctenylene rubber (TOR) reportedly can be blended with NR, BR, SBR, NBR, CR, EPDM, and so forth in minor amounts to improve dimensional stability of the extruded compound at elevated temperatures. GEN: A. Draxler, "A New Rubber: *trans*-Polyoctenamer," Chemische Werke Huels AG, Postfach, Germany.

Oil Effects

Sometimes increasing the part loading of process oil while maintaining a constant carbon black part loading may actually result in an increase in die swell because the actual concentration of carbon black has decreased. GEN: W. Hess, "Characterization of Dispersions," *Rubber Chemistry and Technology*, July–August, 1991, Vol. 64, p. 386.

Vulcanized Vegetable Oil

Using vulcanized vegetable oil (VVO) in NR- or SBR-based compounds will generally reduce die swell and impart better dimensional stability to the extrudate. GEN: S. Botros, F. El-Mohsen, E. Meinecke, "Effect of Brown Vulcanized Vegetable Oil on Ozone Resistance, Aging, and Flow Properties of Rubber Compounds," *Rubber Chemistry and Technology* March–April, 1987, p. 159.

Mixing

Improving the dispersion of carbon black through longer mix cycles and more work history will usually reduce compound extrusion shrinkage after it passes through a maximum point. RT: Chapter 12, "Compounding with Carbon Black and Oil," S. Laube, S. Monthey, M-J. Wang, p. 309. GEN: F. Myers, W. Newell, "Use of Power Integrator and Dynamic Stress Relaxometer to Shorten Mixing Cycles and Establish Scale-up Criteria for Internal Mixers," *Rubber Chemistry and Technology*, May–June, 1978, Vol. 51, p. 180; B. B. Boonstra, A. I. Medalia, "Effect of Carbon Black Dispersion on the Mechanical Properties of Rubber Vulcanizates," *Rubber Chemistry and Technology*, January–March, 1963, Vol. 36, p. 115. RP: J. Stevenson.

Generally as one progresses in the mixing process through the second power peak in the mixing cycle, the extrusion die swell and shrinkage characteristics of the stock start to decrease with additional mixing work history. Hess reported that the cause of this decrease in die swell property is either rubber breakdown or "selective adsorption" of higher weight fractions of the rubber by carbon black. GEN: W. Hess, "Characterization of Dispersions," *Rubber Chemistry and Technology*, July–August, 1991, Vol. 64, p. 386.

Increasing Work History

By increasing the mastication time or milling time for a hot feed extruder, the die swell as measured by a capillary rheometer decreased to a lower constant value. GEN: J. Leblanc, "Factors Affecting the Extrudate Swell and Melt Fracture Phenomena of Rubber Compounds," *Rubber Chemistry and Technology*, November–December, 1981, Vol. 54, p. 905. RP: J. Stevenson.

Use a High-Mastication Screw

To improve extrudate quality and improve dimensional variation, consider using a high-mastication screw. GEN: J. F. Stevenson, J. S. Dick, *Rubber Extrusion Technology Short Course*, Section VI.B.1, University of Wisconsin–Milwaukee, February 12–14, 2003.

Phase Mixing

How rapidly the relaxed die swell increases with a rise in applied shear rate for a carbon black loaded NR/BR compound is dependent on how the compound is mixed. If an NR black masterbatch and BR black masterbatch are mixed together, the rate of increase in die swell with increasing applied shear rate is somewhat more regular (linear). The reason for this more gradual rise in die swell with shear rate increase may be that the carbon black is more evenly dispersed between the NR and BR phases. Mixing in this manner might allow for better control of die swell through control of the screw speed. GEN: J. Leblanc, "Factors Affecting the Extrudate Swell and Melt Fracture Phenomena of Rubber Compounds," *Rubber Chemistry and Technology*, November-December, 1981, Vol. 54, p. 905.

Applied Shear Rate During Extrusion

Lowering the applied shear rate (lowering the screw speed) should reduce the die swell. GEN: J. Leblanc, "Factors Affecting the Extrudate Swell and Melt Fracture Phenomena of Rubber Compounds," *Rubber Chemistry and Technology*, November-December, 1981, Vol. 54, p. 905; R. Kannabrian, "Application of Flow Behavior to Design of Rubber Extrusion Dies," *Rubber Chemistry and Technology*, March-April, 1986, Vol. 59, p. 142. RP: J. Stevenson.

Extrusion Temperature

Extrusion temperature also has an effect on extrusion die swell. GEN: D. Reynolds (Bytewise), "Mistake-Proofing the Tire Manufacturing Process," Presented at a Rubber Seminar in Qingdao, China, August 23, 2011.

Die Land Length

Extrudate swell can be reduced by increasing die land length. GEN: R. Kannabrian, "Application of Flow Behavior to Design of Rubber Extrusion Dies," *Rubber Chemistry and Technology*, March-April, 1986, Vol. 59, p. 142; J. Leblanc, "Factors Affecting the Extrudate Swell and Melt Fracture Phenomena of Rubber Compounds," *Rubber Chemistry and Technology*, November-December, 1981, Vol. 54, p. 905. RP: J. Stevenson.

Porosity

Rubber compounds that tend to develop porosity during extrusion will display higher die swell values. To avoid porosity formation, see Section 5.1 on "Porosity, Bubbles, and Voids." GEN: A. Kasner, E. Meinecke, "Porosity in Rubber: A Review," *Rubber Chemistry and Technology*, July-August, 1996, Vol. 69, p. 424.

5.8 Extrusion: Improving Smoothness of the Extrudate Surface

The smoothness and appearance of many extruded rubber products can be very important to ensure its commercial success.

The following ***experimental*** *ideas may improve the smoothness and appearance of extruded products. Also, all relevant literature sources, including the ones cited below, should be researched and read.* ***Caution:*** *these general experimental ideas may not work in all specific situations. Changes to improve the smoothness of the extrudate surface will certainly affect other properties as well, for better or for worse, and this book does not purport to show how these other properties are affected. Also, this book does not purport to address safety and health issues.*

Molecular Weight

Avoid SBR with very high molecular weight. If these SBRs must be used for their physical properties, use the high molecular weight SBR in an oil-extended masterbatch. RT: Chapter 7, "General Purpose Elastomer and Blends," G. Day, p. 153.

Special Ziegler–Natta Catalyzed EPDM Grades

Through new proprietary Ziegler–Natta catalyst technologies, unique high-temperature crystallinity through ethylene sequencing in EPDM gives better extrudate appearance. With this ethylene sequencing, some crystallinity undergoes multiple crystallinity transitions up to 75 °C (167 °F). GEN: S. Brignac, H. Young, "EPDM with Better Low-Temperature Performance," *Rubber & Plastics News*, August 11, 1997, p. 14.

Gas-Phase EPDM

Try compounding with an ultra-low Mooney viscosity gas-phase polymerized EPDM with very high ethylene content and filler loading (a higher ethylene content reportedly improves the surface smoothness and appearance of an extrudate). GEN: A. Paeglis, "Very Low Mooney Granular Gas-Phase EPDM," Presented at ACS Rubber Div. Meeting, Fall, 2000, Paper No. 12.

IR vs. NR

Generally, a compound based on synthetic IR can be extruded more smoothly than a similar compound based on natural rubber. RT: Chapter 6, "Elastomer Selection," R. School, p. 131.

Avoid BR

Avoid using very high levels of *cis*-BR, which can cause rough extrusions. RT: Chapter 7, "General Purpose Elastomers and Blends," G. Day, p. 145.

CR

The addition of 10 to 20 phr of Neoprene WB® to a polychloroprene compound is reported to improve appearance and reduce die swell in an extrusion operation. RP: L. L. Outzs.

"T"-type Neoprene

When compounding with neoprenes, T-types give smoother extrusions than many other neoprenes. RT: Chapter 8, "Specialty Elastomers," L. Outzs, p. 211.

CR/SBR Blend

Sometimes a small amount of SBR is added to a polychloroprene to improve its processability. GEN: F. Eirich, *Science and Technology of Rubber*, Chapter 9, "The Rubber Compound and Its Composition," M. Studebaker, J. Beatty, Academic Press, 1978, p. 367.

FKM

To improve the extrusion appearance of an FKM compound, consider evaluating 3M's performance additive under the trade name of "Dynamar Rubber Additive RA 5300," which reportedly contains siloxane elastomer and talc. GEN: J. Denham (3M), "Optimizing Performance and Improving Productivity," Presented at the Spring Meeting of the Energy Rubber Group, May 18, 2011, Arlington, TX.

Carbon Black

Compounds have a specific optimal loading for carbon black for optimal smoothness. RT: Chapter 12, "Compounding with Carbon Black and Oil," S. Laube, S. Monthey, M-J. Wang, p. 308.

Compounds loaded with carbon blacks of higher structure tend to have better extrusion smoothness than compounds loaded with carbon blacks of lower structure. RT: Chapter 12, "Compounding with Carbon Black and Oil," S. Laube, S. Monthey, M-J. Wang, pp. 308, 321. GEN: S. Monthey, "The Influence of Carbon Blacks on the Extrusion Operation for Hose Production," *Rubber World*, May, 2000, p. 38.

Consider using a carbon black with a lower surface area because this may improve the compound's extrudate smoothness. GEN: S. Monthey, "The Influence of Carbon Blacks on the Extrusion Operation for Hose Production," *Rubber World*, May, 2000, p. 38.

Ultraclean Carbon Blacks

Consider using the so-called ultraclean carbon blacks, which are reportedly free of large particle foreign impurities such as coke, refractory, or metallic contamination. RT: Chapter 12, "Compounding with Carbon Black and Oil," S. Laube, S. Monthey, M-J. Wang, p. 321. GEN: S. Monthey, "The Influence of Carbon Blacks on the Extrusion Operation for Hose Production," *Rubber World*, May, 2000, p. 38.

Calcium Carbonate

Consider using calcium carbonate as a filler in a rubber compound to possibly improve surface appearance and smoothness. RT: Chapter 13, "Precipitated Silica and Non-black Fillers," W. Waddell, L. Evans, p. 326.

Talc

Consider using talc as a filler in a rubber compound where surface appearance and smoothness are critical. RT: Chapter 13, "Precipitated Silica and Non-black Fillers," W. Waddell, L. Evans, p. 328.

Delaminated Clays

Reportedly, using delaminated clays as compounding additives can give smoother extrusions for a rubber compound. Delaminated kaolin clays are produced from a mechanical process that separates the clay's stacked platelets to form platelets that have high aspect ratios (L/D). GEN: D. Askea (Polymer Valley Chemicals), Paper presented at the Fall Meeting of the Energy Rubber Group, September 15, 2011, Galveston, TX.

Liquid NBR as an Additive

A liquid NBR polymer (such as Hycar 1312) might be used in a nitrile-based compound as a nonextractable plasticizer to improve the appearance of the extrudate. GEN: "A Comparative Evaluation of Hycar Nitrile Polymers," Manual HM-1 Revised, B. F. Goodrich Chemical Co.

Processing Aid

Consider using a processing additive such as a blend of zinc and potassium soaps of fatty acids in a general purpose rubber compound to improve the surface appearance of an extrudate. RT: Chapter 14, "Ester Plasticizers and Processing Additives," C. Stone, pp. 375–376.

Vulcanized Vegetable Oil

Using vulcanized vegetable oil (VVO) in NR- or SBR-based compounds will generally improve the appearance of the extrudate. GEN: S. Botros, F. El-Mohsen, E. Meinecke, "Effect of Brown Vulcanized Vegetable Oil on Ozone Resistance, Aging, and Flow Properties of Rubber Compounds," March–April, 1987, p. 159; J. Sommer, *Elastomer Molding Technology*, Elastech, Hudson, OH, 2003, p. 281.

Bin Storage Times and Bound Rubber Content

Avoid long storage times for mixed stocks because bound rubber content can increase, contributing to high yield stress (high Mooney peak), which can cause melt fracture (occurs when melt strength is lower than yield strength) and appearance problems. GEN: S. Schaal, A. Coran, "The Rheology and Processability of Tire Compounds," *Rubber Chemistry and Technology*, May–June, 2000, Vol. 73, p. 225.

Straining

To improve the appearance of the final extrudate, consider using a gear extruder (a gear pump) for straining the rubber compound before it goes through the final extrusion. GEN: J. F. Stevenson, J. S. Dick, *Rubber Extrusion Technology Short Course*, Section II.A.4, University of Wisconsin-Milwaukee, February 12–14, 2003. RP: J. Stevenson.

Reducing Torn Edges with Heated Dies

Torn edges of some extrudates can be avoided by heating the die to reduce viscosity and increase flow through a given thin section of the die. GEN: J. F. Stevenson, "Die Design for Rubber Extrusion," *Rubber World*, May, 2003, Vol. 228, p. 23; J. F. Stevenson, J. S. Dick, *Rubber Extrusion Technology Short Course*, Section VI.C.7 and III.B.7, University of Wisconsin-Milwaukee, February 12–14, 2003. RP: J. Stevenson.

Use a High-Mastication Screw

To improve extrudate quality and dimensional variation, consider using a high-mastication screw. GEN: J. F. Stevenson, J. S. Dick, *Rubber Extrusion Technology Short Course*, Section VI.B.1, University of Wisconsin-Milwaukee, February 12–14, 2003. RP: J. Stevenson.

Avoid "Starving" the Extruder

If the extruder operation is "starved" (incomplete filling of the screw), then extrudate appearance can be quite poor. To avoid starving the extruder, the operator should do the following:

- Make sure the metered feed rate is fast enough for the extruder pumping capacity.
- Make sure the feed strip cross-section area is not too small.
- Make sure the feed section of the screw is able to provide material at sufficient rate to keep up with the metering requirements (i.e., avoid unbalanced two-stage screw).

GEN: J. F. Stevenson, J. S. Dick, *Rubber Extrusion Technology Short Course,* Section VI.B.6, Section VI.B.2, University of Wisconsin-Milwaukee, February 12-14, 2003. RP: J. Stevenson.

Extrusion Rate and Melt Fracture

For a compound extrudate to have a good appearance, it should be extruded below its critical shear stress ("melt fracture" point), above which extrudate defects and appearance problems can occur. Filler type and loading levels can affect critical shear. An increase in extruder temperature also will reduce the viscosity of the compound in the extruder and increase the critical shear rate, thus preventing appearance problems (provided there are no new scorch problems from running at the higher temperature!). GEN: J. Leblanc, "Factors Affecting the Extrudate Swell and Melt Fracture Phenomena of Rubber Compounds," *Rubber Chemistry and Technology*, November-December, 1981, Vol. 54, p. 905; J. F. Stevenson, J. S. Dick, *Rubber Extrusion Technology Short Course*, Section VI.B.1, University of Wisconsin-Milwaukee, February 12-14, 2003; G. Colbert, "Time Uniformity of Extrudate Melt Temperature," *Rubber World*, July, 1990, Vol. 202, p. 27. RP: J. Stevenson.

Longer Extruder Barrel Length

Using an extruder with a longer barrel length may result in better appearance of the extrudate. A longer barrel may provide more residence time for heat transfer. GEN: G. Colbert, "Time Uniformity of Extrudate Melt Temperature," *Rubber World*, July, 1990, Vol. 202, p. 27. RP: J. Stevenson.

5.9 Extrusion: Increasing Extrusion Rate Without Loss of Quality

Increasing extrusion rate without loss in quality can significantly improve the productivity of an extrusion operation.

The following ***experimental*** *ideas may suggest ways to increase extrusion rates without quality losses. Also, all relevant literature sources, including the ones cited below, should be researched and read.* ***Caution:*** *these general experimental ideas may not work in all specific situations. Changes to increase the maximum extrusion rate without loss in quality will certainly affect other properties as well, for better or for worse, and this book does not purport to show how these other properties are affected. Also, this book does not purport to address safety and health issues.*

Molecular Weight

Avoid SBR with very high molecular weight. If these SBRs must be used for their physical properties, use the high molecular weight SBR in an oil-extended masterbatch. RT: Chapter 7, "General Purpose Elastomer and Blends," G. Day, p. 153.

IR vs. NR

A compound based on synthetic IR generally can be extruded faster than a similar compound based on NR. RT: Chapter 6, "Elastomer Selection," R. School, p. 131.

CV Natural Rubber

Constant viscosity (CV) natural rubber can be substituted for regular TSR natural rubber in order to increase the extrusion speed of the rubber compound. CV grades of natural rubber have had hydroxylamine neutral sulfate (HNS) added in the latex stage before coagulation in order to keep the coagulated NR viscosity from rising too much during shipment in ocean cargo ships. GEN: B. Rodgers, D. Tracey, W. Waddell (ExxonMobil Chemical Co.), "Production, Classification, and Properties of Natural Rubber," Paper No. 37 presented at the Spring Meeting of the Rubber Division, ACS, May 16–18, 2005, San Antonio, TX.

Liquid IR as an Additive

Consider using a liquid polyisoprene as a processing aid. On vulcanization, it crosslinks into the compound and cannot be extracted. RT: Chapter 7, "General Purpose Elastomers and Blends," G. Day, p. 143.

Liquid EPDM as an Additive

Use a liquid EPDM with solid grades of EPDM to increase extrusion rates while minimizing losses in cured hardness. GEN: W. Sigworth, "Liquid EP(D)M Polymers in Mechanical Goods Applications," Presented at ACS Rubber Div. Meeting, Fall, 2000, Paper No. 9.

TOR as an Additive

Reportedly polyoctenamer or *trans*-polyoctenylene rubber (TOR) can be blended with NR, BR, SBR, NBR, CR, EPDM, and so forth in minor amounts to improve extrusion rates. GEN: A. Draxler, "A New Rubber: *trans*-Polyoctenamer," Chemische Werke Huels AG, Postfach, Germany.

EPDM

Highly crystalline EPDM-based compounds may possess a faster extrusion rate. GEN: S. Brignac, H. Young, "EPDM with Better Low-Temperature Performance," *Rubber & Plastics News*, August 11, 1997, p. 14.

Special Metallocene-Catalyzed EPDM Grades

Through the new metallocene catalyst for commercial EPDM polymerization, it is possible to specifically control independently both the molecular weight distribution and long-chain branching to produce commercial grades of EPDM with greater shear thinning behavior and greater extruder output. GEN: D. Parikh, M. Hughes, M. Laughner, L. Meiske, R. Vara, "Next Generation of Ethylene Elastomers," Presented at ACS Rubber Div. Meeting, Fall, 2000.

HNBR

Traditionally, HNBR grades that were available earlier had relatively high Mooney viscosity values. Now there are new grades of HNBR that are available with lower Mooney viscosity values that will increase the extrusion speed of their respective compounds. GEN: F. Guerin, S. Guo (Lanxess), "Improving the Processibility of HNBR," Paper No. 62 presented at the Fall Meeting of the Rubber Division, ACS, October 5–8, 2004, Columbus, OH; E. Campomizzi, L. Ferrari, R. Pazur (Lanxess), "Enhancing Compound Properties and Aging Resistance by Using Low Viscosity HNBR," Paper No. 69 presented at the Spring Meeting of the Rubber Division, ACS, May 16–18, 2005, San Antonio, TX.

Reclaim Rubber

Partially substituting virgin rubber with reclaim rubber in many cases will enable a given compound to be extruded faster. GEN: H. Gandhi, A. Barvey

(Gujarat Reclaim and Rubber Products Ltd.), Paper No. 10 presented at the Fall Meeting of the Rubber Division, ACS, October 10–12, 2006, Cincinnati, OH.

Block Polymer

Small amounts of block styrene in what is normally random SBR may increase the extrusion rate. RT: Chapter 7, "General Purpose Elastomer and Blends," G. Day, p. 148.

"T"-type Neoprenes

When compounding with neoprenes, T-types give a faster extrusion than other neoprenes. RT: Chapter 8, "Specialty Elastomers," L. L. Outzs, p. 211.

CM

Selecting grades of chlorinated polyethylene with a lower percentage of chlorine will allow a greater acceptance of plasticizer for the compound to have faster extrusions. RT: Chapter 8, "Specialty Elastomers," L. Weaver, p. 213.

Carbon Black

Selecting a carbon black with a low surface area should increase a rubber compound's extrusion rate. RT: Chapter 12, "Compounding with Carbon Black and Oil," S. Laube, S. Monthey, M-J. Wang, p. 321.

Silica

Use an organosilane on a silica dispersion aid (SDA from Flow Polymers Inc.) instead of on an N330 carbon black (such as with X50-S). By using the organosilane on the SDA carrier in place of the traditional N330 carrier, reportedly a reduction in extrusion time may be achieved. GEN: J. Fusco, J. Hoover, "Using a Dispersion Aid to Facilitate Application of Silica at High Loadings," ITEC '98 Select, p. 78.

Talc

The use of special talcs in a rubber compound reportedly can allow the compound to run faster in an extruder without loss in quality. GEN: O. Noel, Education Symposium on Fillers, "Talc: A Functional Mineral for Rubber," Presented at ACS Rubber Div. Meeting, Fall, 1995. RP: O. Noel.

Extruder Output, Viscous Heating, and Heat Transfer

Extruder output is proportional to screw speed (N). However, viscous heating or heat generation is equal to the square of screw speed (N^2). Therefore, viscous heating is a limiting factor and is dependent on the overall heat transfer rate in the extruder, which is related to the extruder's design. If the rubber can be cooled adequately, then a greater output can be achieved without encountering excessive heat generation. GEN: P. Johnson, "Developments in Extrusion Science and Technology," *Rubber Chemistry and Technology*, July-August, 1983, p. 575.

Gear Pump

Use a gear pump at the head of an extruder that is better at the metering and pressurizing functions. The gear pump thus allows the extruder to do a better job of heating and masticating. A gear pump will allow higher extruder output with better uniformity at lower temperatures. GEN: D. Eckenberg and G. Folie, "Continuous Production of Rubber Profiles: State of Extrusion Line Technology," Paper No. 43, ACS Rubber Div., October 17-19, 1995; R. Uphus, "Extruder/Gear Pump Combinations for Processing," *Rubber World*, July, 2001, Vol. 224, p. 23; J. F. Stevenson, J. S. Dick, Rubber Extrusion Technology Short Course, Section II.C.3, University of Wisconsin-Milwaukee, February 12-14, 2003. RP: J. Stevenson.

Pin Barrel Extruder

While the output rate for the pin barrel extruder is actually lower than for a conventional extruder of the same size, the significant improvement in quality from the pin barrel, when needed, yields higher productivity and quality output. Also, by removing the pins, one can adjust temperature and output rate. GEN: K. C. Shin, J. L. White, "Basic Studies of Extrusion of Rubber Compounds in a Pin Barrel Extruder," *Rubber Chemistry and Technology*, March-April, 1993, Vol. 66, p. 121. RP: J. Stevenson.

Ring Extruder

Consider going to a continuous mixing operation, such as with a ring extruder, to improve productivity and reduce manufacturing costs. GEN: Gerard Nijman (Vredestein Banden BV), "Continuous Mixing: A Challenging Opportunity?," Paper No. 72 presented at the Fall Meeting of the Rubber Division, ACS, October 8-11, 2002, Pittsburgh, PA.

Multicut Transfermix

Consider using a transfermix extruder for better output and quality. GEN: F. W. Fischer, M. W. Hohl, "MCTD Extruders: From Theory to Practice," *Rubber World*, July, 2000, Vol. 222, p. 25. RP: J. Stevenson.

Preheat and Masticate Feed

Consider applying a hot feed to a cold feed extruder. GEN: J. F. Stevenson, J. S. Dick, Rubber Extrusion Technology Short Course, Section V.B.I, University of Wisconsin-Milwaukee, February 12-14, 2003. RP: J. F. Stevenson.

Screw Temperature

For extruding NR stock in a pin barrel, increasing screw temperature also increases output per screw revolution with only small increases in stock temperature. By contrast, barrel temperature strongly affects stock temperature but has much less effect on rate of output. Therefore, usually a combination of increasing screw temperature and decreasing barrel temperature (within limits) will usually give an increase in output with a lower stock temperature. GEN: J. F. Stevenson, J. S. Dick, Rubber Extrusion Technology Short Course, Section VI.A.I, University of Wisconsin-Milwaukee, February 12-14, 2003. RP: J. Stevenson.

Extruder Temperature vs. Feed Temperature

It is sometimes believed to be good practice to try to reduce the temperature differences between the extruder and feed to achieve better quality output. GEN: G. Colbert, "Time Uniformity of Extrudate Melt Temperature," *Rubber World*, July, 1990, Vol. 202, p. 27. RP: J. Stevenson.

Critical Shear Stress

Achieving high output with good quality is dependent on not exceeding the critical shear stress value. If the extruder operates above this critical value (which is compound dependent), the quality of the output will deteriorate. GEN: J. Leblanc, "Factors Affecting the Extrudate Swell and Melt Fracture Phenomena of Rubber Compounds," *Rubber Chemistry and Technology*, November-December, 1981, Vol. 54, p. 905. RP: J. Stevenson.

Critical Shear Stress for Natural Rubber Based Stock

NR compounds extruded above their critical shear stress result in a very rough extrudate because of the onset of strain-induced crystallization in the extruder. This critical shear stress for NR compounds can be moved to a higher output rate through additional mastication or increasing the temperature of the rubber. GEN: V. L. Folt, R. W. Smith, C. E. Wilkes, "Crystallization of *cis*-Polyisoprenes in a Capillary Rheometer," *Rubber Chemistry and Technology*, March, 1971, Vol. 44, p. 1; J. F. Stevenson, J. S. Dick, Rubber Extrusion Technology Short Course, Section VI.B1, (Data of E. L. Ong), University of Wisconsin-Milwaukee, February 12-14, 2003. RP: J. Stevenson.

Extrusion Grades of EPDM

A study reported that the optimal output of an EPDM-based compound at a consistently acceptable surface appearance was reported to occur when the base EPDM has a relatively low Mooney viscosity. GEN: N. P. Cheremisinoff, C. Shulman, *Automotive Polymers and Design*, June, 1989, p. 82. RP: J. Stevenson.

Ultrasonically Aided Extrusion

Normally, carbon nanotube loaded natural rubber compounds are very difficult to extrude. However, using the experimental ultrasonically aided extrusion process, these high-viscosity CNT compounds could be processed with significantly less difficulty. GEN: J. Choi, A. Isayev (University of Akron), "Natural Rubber/Carbon Nanotube Nanocomposites Prepared by Ultrasonically Aided Extrusion," Paper No. 77 presented at the Fall Meeting of the Rubber Division, ACS, October 12-14, 2011, Cleveland, OH.

5.10 Calendering: Eliminating Blisters

Blistering from a calendering operation in the factory can be a major quality problem that could be very costly.

The following ***experimental*** *ideas may help to eliminate blisters. Also, all relevant literature sources, including the ones cited below, should be researched and read.* ***Caution:*** *these general experimental ideas may not work in all specific situations. Changes to eliminate blisters will certainly affect other properties as well, for better or for worse, and this book does not purport to show how these other properties are affected. Also, this book does not purport to address safety and health issues.*

Viscosity

Keeping the compound viscosity on the high side may help reduce blistering. GEN: B. Shama, B. Rodgers, D. Tracey, N. Tambe, D. Rouckhout (ExxonMobil), “Tire Halobutyl Rubber Innerliner,” Paper No. 94 presented at the Fall Meeting of the Rubber Division, ACS, October 10–12, 2006, Cincinnati, OH.

Star-Branched Polymers

Use star-branched halobutyl in place of regular halobutyl to increase the green strength of calendered sheets, which decreases pooling of entrapped air, which in turn reduces blisters. RT: Chapter 8, “Specialty Elastomers,” G. Jones, D. Tracey, A. Tisler, p. 182.

AEM

When using ethylene-acrylic elastomers (AEM), reportedly blisters can be reduced by selecting the terpolymer instead of the copolymer (dipolymer) grades of AEM that require a peroxide cure. RT: Chapter 8, “Specialty Elastomers,” T. Dobel, p. 224.

Avoid Butyl

If possible, avoid a butyl rubber base because butyl rubber has a particular problem with blisters due to its low air permeability. GEN: M. Chase, “Roll Coverings Past, Present, and Future,” Presented at Rubber Roller Group Meeting, New Orleans, May 15–17, 1996, p. 8.

Milling

Maintain a “minimum rolling bank” while milling. GEN: “Polysar Halobutyl Innerliner Problem Solving Guide,” Processing Problem No. 6.

Reduce the gage of the feed to the nip of the mill. GEN: "Polysar Halobutyl Innerliner Problem Solving Guide," Processing Problem No. 6.

Try a small friction ratio of perhaps 1.05 : 1.0. GEN: "Polysar Halobutyl Innerliner Problem Solving Guide," Processing Problem No. 6.

Moisture Effects

Check to make sure that all compounding ingredients used in the formulation are free of volatiles and moisture. GEN: "Polysar Halobutyl Innerliner Problem Solving Guide," Processing Problem No. 6.

Eliminate exposure to any water after completing mixing. GEN: "Polysar Halobutyl Innerliner Problem Solving Guide," Processing Problem No. 6.

Tire Innerliner

To prevent blisters from forming in the tire innerliner production process, consider minimizing the compound waiting time between the calender or extruder and the tire building operation and the curing stage. Also, blisters in the innerliner can be prevented by using a squeeze roll with uniform contact pressure across the liner while it is plied up with a barrier or squeegee. In addition, using a higher viscosity halobutyl rubber or increasing the ratio of carbon black to oil used in the compound may reduce the number of blisters. Also, using higher molecular weight processing oils or oils that contain only a very small fraction of low molecular weight components may also help minimize blister occurrence. GEN: B. Shama, B. Rodgers, D. Tracey, N. Tambe, D. Rouckhout (ExxonMobil), "Tire Halobutyl Rubber Innerliner," Paper No. 94 presented at the Fall Meeting of the Rubber Division, ACS, October 10–12, 2006, Cincinnati, OH.

5.11 Calendering: Improving Calender Release

Poor release from the calender can be a source of factory quality problems.

The following ***experimental*** *ideas may improve calender release. Also, all relevant literature sources, including the ones cited below, should be researched and read.* ***Caution:*** *these general experimental ideas may not work in all specific situations. Changes to improve calendar release will certainly affect other properties as well, for better or for worse, and this book does not purport to show how these other properties are affected. Also, this book does not purport to address safety and health issues.*

Star Polymer

Using star-branched halobutyl in place of regular halobutyl will result in faster stress relaxation of calendered sheets and easier roll release. RT: Chapter 8, "Specialty Elastomers," G. Jones, D. Tracey, A. Tisler, p. 182.

CR

Adding 5 to 10 phr of a high *cis*-polybutadiene to a polychloroprene compound should significantly improve the compound's roll release in a calender operation. RP: L. L. Outzs.

The use of 2 to 3 phr of a polyethylene wax as a compound additive will also improve calender release. RP: L. L. Outzs.

5.12 Raw Material and Mixed Stock Bin Storage: Lengthening Useful Shelf Life

Sometimes if raw materials are stored too long in the plant under adverse conditions before use, they may impart poor properties to the mixed stock. Sometimes the mixed batches themselves can be stored too long as well. Excessive storage time can affect a batch's processability as well as the ultimate cured physical properties.

*The following **experimental** ideas may help prevent quality problems with mixed batches. Also, all relevant literature sources, including the ones cited below, should be researched and read. **Caution:** these general experimental ideas may not work in all specific situations. Changes to lengthen bin storage shelf life will certainly affect other properties as well, for better or for worse, and this book does not purport to show how these other properties are affected. Also, this book does not purport to address safety and health issues.*

Preweighed Blends

When using preweighed and powder blended curative packages, avoid blending curatives that may react with each other, such as DPG mixed with sulfenamide accelerators and so forth. This will shorten the powder preblend's shelf life. GEN: J. Sommer, "Stabilized Curative Blends for Rubber," *Rubber Chemistry and Technology*, March-April, 1988, Vol. 61, p. 149.

5.13 Mixed Stock Bin Storage: Reducing Bloom

Bloom is the exudation of one or more compounding ingredients to the surface of the compound stock after mixing and cooling. This separation of compounding ingredients is usually due to chemical insolubility or incompatibility. It can occur over a given time period for both uncured and cured rubber compounds. Bloom in the uncured state can reduce a compound's building tack and can cause adhesion problems. Bloom can also present an appearance problem for cured rubber in a product application.

The following ***experimental*** *ideas may reduce the occurrence of bloom. Also, all relevant literature sources, including the ones cited below, should be researched and read.* ***Caution:*** *these general experimental ideas may not work in all specific situations. Changes to reduce bloom will certainly affect other properties as well, for better or for worse, and this book does not purport to show how these other properties are affected. Also, this book does not purport to address safety and health issues.*

Oils

Reducing the level of processing oil used in the recipe may reduce the tendency for bloom to occur. RT: Chapter 12, "Compounding with Carbon Black and Oil," S. Laube, S. Monthey, M-J. Wang, p. 311.

Select oils that are more compatible with the base elastomer(s). Oils with higher viscosity gravity constants (VGCs) generally tend to be more compatible with SBR and BR. Oils with lower VGC and lower aromaticity tend to be more compatible with elastomers such as EPDM and butyl rubber. RT: Chapter 12, "Compounding with Carbon Black and Oil," S. Laube, S. Monthey, M-J. Wang, p. 312. GEN: J. Dick, "Oils, Plasticizers and Other Rubber Chemicals," *Basic Rubber Testing,* ASTM, 2003, p. 124.

Plasticizers

Be careful to avoid using too high a concentration of polymeric plasticizers because these plasticizers can be too polar and bloom from a cured rubber compound. RT: Chapter 14, "Ester Plasticizers and Processing Additives," W. Whittington, p. 356.

Consider trying a synthetic ester plasticizer with fewer carbons in its linear alcohol segment for use in compounds based on polar specialty elastomers to avoid exudation or blooming out from the compound. RT: Chapter 14, "Ester Plasticizers and Processing Additives," W. Whittington, p. 363.

For polar elastomer based compounds, consider selecting glutarate polymeric plasticizers for reportedly very good compatibility and migration resistance. GEN: S. O'Rourke (CP Hall), "High Performance Ester Plasticizers," Paper No. 61 presented at the Fall Meeting of the Rubber Division, ACS, October 16–19, 2001, Cleveland, OH.

Accelerators

Some accelerators used as curatives for EPDM compounds may be somewhat insoluble in this polymer at higher concentrations and can sometimes present a bloom problem. To reduce the risk of bloom in curing an EPDM compound, one may consider selecting the "triple 8" cure system, which consists of sulfur at 2 phr, MBT at 1.5 phr, TeDEC at 0.8 phr, DPTT at 0.8 phr, and TMTD at 0.8 phr. Another nonbloom cure system for EPDM is called "general purpose." It consists of sulfur at 2.0 phr, MBTS at 1.5 phr, ZBDC at 2.5 phr, and TMTD at 0.8 phr. RT: Chapter 16, "Cures for Specialty Elastomers," B. H. To, p. 395. GEN: *Compounders Pocket Book*, Flexsys, 2002, pp. 122–123.

Insoluble Sulfur

Use insoluble sulfur in place of regular rhombic crystalline sulfur to avoid sulfur bloom. GEN: S. Tobing, "Covulcanization in NR/EPDM Blends," *Rubber World*, February, 1988, p. 33.

Take steps to ensure that the insoluble sulfur does not revert to crystalline rhombic sulfur before it is used as a curative. Otherwise sulfur bloom may occur during processing before cure. For example, avoid storing insoluble sulfur as a raw material at temperatures higher than 40 °C. Avoid storing insoluble sulfur near alkaline materials. Also, avoid excessively high mixing temperatures, especially when alkaline ingredients are present in the formulation. When insoluble sulfur is used in preblend packages, make sure the sulfenamide accelerator that it might be blending with is of good quality and not aged. Also, it is a good idea to use an inventory system of first-in/first-out in order to minimize total storage time in the warehouse for insoluble sulfur. GEN: B. To (Flexsys), "Insoluble Sulfur Compounding Technology," Presented at the ITEC Meeting, September, 2002, Akron, OH.

Peroxides

For peroxide cures, consider selecting a peroxide such as Luperox TBEC, which normally does not generate decomposition byproducts on curing that might lead to bloom. RT: Chapter 17, "Peroxide Cure Systems," L. Palys, p. 422.

For peroxide cures, consider selecting certain peroxides of the peroxyketal class, which normally do not generate decomposition byproducts on curing that might lead to bloom. RT: Chapter 17, "Peroxide Cure Systems," L. Palys, pp. 422–424.

Avoid curing with either DCP or BBPIB (peroxides) because they decompose into byproducts such as acetophenone or bis(hydroxyisopropyl)benzene and 1-acetyl-4-(hydroxyisopropyl) benzene that bloom. RT: Chapter 17, "Peroxide Cure Systems," L. Palys, pp. 426–427.

Peroxide Coagents

Consider using low molecular weight (liquid) high-vinyl 1,2-polybutadiene resins (such as Ricon® grades) as a coagent in a peroxide cure of an EPDM compound to prevent surface bloom when compared to some other traditional coagents. GEN: R. Drake, "Using Liquid Polybutadiene Resin to Modify Elastomeric Properties," *Rubber & Plastics News*, February 28 and March 14, 1983.

AOs

If possible, avoid the use of diaryl PPDs because members of this class of antiozonants are usually the most susceptible to bloom problems. If possible use other types of PPDs instead. RT: Chapter 19, F. Ignatz-Hoover, p. 458.

BIMS

Brominated isobutylene-co-*p*-methylstyrene (BIMS)-based compounds may have less tendency to bloom than some other rubber compounds because the BIMS compounds do not need to use antiozonants that bloom or exude to the surface of the compound, causing appearance problems. This is because compounds based on BIMS have an inherent resistance to ozone attack (possessing a saturated backbone). GEN: A. Tisler, K. McElrath, D. Tracey, M. Tse, "New Grades of BIMS for Non-stain Tire Sidewalls," Presented at ACS Rubber Div. Meeting, Fall, 1997, Paper No. 66.

EPDM Oil Bloom

Be careful about high levels of paraffinic oil blooming from a high-ethylene (greater than 68%) EPDM-based compound. Consider possibly substituting a blend of naphthenic and paraffinic oils in place of the loading of all paraffinic oil. Also consider substituting an EPDM polymer with lower ethylene content. RP: L. L. Outzs.

Methylene Donors

When using phenolic or resorcinol resins, minimize the use of hexamethylenetetramine (HMTA) as the methylene donor. Excessive levels of HMTA are prone to blooming. Consider using another methylene donor, such as HMMM, which might have less tendency to bloom. RP: M. A. Lawrence.

Mixing

Use an internal mixer that has an effective cooling system so that the batch temperature does not rise too quickly. Rising batch temperature adds to the batch's heat history (thus affecting the propensity for bloom in some cases) and causes the compound viscosity to decrease, which causes lower shear mixing forces and less effective dispersion at the higher temperature range. Therefore, keeping the batch at a lower temperature range longer should improve the ultimate dispersion and state of mix and possibly reduce the likelihood of bloom. Besides assuming that the optimum batch size has been selected and working with an effective cooling system, one can also gain some additional control of mixing temperature by adjusting either ram pressure or rotor speed. RT: Chapter 23, "Rubber Mixing," W. Hacker, p. 514.

Talc

The addition of fine particle size talc as a compounding ingredient can sometimes reduce iridescent bloom. RP: O. Noel.

5.14 Compression/Transfer/Injection Molding: Improving Mold Release

Poor mold release can slow down the demolding process and sometimes even damage the rubber product.

*The following **experimental** suggestions may improve mold release. Also, all relevant literature sources, including the ones cited below, should be researched and read. **Caution:** these general experimental ideas may not work in all specific situations. Changes to improve mold release will certainly affect other properties as well, for better or for worse, and this book does not purport to show how these other properties are affected. Also, this book does not purport to address safety and health issues.*

Polyacrylates

Increasing the stearic acid levels in a polyacrylate compound may improve mold release. RT: Chapter 8, "Specialty Elastomers," P. Manley, C. Smith, p. 206.

Fluoroelastomers

Carnauba wax as a compounding ingredient in fluoroelastomer (FKM) compounds reportedly improves mold release properties. GEN: J. Sommer, *Elastomer Molding Technology*, Elastech, Hudson, OH, 2003, p. 276.

Also, selecting either a bisphenol cure or a peroxide cure for an FKM compound will probably give better mold release than using a diamine cure, which may give poor mold release. GEN: J. Denham (3M), "Basic Fluoroelastomer Technology," Presented at a meeting of the Energy Rubber Group, September 13, 2011, Galveston, TX.

Also, to improve the mold release of an FKM compound, consider evaluating 3M's performance additive under the trade name of "Dynamar Rubber Additive RA 5300," which reportedly contains siloxane elastomer and talc. GEN: J. Denham (3M), "Optimizing Performance and Improving Productivity," Presented at the Spring Meeting of the Energy Rubber Group, May 18, 2011, Arlington, TX.

External Release Agents

There are many proprietary external release agents available for use to achieve good mold release. GEN: M. Kuschnerus, M. Hensel, R. Mille (Schill + Seilacher Strukol GmbH), "Release Agent Systems for Moulding and Shaped Hose Application, Ways to Reduce Production Downtime," Paper No. 39 presented at the Indian Rubber Expo 2011, January 19, 2011, Chennai, India.

Polychloroprenes

It is recommended that an effective low molecular weight polyethylene mold release agent be considered for polychloroprene (CR) compounds. GEN: J. C. Bament, *Neoprene Synthetic Rubber*, "A Guide to Grades, Compounding and Processing," DuPont, p. 24; J. Sommer, *Elastomer Molding Technology*, Elastech, Hudson, OH, 2003, pp. 276–277.

Consider evaluating Struktol WB16 (a mixture of calcium soaps and amides of saturated fatty acids) for use in a CR compound. Reportedly, the calcium soaps in this blend have an affinity for the metal surfaces while the polarity of the amide additives promote surface activity. This combination may promote mold release as well as mold flow. GEN: C. Clarke, M. Hensel (Struktol), "High Technology Process Additives Developed for Use with Different Mixing Techniques for Improved Performance," Paper No. 36 presented at the Indian Rubber Expo 2011, January 19, 2011, Chennai, India.

Silicone Rubber

It has been reported that diatomaceous earth, when used as a filler in silicone rubber, can impart better mold release to the compound than fumed silica can exhibit. GEN: R. Singh (World Minerals), "Diatomaceous Earth: Filler For Rubber Reinforcement," Paper No. 94 presented at the Fall Meeting of the Rubber Division, ACS, October 13–15, 2009, Pittsburgh, PA.

Mold Coating

Consider coating the mold with Teflon®. GEN: J. Sommer, *Elastomer Molding Technology*, Elastech, Hudson, OH, 2003, p. 282.

5.15 Compression/Transfer/Injection Molding: Reducing or Eliminating Mold Fouling

Mold fouling or "buildup" from repeated molding can be a problem. High levels of fouling can result in less heat transfer to the rubber stock from the mold. In severe cases of mold fouling, sometimes the "fouling material" may adhere itself to the surface of the molded part and cause appearance problems or worse.

*The following **experimental** ideas might reduce or eliminate mold fouling. Also, all relevant literature sources, including the ones cited below, should be researched and read. **Caution:** these general experimental ideas may not work in all specific situations. Changes to reduce or eliminate mold fouling will certainly affect other properties as well, for better or for worse, and this book does not purport to show how these other properties are affected. Also, this book does not purport to address safety and health issues.*

Hot Tear Resistance

Parts made of compounds with better hot tear resistance are less likely to increase the difficulties in demolding that may result in "fouled residues" in certain mold surface areas. One might consider cutting back on high loadings of large particle size mineral fillers because many times too high a loading of these inexpensive fillers may result in poor hot tear properties. (See Sections 2.7 and 2.8 for ideas for improving tear resistance and hot tear resistance, respectively.) GEN: J. Sommer, *Elastomer Molding Technology*, Elastech, Hudson, OH, 2003, p. 270.

Bloom

Some forms of bloom will increase mold fouling. For example, tetramethylthiuram disulfide (TMTD) accelerator tends to bloom from some general purpose elastomer based compounds. If a cure system can be substituted based on tetrabutylthiuram disulfide (TBTD) instead of TMTD, then less bloom may occur because TBTD may be more soluble in some general purpose elastomers than TMTD. Thus the use of TBTD may result in less mold fouling. (See Section 5.13 for ways to reduce bloom.) GEN: J. Sommer, *Elastomer Molding Technology*, Elastech, Hudson, OH, 2003, p. 271.

Cure Temperature

In injection molding, if a compound can be cured at a lower temperature for a longer time, there may be less mold fouling. GEN: J. Sommer, *Elastomer Molding Technology*, Elastech, Hudson, OH, 2003, pp. 271–273.

Molds

Using molds made of a better grade of steel reportedly will reduce fouling. GEN: J. Sommer, *Elastomer Molding Technology*, Elastech, Hudson, OH, 2003, p. 271.

Diene Rubber Base

Compounds based on diene elastomers can cause high mold fouling. Reportedly, natural rubber is the worst, followed by polyisoprene, followed by nitrile rubber, followed by SBR and BR, which are not as detrimental.

NR (high fouling) > IR > NBR > SBR ~ BR (less fouling)

Generally speaking, these diene elastomers are very susceptible to oxidation, which leads to mold fouling. Therefore, an effective use of antioxidants may help slow down the formation of mold fouling residues. GEN: J. Sommer, *Elastomer Molding Technology*, Elastech, Hudson, OH, 2003, pp. 271–274.

FKMs

There are problems with mold fouling using fluoroelastomers (FKMs). S. Bowers and E. Thomas reported on a new class of FKMs that do not generate as much mold fouling. Also, using a mold made of stainless steel may help. GEN: S. Bowers, E. Thomas, *Rubber & Plastics News*, February 5, 2001, Vol. 30, p. 23; *Rubber & Plastics News*, February 19, 2001, Vol. 30, p. 14; J. Sommer, *Elastomer Molding Technology*, Elastech, Hudson, OH, 2003, p. 276.

Also, to reduce mold fouling of an FKM compound, consider evaluating 3M's performance additive under the trade name of "Dynamar Rubber Additive RA 5300," which reportedly contains siloxane elastomer and talc. GEN: J. Denham (3M), "Optimizing Performance and Improving Productivity," Presented at the Spring Meeting of the Energy Rubber Group, May 18, 2011, Arlington, TX.

Also, a processing additive such as Struktol HT 290 (reportedly a blend of fatty acid derivatives and waxes) might be considered for FKM compounds in order to reduce mold fouling in injection molding. GEN: J. Bruhn, J. Bertrand, C. Clarke, M. Hensel (Struktol), "Hi-tech Fluoro Polymers, Now Used in Increasing Volume Production, Gain Important Benefit from Hi-tech Process Additives," Paper No. 21 presented at the Spring Meeting of the Rubber Division, ACS, May 17–19, 2004, Grand Rapids, MI.

Polychloroprenes (CR)

Compounds based on CR can also cause significant mold fouling. It is suggested that molds made of nickel-chrome steels may have less mold fouling from compounds based on CR. For less mold fouling from CR compounds, select the new low-mold-fouling grades of CR such as the M grades. Also consider using a concentration greater than 4 phr of a good high-activity magnesium oxide in the CR formulation to reduce mold fouling. Also, it is recommended that an effective low molecular weight polyethylene mold release agent be used. In addition, efforts should be made to keep the molding temperatures low, such as between 180 and 185°C (356 to 365°F). Lastly, avoid plasticizers that contain free acidic impurities. GEN: J. C. Bament, *Neoprene Synthetic Rubber*, "A Guide to Grades, Compounding and Processing," DuPont, p. 24; J. Sommer, *Elastomer Molding Technology*, Elastech, Hudson, OH, 2003, pp. 276–277.

Nitrile Rubber

Emulsion-polymerized nitrile rubbers (NBRs) tend to have differing tendencies to impart mold fouling. Select a low-mold-fouling grade of NBR such as Perbunan® NT. In addition, two 30-Mooney NBR grades that are reported to have relatively low mold fouling characteristics are Chemigum® N683B and Paracril® X3684. GEN: J. Sommer, *Elastomer Molding Technology*, Elastech, Hudson, OH, 2003, pp. 278–280.

ECO

The chlorine content in epichlorohydrin copolymer (ECO) is reported to cause mold fouling. ECO compounds cured with ethylene thiourea can lead to mold fouling after several mold cycles. On the other hand, triazine derivative curing systems with the appropriate acid acceptors (magnesium oxide) may be better suited to decrease mold fouling. GEN: J. Sommer, *Elastomer Molding Technology*, Elastech, Hudson, OH, 2003, p. 280.

CSM

When cured, chlorosulfonated polyethylene elastomers (CSMs) give off sulfurous acid, which causes corrosion of the steel surfaces of a mold. It is recommended that stainless steel molds be considered when curing CSM compounds GEN: J. Sommer, *Elastomer Molding Technology*, Elastech, Hudson, OH, 2003, p. 280.

EPDM Compounds

Sometimes, multiple accelerators and curatives are used for EPDM compounds. Under certain circumstances, one of the accelerators, such as TMTD, can contribute to mold fouling. One should develop a cure system that does not promote excessive mold fouling. In addition, often EPDM compounds may contain high filler-oil loadings. Sometimes these oils may contain high-volatility components that can promote mold fouling. This type of problem might be solved by either (1) substituting low-volatility oils in place of those high-volatility oils, (2) reducing the concentration of the oils with high volatility, or (3) substituting factice for these high-volatility oils. Perhaps some combination of all three of these suggested changes could also be worked out. GEN: J. Sommer, *Elastomer Molding Technology*, Elastech, Hudson, OH, 2003, p. 281.

Special Additives

Special compounding additives such as PPA-790® have been reported to reduce mold fouling in some situations. GEN: J. Sommer, *Elastomer Molding Technology*, Elastech, Hudson, OH, 2003, p. 282.

Caution with Mold Release Agents

Be careful with mold release agents such as emulsified silicones, which have been reported to cause mold fouling. GEN: J. Sommer, *Elastomer Molding Technology*, Elastech, Hudson, OH, 2003, p. 280.

Synthetic-Based Emulsification System

Use an emulsion NBR with a synthetic-based emulsification system instead of a rosin acid (worst), fatty acid, or mixed system. RT: Chapter 8, "Specialty Elastomers," M. Gozdiff, p. 195.

Processing Oil

Reducing the level of processing oil used in the recipe may reduce the tendency of a compound to foul molds. RT: Chapter 12, "Compounding with Carbon Black and Oil," S. Laube, S. Monthey, M-J. Wang, p. 311.

Carbon Black

Use a new ultra-low-structure, semireinforcing carbon black instead of processing aids to keep compound viscosity low and maintain good mold flow without mold fouling. GEN: S. Bussolari, S. Laube, "A New Cabot Carbon Black for Improved Performance in Peroxide Cured Injection Molded Compounds," Presented at ACS Rubber Div. Meeting, Fall, 2000, Paper No. 98.

Mold Coating

Consider coating the mold with Teflon®. GEN: J. Sommer, *Elastomer Molding Technology*, Elastech, Hudson, OH, 2003, p. 282.

■ 5.16 Compression/Transfer/Injection Molding: Improving Mold Flow

Improving mold flow is important to prevent underfills or nonfills from occurring.

The following ***experimental*** *ideas might help to improve mold flow. Also, all relevant literature sources, including the ones cited below, should be researched and read.* ***Caution:*** *these general experimental ideas may not work in all specific situations. Changes to improve mold flow will certainly affect other properties as well, for better or for worse, and this book does not purport to show how these other properties are affected. Also, this book does not purport to address safety and health issues.*

Use an emulsion NBR with fatty acid based emulsification system instead of a rosin acid, mixed system, or synthetic (worst). RT: Chapter 8, "Specialty Elastomers," M. Gozdiff, p. 195.

Use a liquid EPDM with solid grades of EPDM to increase injection mold flow rates while minimizing losses in cured hardness. GEN: W. Sigworth, "Liquid EP(D)M Polymers in Mechanical Goods Applications," Presented at ACS Rubber Div. Meeting, Fall, 2000, Paper No. 9.

The addition of fine particle size talc as a compounding ingredient reportedly can improve mold flow compared to some other fillers such as calcined clay. GEN: O. Noel, "Talc for Injection Molding of Rubber," Technical Seminar on Injection Molding of Rubber for the Canadian Society of Chemistry, Sponsored by the Ontario Rubber Group, May, 2000. RP: O. Noel.

Consider evaluating Struktol WB16 (a mixture of calcium soaps and amides of saturated fatty acids) in CR compounds. Reportedly, the calcium soaps in this blend have an affinity for the metal surfaces while the polarity of the amide additives promotes surface activity. This combination may promote mold flow. GEN: C. Clarke, M. Hensel (Struktol), "High Technology Process Additives Developed for Use with Different Mixing Techniques for Improved Performance," Paper No. 36 presented at the Indian Rubber Expo 2011, January 19, 2011, Chennai, India.

Traditionally, HNBR grades that were available earlier had relatively high Mooney viscosity values. Now there are new grades of HNBR that are available with lower Mooney viscosity values that will improve mold flow in injection molding. GEN: F. Guerin, S. Guo (Lanxess), "Improving the Processibility of HNBR," Paper No. 62 presented at the Fall Meeting of the Rubber Division, ACS, October 5-8, 2004, Columbus, OH; E. Campomizzi, L. Ferrari, R. Pazur (Lanxess), "Enhancing Compound Properties and Aging Resistance by Using

Low Viscosity HNBR," Paper No. 69 presented at the Spring Meeting of the Rubber Division, ACS, May 16–18, 2005, San Antonio, TX.

In tire curing, special small vents (spews) are usually required to allow trapped air to escape in order to not distort the tire or result in an unacceptable tire surface appearance. GEN: M. Stefanidis, G. Alloys, R. Coleman, "Spring Vent Technical Paper," Paper No. 9A presented at a meeting of ITEC, September 15–17, 2008, Akron, OH.

In addition, it should be noted that in general mold flow can relate to compound viscosity, nerve, and scorch time. In the special case of injection molding, mold flow also relates to shear thinning as well. The reader may wish to reference the following alternative book sections to consider additional experimental ideas for improving mold flow:

Section 4.1 Reducing Viscosity
Section 4.2 Increasing Shear Thinning (for Injection Molding)
Section 4.3 Reducing Elasticity (Nerve)
Section 4.8 Increasing Scorch Safety Time

5.17 Compression/Transfer/Injection Molding: Reducing Shrinkage of Part

Excessive shrinkage of a rubber part after molding can lead to problems in meeting dimensional specifications.

The following ***experimental*** *ideas may reduce or eliminate this shrinkage problem. Also, all relevant literature sources, including the ones cited below, should be researched and read.* ***Caution:*** *these general experimental ideas may not work in all specific situations. Changes to reduce shrinkage of mold-cured rubber compounds will certainly affect other properties as well, for better or for worse, and this book does not purport to show how these other properties are affected. Also, this book does not purport to address safety and health issues.*

Devolatilization

Using devolatilized rubber and compounding ingredients helps reduce or eliminate any porosity or microvoids (which cannot be seen with the naked eye), which results in a higher cured density, leads to a decrease in bulk compressibility, and results in less shrinkage on demolding. GEN: A. Kasner, E. Meinecke, "Porosity in Rubber: A Review," *Rubber Chemistry and Technology*, July-August, 1996, Vol. 69, p. 424.

Synthetic Ester Plasticizer

Sometimes reducing the concentration of synthetic ester plasticizer in a nitrile formulation may reduce shrinkage of cured rubber parts. GEN: J. R. Beatty, "Effect of Composition on Shrinkage of Mold Cured Elastomeric Compounds," *Rubber Chemistry and Technology*, November-December, 1978, Vol. 51, p. 1044.

Filler Loading

Higher loadings of filler in general will reduce the shrinkage of cured molded parts. However, certain fillers are better at retarding shrinkage than others. For example, clays are reportedly significantly better at retarding compound mold shrinkage than other fillers such as whiting, carbon black, or zinc oxide. GEN: J. R. Beatty, "Effect of Composition on Shrinkage of Mold Cured Elastomeric Compounds," *Rubber Chemistry and Technology*, November-December, 1978, Vol. 51, p. 1044.

RT: *Rubber Technology, Compounding and Testing for Performance*. 2nd ed. Edited by J. S. Dick, Hanser Publications, 2009 ■ GEN: General References from the Literature ■ RP: Review Panel (Advisors for this book) ■ IMPORTANT: The information contained herein is experimental in nature and is meant for individuals with advanced scientific training and rubber compounding experience. All safety precautions should be taken when replicating or applying this information. The author makes no representation or warranty, either expressed or implied.

Clay Loading

Increasing the loading of clay in a rubber compound can significantly reduce the compound's mold shrinkage. GEN: "A Comparative Evaluation of Hycar Nitrile Polymers," Manual HM-1 Revised, B. F. Goodrich Chemical Co.

Talcs

Reportedly, talc as a compounding ingredient may help reduce cured shrinkage. RP: O. Noel.

Sulfur Concentration

With increasing loadings of sulfur to concentrations higher than 3 phr, cured mold shrinkage of the rubber part can become greater. However, adjusting sulfur loads within concentration ranges below 3 phr usually has only a minor effect on mold shrinkage. GEN: J. R. Beatty, "Effect of Composition on Shrinkage of Mold Cured Elastomeric Compounds," *Rubber Chemistry and Technology*, November-December, 1978, Vol. 51, p. 1044.

Cure Pressure

Curing with very high pressure applied to the rubber compound will reduce or eliminate any porosity or microvoids (which cannot be seen with the naked eye), which results in a higher cured density, leads to a decrease in bulk compressibility, and results in less shrinkage on demolding. GEN: A. Kasner, E. Meinecke, "Porosity in Rubber: A Review," *Rubber Chemistry and Technology*, July-August, 1996, Vol. 69, p. 424.

Molding Temperature

Molding temperature can have an effect on mold shrinkage after cure. GEN: J. R. Beatty, "Effect of Composition on Shrinkage of Mold Cured Elastomeric Compounds," *Rubber Chemistry and Technology*, November-December, 1978, Vol. 51, p. 1044. For injection molding, significantly higher rates of shrinkage resulted when the molding temperature of an injection molding operation was increased. The counterpart compression molding did not show quite as much increase in shrinkage with the same rise in molding temperature. GEN: J. Sommer, *Elastomer Molding Technology*, Elastech, 2003, p. 185.

Compound Orientation

Rubber compounds may have a certain degree of orientation (grain) from calendering or extrusion. This will affect the shrinkage of the molded rubber part. Shrinkage can be greater in the transverse direction compared to the longitudinal direction. GEN: J. R. Beatty, "Effect of Composition on Shrinkage of Mold Cured Elastomeric Compounds," *Rubber Chemistry and Technology*, November-December, 1978, Vol. 51, p. 1044.

Post-cure

If possible, avoid base elastomers that require a post-cure because the post-cure tends to increase shrinkage of the molded rubber part. GEN: J. R. Beatty, "Effect of Composition on Shrinkage of Mold Cured Elastomeric Compounds," *Rubber Chemistry and Technology*, November-December, 1978, Vol. 51, p. 1044.

5.18 Compression/Transfer/Injection Molding: Improving Surface Appearance

Improving surface appearance for a molded part can be very important commercially.

The following ***experimental*** *ideas can help improve surface appearance. Also, all relevant literature sources, including the ones cited below, should be researched and read.* ***Caution:*** *these general experimental ideas may not work in all specific situations. Changes to improve surface appearance will certainly affect other properties as well, for better or for worse, and this book does not purport to show how these other properties are affected. Also, this book does not purport to address safety and health issues.*

Consider using a processing additive such as a blend of zinc and potassium soaps of fatty acids in a general purpose rubber compound to improve the surface appearance of an injection-molded part. RT: Chapter 14, "Ester Plasticizers and Processing Additives," C. Stone, pp. 375-376.

5.19 Increasing the Blow Rate During the Curing of a Sponge Compound

Maintaining a good "cure-blow" balance for a sponge compound is very important. For example, if the rate of blow (from the decomposition of the blowing agents) is too slow compared to the cure rate, then the cells will not form well or will be too small. If the blow reaction is too fast relative to the cure, then the opposite happens with large open cells forming.

The following ***experimental*** *ideas may increase the blow reaction rate during the curing of the compound. Also, all relevant literature sources, including the ones cited below, should be researched and read.* ***Caution:*** *these general experimental ideas may not work in all specific situations. Changes to increase the blow rate during the curing of a sponge compound will certainly affect other properties as well, for better or for worse, and this book does not purport to show how these other properties are affected. Also, this book does not purport to address safety and health issues.*

Inorganic vs. Organic Blowing Agents

When an inorganic blowing agent such as sodium bicarbonate is used in rubber compounding, usually an open cell structure results. On the other hand, when organic blowing agents such as ADC or OBSH are used, then a closed cell structure may result. RT: Chapter 21, "Chemical Blowing Agents," R. Annicelli, p. 476.

ADC vs. OBSH

The blowing agent *p,p*-Oxybis(benzenesulfonylhydrazide) (OBSH) has a lower decomposition temperature and a much faster decomposition rate than the blowing agent azodicarbonamide (ADC). In some compounds OBSH and ADC are used together to achieve a greater temperature range in manufacturing. RT: Chapter 21, "Chemical Blowing Agents," R. Annicelli, pp. 479–485.

Particle Size of Blowing Agent

For faster decomposition of ADC blowing agent, use the fine particle size grade. RT: Chapter 21, "Chemical Blowing Agents," R. Annicelli, pp. 481–483. GEN: J. S. Dick, R. Annicelli, "Compound Changes to Balance the Cure and Blow Reactions Using the MDR-P to Control Cellular Density and Structure," *Rubber & Plastics News*, November 16, 1998; *Gummi Fasern Kunststoffe*, April, 1999.

Blow Activators

For ADC blowing agent, the actual decomposition temperature can be lowered by using the appropriate "blow activator" for ADC. Zinc oxide (usually already present as a cure activator) is also a very effective blow activator for ADC. Urea, triethanolamine, barium/calcium stearates, and polyethylene glycol can also be used in a formulation as blow activators for ADC. RT: Chapter 21, "Chemical Blowing Agents," R. Annicelli, pp. 480–483.

For OBSH blowing agent, the blow activators that work best at lowering the decomposition temperature of OBSH are urea and triethanol amine (both very strong), as well as diphenyl guanidine, barium/calcium stearates, and stearic acid. RT: Chapter 21, "Chemical Blowing Agents," R. Annicelli, pp. 480–483.

Lower Compound Viscosity

Sponge rubber compounds should be formulated to have a relatively low Mooney viscosity. The compound viscosity must be low enough to allow good cell formation. Many times these formulations will contain significant levels of process oil(s) and large particle size fillers, which will result in a lower compound viscosity. GEN: J. S. Dick, R. Annicelli, "Compound Changes to Balance the Cure and Blow Reactions Using the MDR-P to Control Cellular Density and Structure," *Rubber & Plastics News*, November 16, 1998; *Gummi Fasern Kunststoffe*, April, 1999.

EPDM Grades

Select a EPDM grade with the highest level of diene (for a fast cure rate) to help match a fast blowing rate. RT: Chapter 8, "Specialty Elastomers," R. Vara, J. Laird, p. 193.

Cure–Blow Balance

To achieve a fine, closed cell structure for a sponge product, it is very important to establish and control the so-called "cure–blow balance." If the cure is too slow in relation to the blow reaction, you may lose gas and have an open cell structure. On the other hand, if the cure reaction is too fast in relation to the blow (decomposition) reaction, then the compound modulus will already be too high as the gas decomposition product is forming, and thus the cell formation, if any, will be too small in size. It is very important to keep the cure and the decomposition reaction rates in balance. RT: Chapter 21, "Chemical Blowing Agents," R. Annicelli, pp. 480, 487.

Activators for Inorganic Blowing Agent

Sodium bicarbonate blowing agent is usually activated with fatty acids such as stearic acid or even oleic acid to give higher gas yields. RT: Chapter 21, "Chemical Blowing Agents," R. Annicelli, p. 478.

Dispersion of Sodium Bicarbonate

It is normally difficult to achieve good dispersion of sodium bicarbonate as a blowing agent. For better dispersion, one should probably use sodium bicarbonate that has been coated with an oil or wax to improve dispersion. RT: Chapter 21, "Chemical Blowing Agents," R. Annicelli, p. 478.

5.20 Eliminating Trapped Air During Molding

Entrapped air during molding can greatly increase the scrap rate.

The following ***experimental*** *ideas may prevent entrapment of air during molding. Also, all relevant literature sources, including the ones cited below, should be researched and read.* ***Caution:*** *these general experimental ideas may not work in all specific situations. Changes designed to prevent entrapped air during molding will certainly affect other properties as well, for better or for worse, and this book does not purport to show how these other properties are affected. Also, this book does not purport to address safety and health issues.*

Bumping of Presses

Entrapped air can be eliminated during the curing process by "bumping" the presses during the earlier stages of the molding cycle. RT: Chapter 8, "Specialty Elastomers," G. Jones, D. Tracey, A. Tisler, p. 182.

Gelled Polymer

Consider using hot NBRs that are crosslinked with a difunctional monomer (gelled), and use at about 10 to 25 phr with other elastomers, such as SBR, XNBR, or cold NBR, to provide sufficient mold forces and "back pressure" to avoid trapped air. RT: Chapter 8, "Specialty Elastomers," M. Gozdiff, p. 198.

Use Ground Rubber

The use of ground rubber as a compounding ingredient may reduce trapped air. RP: J. M. Long.

Increase Compound Viscosity for Compression Molding

Epichlorohydrin compounds compression mold better if they are compounded to have higher compound viscosity, for example, 90 to 120 Mooney units. This higher viscosity is usually achieved with higher filler loadings and perhaps 2 to 5 phr of plasticizer to help flow. If the compound is too soft, then this could lead to air entrapment during compression molding. RT: Chapter 8, "Specialty Elastomers," C. Cable, p. 219.

Maybe Decreasing Compound Viscosity for Injection Molding

Epichlorohydrin compounds usually injection mold better if they are compounded to have lower viscosity than they might have for compression molding. However, a lower viscosity may not neccessarily be needed because of the heating and mastication of the compound by the barrel and screw in injection molding. RT: Chapter 8, "Specialty Elastomers," C. Cable, p. 221.

5.21 Reducing Backrinding

Backrinding from molding of a rubber compound may occur when the mold is opened after cure. Backrinding may result when the rubber vulcanizate expands from the opening of the mold after cure, resulting in a rupture at the part line. The backrinding occurs when the expansion of the cured rubber is concentrated in a limited area and is greater than the ultimate cured elongation of the compound, resulting in tearing of the rubber vulcanizate. The consequence of this backrinding is a "torn" or "gouged" appearance at the part line for the rubber part or article. Backrinding is most commonly associated with compression molding; however, it can also occur with transfer and injection molding.

The following are some ***experimental*** *ideas, which may help prevent or eliminate backrinding. Also, all relevant literature sources, including the ones cited below, should be researched and read.* ***Caution:*** *these general experimental ideas may not work in all specific situations. Changes to reduce backrinding will certainly affect other properties as well, for better or for worse, and this book does not purport to show how these other properties are affected. Also, this book does not purport to address safety and health issues.*

Mold Temperature

Curing a compound at a lower temperature for a longer period of time will allow more rubber to exit the mold from thermal expansion before scorch and cure are reached. The rubber compound will most likely have better tear resistance at lower temperatures as well. GEN: J. Sommer, *Elastomer Molding Technology*, Elastech, Hudson, OH, 2003, p. 80; R. M. Murray, D. C. Thompson, *The Neoprenes*, DuPont Inc., 1963, p. 15.

Mold Design

In designing the mold, place the part line in a noncritical area that does not affect product performance. GEN: R. M. Murray, D. C. Thompson, *The Neoprenes*, DuPont Inc., 1963, p. 15.

Backrinding can be severe in the molding of large, bulky articles with a large amount of mass. In designing the rubber mold, if the ratio of surface area to volume can be increased, then backrinding may be less of a problem. GEN: R. M. Murray, D. C. Thompson, *The Neoprenes*, DuPont Inc., 1963, p. 15; J. Sommer, *Elastomer Molding Technology*, Elastech, Hudson, OH, 2003, p. 80.

Cooling the Mold

Cooling the mold to lower the pressure that will be released on demolding is another option to consider to reduce or eliminate backrinding. GEN: R. M. Murray, D. C. Thompson, *The Neoprenes*, DuPont Inc., 1963, p. 15.

Preheating the Preform

Preheating the stock before placing it in the mold is another way to reduce backrinding. By preheating it beforehand, you also expand it before placing it in the mold. GEN: R. M. Murray, D. C. Thompson, *The Neoprenes*, DuPont Inc., 1963, p. 15; J. Sommer, *Elastomer Molding Technology*, Elastech, Hudson, OH, 2003, p. 80.

Extending Scorch Time

Extending scorch time through compound modifications (see Section 4.8 for ways to increase scorch time) should favor reduced backrinding because it provides more time for rubber to exit the mold cavity before curing occurs. J. Sommer, *Elastomer Molding Technology*, Elastech, Hudson, OH, 2003, p. 81.

Improving Tear Resistance

Explore ways of improving the compound's cured tear resistance (Section 2.7) and especially hot tear resistance (Section 2.8) to reduce backrinding. GEN: R. M. Murray, D. C. Thompson, *The Neoprenes*, DuPont Inc., 1963, p. 15.

6 Tire Performance Properties

6.1 Improving Tire Wet Traction

Improving the wet traction of a tire is very important.

*The following **experimental** ideas may improve this tire performance characteristic. Also, all relevant literature sources, including the ones cited below, should be researched and read. **Caution:** these general experimental ideas may not work in all specific situations. Changes to improve tire wet traction will certainly affect other properties as well, for better or for worse, and this book does not purport to show how these other properties are affected. Also, this book does not purport to address safety and health issues.*

BR Microstructure

Use solution BR with a higher 1,2 microstructure in the tread to increase the glass transition temperature (T_g) of the tread and reportedly increase the imparted wet traction without significantly hurting tire rolling loss. RT: Chapter 7, "General Elastomers and Blends," G. Day, p. 157.

E-SBR/BR Blend

For a tire tread based on a classic blend of E-SBR and 1,4-*cis*-BR, increasing the amount of E-SBR loading in relation to the BR will usually improve the wet traction. GEN: J. Palombo (Kumho), "Compounding for Rolling Resistance, Traction and Tread Wear: A Review of the Fundamentals," Paper No. 36 presented at the Spring Meeting of the Rubber Division, ACS, April 29–May 1, 2002, Savannah, GA.

ENR

Use epoxidized natural rubber (ENR) to increase T_g and improve wet traction of tire treads. RT: Chapter 7, "General Elastomers and Blends," G. Day, p. 144.

Normally, epoxidized NR is reportedly not "compatible" with technically specified NR such as SIR10, and it gives a higher damping response. However, by basing a tread compound on an ENR50 and SIR10 blend, an improvement in wet traction might be achieved. RT: Chapter 7, "General Elastomers and Blends," G. Day, p. 169.

BR/BIIR

Adding bromobutyl rubber to a tread based on BR will greatly improve wet traction performance. Unfortunately, because of the weak van der Waals interaction between the BIIR and the carbon black, the tread abrasion resistance is drastically reduced. However, it is reported that some of this "trade-off" problem can be overcome through the introduction of precipitated hydrated silica with a suitable silane coupling agent. GEN: R. Resendes, K. Kulbaba, A. Nizioiek (Lanxess), "BIIR in Treads: Preparation of High Traction Tread Compounds Through Polymer-Filler Modification," Paper No. 15 presented at the Fall Meeting of the Rubber Division, ACS, October 5–8, 2004, Columbus, OH.

Other Wet Grip Enhancing Polymers for Treads

Consider using between 5 to 20 phr of acrylonitrile butadiene rubber (NBR) as a wet grip enhancer in a tread, possibly with a small loading of dioctyl phthalate (DOP). While NBR has poor compatibility with SBR, it has better compatibility with precipitated hydrated silica, which is commonly used in "green" tread formulations to improve wet grip (wet traction), which is usually measured as $\tan\delta$ at 0 °C. GEN: R. Engehausen, A. Rawlinson (Bayer), "A Comparison of Wet Grip Enhancing Polymers in Tire Tread Compounds," Paper No. 5A presented at the ITEC Meeting, September 10–12, 2002, Akron, OH.

Consider using between 5 to 20 phr of acrylonitrile styrene butadiene rubber (NSBR) as a wet grip enhancer in a tread. NSBR has somewhat better compatibility with SBR compared to NBR; it also has good compatibility with precipitated hydrated silica, which is commonly used in "green" tread formulations as well to improve wet grip (wet traction), which is usually measured as $\tan\delta$ at 0 °C. GEN: R. Engehausen, A. Rawlinson (Bayer), "A Comparison of Wet Grip Enhancing Polymers in Tire Tread Compounds," Paper No. 5A presented at the ITEC Meeting, September 10–12, 2002, Akron, OH.

Consider using between 5 to 20 phr of 3,4-polyisoprene (3,4-IR). This IR polymer has a relatively high glass transition value and is also incompatible with SBR and BR. However, it has been used in high-performance tire treads to improve wet grip even though it may adversely affect wear resistance. GEN: R. Engehausen, A. Rawlinson (Bayer), "A Comparison of Wet Grip Enhancing Polymers in Tire Tread Compounds," Paper No. 5A presented at the ITEC Meeting, September 10–12, 2002, Akron, OH.

Consider using about 15 phr of high-vinyl polyisoprene (under the Goodyear trade name of "Traxsyn") in a tire tread formulation in order to improve wet traction. GEN: D. Benko (Goodyear), "New Polymers for Tires," Paper No. 3A presented at the ITEC Meeting, September 10–12, 2002, Akron, OH.

High Styrene Content

Normally SBR and NR are fairly compatible. However, SBR 1013 with 40% bound styrene is not very compatible with NR because of the high styrene content. However, a tread compound based on a blend of SBR 1013 and NR might increase wet traction. RT: Chapter 7, "General Elastomers and Blends," G. Day, p. 169. SBR 1721 (an oil-extended SBR with 40% bound styrene) is commonly used in high-performance tires. RP: J. M. Long.

Increasing the bound styrene content in emulsion SBR may improve wet traction. GEN: E. McDonel, K. Baranwal, J. Andries, *Polymer Blends*, Vol. 2, Chapter 19, "Elastomer Blends in Tires," Academic Press, 1978, p. 283.

It has been reported that treads containing SBR 1721 (based on an emulsion SBR with 40% bound styrene and extended with aromatic oil) give better wet traction. However, aromatic oils were banned in Europe in 2010. GEN: R. Engehausen, G. Marwede (Bayer AG), "The Influence of Rubber/Filler Systems on the Wet Traction of Radial Tires," Paper No. 22A presented at ITEC, September, 1998, Akron, OH.

High-Vinyl SBR

It is reported that using high-vinyl SBR in a tread formulation will improve a tire's wet grip performance. GEN: R. Engehausen, G. Marwede (Bayer AG), "The Influence of Rubber/Filler Systems on the Wet Traction of Radial Tires," Paper No. 22A presented at ITEC, September, 1998, Akron, OH; H. Colvin (Solvay Engineered Polymers), "Effect of SBR Structure on Compounded Properties," Paper D presented at the Spring Meeting of the Rubber Division, ACS, April 29–May 1, 2002, Savannah, GA.

IBR

Consider the use of isoprene butadiene rubber (IBR) for better wet traction and ice traction. RP: R. Dailey.

CIIR/BR

Treads based on 10 to 30% BIIR or CIIR with BR are reported to give significantly better wet traction and possibly not too much increase in rolling resistance. GEN: J. Fusco, "New Isobutylene Polymers for Improved Tire Processing," Presented at Akron Rubber Group Meeting, January 24, 1991.

A chlorobutyl/*cis*-BR blend in a tread reportedly gave better wet traction. GEN: L. Chang, J. Shackleton, "An Overview of Rolling Resistance," *Elastomerics*, March, 1983, p. 23.

BIMS

It is claimed that the wet tire traction of a tire tread can be improved if 25 to 50 phr of brominated isobutylene-co-*p*-methylstyrene (BIMS) is substituted for either NR or solution SBR in a tread formulation normally composed of 25 phr of BR, 50 phr of S-SBR, and 25 phr of NR with 75 phr of silane-coupled silica. GEN: R. Poulter, Presented at ITEC 2000, Paper No. 12C.

Reportedly, with a four-step phase mixing sequence involving brominated isobutylene-co-paramethylstyrene (BIMSM), *cis*-polybutadiene, and solution-polymerized styrene-butadiene rubber with both carbon black and precipitated hydrated silica, wet traction could be improved through the selective masterbatching of BIMSM with precipitated hydrated silica. GEN: W. Waddell, R. Poulter (Exxon), "Phase Mixing of Brominated Isobutylene-co-para-methylstyrene with Precipitated Silica to Enhance the Properties of a Tire Tread Compound," Paper No. 32 presented at the Spring Meeting of the Rubber Division, ACS, April 13–16, 1999, Chicago, IL.

IIR and Vinyl BR

It is commonly known that butyl rubber (IIR) and vinyl 1,2-polybutadiene rubber can impart very high wet skid resistance to a tire tread, if wet traction is the only property that concerns you. GEN: R. Rahalkar, "Dependence of Wet Skid Resistance upon the Entanglement Density and Chain Mobility According to the Rouse Theory of Viscoelasticity," *Rubber Chemistry and Technology*, May–June, 1989, Vol. 62, p. 246.

SIBR

SIBR can have two glass transition temperature (T_g) values. The lower T_g for SIBR relates to tire rolling resistance and fuel economy. The higher T_g value can relate to the wet traction characteristics that the SIBR polymer can also impart to a tread compound. The proper selection of an SIBR polymer can provide a better trade-off between wet traction vs. rolling resistance. GEN: A. Halasa, B. Gross, W. Hsu (Goodyear Tire and Rubber Company), "Multiple Glass Transition Terpolymers of Isoprene, Butadiene, and Styrene," Paper No. 91 presented at the Fall Meeting of the Rubber Division, ACS, October, 2009, Cleveland, OH.

LL Carbon Black

The use of a high-structure "LL carbon black" (long linkage carbon blacks that are crush resistant during mixing) in place of regular high-structure carbon black reportedly can be used to impart better wet traction to a tread. According to one study, high-structure carbon blacks are commonly susceptible to "crush" during mixing and convert to ordinary carbon blacks. These special high-structure LL carbon black grades are reported to be crush resistant. GEN: H. Mouri, K. Akutagawa, "Reducing Energy Loss to Improve Tire Rolling Resistance," Presented at ACS Rubber Div. Meeting, Spring, 1997, Paper No. 14.

Using a carbon black with a higher structure and/or a higher surface area and/or a higher loading level may improve the wet traction of a given tire tread. GEN: J. Palombo (Kumho), "Compounding for Rolling Resistance, Traction and Tread Wear: A Review of the Fundamentals," Paper No. 36 presented at the Spring Meeting of the Rubber Division, ACS, April 29–May 1, 2002, Savannah, GA.

Carbon Black Coupling Agents

Use of a carbon black–rubber coupling agent or "chemical promotor" in a carbon black loaded rubber compound may increase resilience and modulus as well as improve abrasion loss. It may also improve tire rolling resistance and wet grip. In the past, such coupling agents as *N*-(2-methyl-2-nitropropyl)-4-nitrosoaniline, *N*-4-dinitroso-*N*-methylaniline, *p*-nitro-sodiphenylamine, and *p*-nitroso-*N*,*N*-dimethylaniline were used. These nitroso compounds are no longer used because they give off nitrosoamines, which are suspected carcinogens. Therefore other coupling agents have been tried. A new coupling agent, *p*-aminobenzenesulfonyl azide (or amine-BSA), has recently been investigated that may impart these performance advantages to a compound, including improving wet traction and rolling resistance. GEN: L. Gonzalez, A. Rodriguez, J. deBenito, A. Marcos, "A New Carbon Black–Rubber Coupling Agent to Improve Wet Grip and Rolling Resistance of Tires," *Rubber Chemistry and Technology*, May–June, 1996, Vol. 69, p. 266.

Carbon Black–Silica Hydrid

Using a special carbon black–silica hybrid (dual-phase filler CRX4210A from Cabot Corporation) with Vulcan 1436 broad aggregate size carbon black and Si69 silane coupling agent in a tire tread gave very good rolling resistance and good wet traction. GEN: C. Flanigan, L. Beyer, D. Klekamp, D. Rohweder (Ford Motor Co.), B. Stuck, E. Terrill (ARDL), "Comparative Study of Silica, Carbon Black and Novel Fillers in Tread Compounds," Paper No. 34 presented at the Fall Meeting of the Rubber Division, ACS, October 11–13, 2011, Cleveland, OH.

Treated Aramid Fiber Granule

Using only 2 phr of treated aramid fiber granule (Sulfron 3001) in a tire tread did improve wet traction without hurting rolling resistance. GEN: C. Flanigan, L. Beyer, D. Klekamp, D. Rohweder (Ford Motor Co.), B. Stuck, E. Terrill (ARDL), "Comparative Study of Silica, Carbon Black and Novel Fillers in Tread Compounds," Paper No. 34 presented at the Fall Meeting of the Rubber Division, ACS, October 11–13, 2011, Cleveland, OH.

Oil and Carbon Black Loading

Increasing oil loading at fixed black levels may increase wet traction significantly more than the resulting increase in rolling resistance. GEN: W. Hess, W. Klamp, "The Effects of Carbon Black and Other Compounding Variables on the Tire Rolling Resistance and Traction," *Rubber Chemistry and Technology*, Vol. 56, p. 390.

Tread Hardness and Hysteresis

Increasing the hardness of treads made of IIR, NR, SBR, or BR reportedly caused the wet cornering coefficients at high speeds to also increase. Good wet traction at high speeds reportedly requires a combination of relatively high tread hardness and high hysteresis. GEN: W. Hess, W. Klamp, "The Effects of Carbon Black and Other Compounding Variables on the Tire Rolling Resistance and Traction," *Rubber Chemistry and Technology*, Vol. 56, p. 390.

Tread Hardness

A tire tread with too low a durometer hardness may have poor wet traction. RT: Chapter 12, "Compounding with Carbon Black and Oil," S. Laube, S. Monthey, M-J. Wang, p. 310.

S-SBR and Silica

Chemically modified solution SBR used with silica reportedly gives a better balance between wet skid resistance and rolling resistance. In addition, reportedly NS-116 gives a better rolling resistance vs. wet traction performance trade-off when used with silica. GEN: F. Suzuki (Nippon Zeon), "Rubbers for Low Rolling Resistance," *Tire Technology International*, 1997, p. 87.

Reportedly, an improvement in wet traction of a tire tread was achieved by using a special emulsion SBR that was functionalized with acrylonitrile (ACN) during its emulsion polymerization process. The final terpolymer produced was about 1.9% ACN in composition. The inclusion of this third monomer into the polymer raised the polymer's glass transition temperature (T_g). Organosilane is the more expensive compounding additive that was also used in this silica-loaded compound. GEN: V. Monroy, S. Hofmann, R. Tietz (Dow Chemical), "Effects of Chemical Functionalization of Polymers in Tire Silica Compounds: Emulsion Polymers," Paper No. 63A presented at the Fall Meeting of the Rubber Division, ACS, October 16–19, 2001, Cleveland, OH.

Silica

Use a dispersible silica silane system with a specially manufactured solution SBR and allow an effective silanization reaction to occur during mixing to establish better wet traction. GEN: A. McNeish, "Nanoblacks for Rolling Resistance," Presented at ITEC Meeting, Fall 2000, Paper No. 23A.

Silica/Silane Tread

For the common silica/silane tread used for better rolling resistance and good wet traction, using a precipitated hydrated silica around 160 square meters per gram of surface area instead of 125 square meters per gram of surface area may improve the imparted wet traction. GEN: H. Luginsland, W. Niedermeier (Degussa), "New Reinforcing Materials for Rising Tire Performance Demands," Paper No. 48 presented at the Spring Meeting of the Rubber Division, ACS, April 28–30, 2003, San Francisco, CA.

Also, for the common silica/silane tread compounds, using a higher loading of the precipitated hydrated silica may also improve the wet traction performance of the tire. GEN: H. Luginsland, W. Niedermeier (Degussa), "New Reinforcing Materials for Rising Tire Performance Demands," Paper No. 48 presented at the Spring Meeting of the Rubber Division, ACS, April 28–30, 2003, San Francisco, CA.

Mix long enough at a sufficiently high mixing temperature in order to complete the *in situ* silanization reaction to ensure not only low rolling resistance but also good wet traction for a tire tread stock. GEN: H. Luginsland, A. Hasse (Degussa), "Processing of Silica/Silane Filled Tread Compounds," Paper No. 34 presented at the Spring Meeting of the Rubber Division, ACS, April 4–6, 2000, Dallas, TX.

Dispersible Silica

Using the new dispersible silicas in the new silica/silane tread stocks enables one to better achieve the wet skid targets. GEN: S. Uhrlandt, A. Blume (Degussa), "Unique Production Process, Unique Silica Structure," Paper No. 15 presented at the Spring Meeting of the Rubber Division, ACS, April 29–May 1, 2002, Savannah, GA.

ATH Filler

Reportedly, alumina trihydrate (ATH) was used as a partial substitute for reinforcing carbon black in a tread formulation to achieve an improvement in wet traction and rolling resistance with only a small loss in tread wear. GEN: H. Mouri, K. Akutagawa, "Improved Tire Wet Traction Through the Use of Mineral Fillers," *Rubber Chemistry and Technology*, November-December, 1999, Vol. 72, p. 960.

Sepiolite as a Filler

Sepiolite, a mineral filler that is a form of magnesium silicate from Spain, can reportedly be used as a partial substitute for carbon black (up to 30%) in tire treads to reduce compound costs. It can reportedly improve wet traction. If used with silane coupling agents, it can also possibly improve tire rolling resistance. GEN: L. Hernandez, L. Rueda, C. Anton, "Magnesium Silicate Filler in Rubber Tread Compounds," *Rubber Chemistry and Technology*, September-October, 1987, Vol. 60, p. 606.

High-Viscosity Oil/Resins

Replacing conventional extender oil with either a high-viscosity oil or a resin may improve tire wet traction. GEN: W. Hess, W. Klamp, "The Effects of Carbon Black and Other Compounding Variables on the Tire Rolling Resistance and Traction," *Rubber Chemistry and Technology*, Vol. 56, p. 390.

Consider adding 20 phr of a C9 hydrocarbon resin, with a softening point of 90 to 100°C (194 to 212°F) (such as Rhenosin TT100®) to an SBR/BR tread compound in order to improve wet traction. GEN L. Steger (Rhein Chemie), K. Hillner, S. Schroter, "Resins in Tire Compounds,"

Tread Design

According to A. Veith, wet traction is a function of tread design that includes "groove void volume" (which normal tire wear will decrease) and "developed tread width" (defined as the tire footprint width, a function of the tire's aspect ratio). According to Veith, much of a tire's wet traction has to do with its "ability to discharge water from the front of the leading edge of the tire's footprint," as well as its ability to remove water from the footprint's interface. GEN: A. Veith, "Tread Groove Void and Developed Tread Width (Aspect Ratio): Their Joint Influence on Wet Traction," *Rubber Chemistry and Technology*, September-October, 1999, Vol. 72, p. 684.

6.2 Improving Tire Dry Traction

Improving dry traction is important to ensure the satisfactory performance of a tire.

The following ***experimental*** *ideas may help to improve this tire performance characteristic. Also, all relevant literature sources, including the ones cited below, should be researched and read.* ***Caution:*** *these general experimental ideas may not work in all specific situations. Changes to improve tire dry traction will certainly affect other properties as well, for better or for worse, and this book does not purport to show how these other properties are affected. Also, this book does not purport to address safety and health issues.*

Tread Hardness

A tire tread that has too high a durometer hardness may have poor dry traction. RT: Chapter 12, "Compounding with Carbon Black and Oil," S. Laube, S. Monthey, M-J. Wang, p. 310.

Vulcanized vegetable oil (VVO) at 10 phr as a partial replacement for TDAE processing oil in a tire tread stock imparted better dry handling characteristics than the control. GEN: C. Flanigan, F. Beyer, D. Klekamp, D. Rohweder (Ford Motor Co.), B. Stuck, E. Terrill (ARDL), "Comparative Study of Silica, Carbon Black, and Novel Fillers in Tread Compounds," Paper No. 34 presented at the Fall Meeting of the Rubber Division, ACS, October 11–13, 2011, Cleveland, OH.

High-Styrene SBR

Commonly, high-styrene SBR is used as the base polymer in the treads for high-performance tires to achieve high dry grip properties. GEN: H. Takino, S. Iwama, Y. Yamada, S. Kohjiya, "Carbon Black Dispersion and Grip Property of High-Performance Tire Tread Compound," Presented at ACS Rubber Div. Meeting, Spring, 1996, Paper No. 2.

Silica treads that are mixed at higher temperatures for longer mix times may have higher dry traction and better rolling resistance. GEN: P. Van Raepenbusch, R. Pessina, D. Nichetti (Pirelli), "On the Dependency of Tire Performance and the Degree of Mixing," Paper No. 14 presented at the Spring Meeting of the Rubber Division, ACS, April 4–6, 2000, Dallas, TX.

High Reinforced Carbon Black/Aromatic Oil

Typically a "super SAF" or ultrafine particle size carbon black at 100 phr with a high loading of an aromatic-type oil is used to achieve high dry grip properties for high-performance tires. GEN: H. Takino, S. Iwama, Y. Yamada, S. Kohjiya, "Carbon Black Dispersion and Grip Property of High-Performance

Tire Tread Compound," Presented at ACS Rubber Div. Meeting, Spring, 1996, Paper No. 2.

Carbon Black vs. Silica

Carbon blacks reportedly have an advantage over silica with compounding for the best dry grip or "lock braking under dry conditions." GEN: A. McNeish, "Nanoblacks for Rolling Resistance," Presented at ITEC Meeting, Fall, 2000, Paper No. 23A.

Vinyl Acetate and Dispersion

A vinyl acetate wax can be used in the mixing of high-performance treads with high loadings of ultrafine carbon black in order to achieve improved dispersion and superior dry grip properties at higher temperatures. GEN: H. Takino, S. Iwama, Y. Yamada, S. Kohjiya, "Carbon Black Dispersion and Grip Property of High-Performance Tire Tread Compound," Presented at ACS Rubber Div. Meeting, Spring, 1996, Paper No. 2.

Tire Temperature

The tire temperature has a great effect on the dry grip property of a high-performance tire. As a tire heats up from continuous running at high speeds, its dry grip properties decrease. GEN: H. Takino, S. Iwama, Y. Yamada, S. Kohjiya, "Carbon Black Dispersion and Grip Property of High-Performance Tire Tread Compound," Presented at ACS Rubber Div. Meeting, Spring, 1996, Paper No. 2.

Footprint

Increasing the contact width (having a wider footprint) or a high width-to-diameter for a tire will generally improve the braking friction of a tire. GEN: C. Ettles, J. Shen, "The Influence of Frictional Heating on the Sliding Friction of Elastomers and Polymers," *Rubber Chemistry and Technology*, March–April, 1988, Vol. 61, p. 119.

6.3 Improving Tire Winter Traction or Ice Traction

Traction on ice or snow is very important.

The following ***experimental*** *ideas may provide guidance on ways to improve this tire performance characteristic. Also, all relevant literature sources, including the ones cited below, should be researched and read.* ***Caution:*** *these general experimental ideas may not work in all specific situations. Changes to improve tire winter traction or ice traction will certainly affect other properties as well, for better or for worse, and this book does not purport to show how these other properties are affected. Also, this book does not purport to address safety and health issues.*

Low Glass Transition Temperature Polymers

Generally, for good ice traction, the base rubber should have a relatively low glass transition temperature. Polymers such as natural rubber and *cis*-polybutadiene are known to impart good ice traction. It has been found that tread compounds that have softness and resiliency at low temperatures usually impart better ice traction. GEN: A. Ahagon, T. Kobayashi, M. Misawa, "Friction on Ice," *Rubber Chemistry and Technology*, March-April, 1988, p. 14.

BIMS

It is claimed that the winter traction of a tire tread can be improved if 25 to 50 phr of brominated isobutylene-co-*p*-methylstyrene (BIMS) is substituted for either NR or solution SBR in a tread formulation normally composed of 25 phr BR, 50 phr of S-SBR, and 25 phr of NR with 75 phr of silane-coupled silica. GEN: R. Poulter, Presented at ITEC 2000, Paper No. 12C.

Vulcanized Vegetable Oil (VVO)

Vulcanized vegetable oil (VVO) at 10 phr as a partial replacement for TDAE processing oil in a tire tread stock imparted better winter traction characteristics than the control. GEN: C. Flanigan, F. Beyer, D. Klekamp, D. Rohweder (Ford Motor Co.), B. Stuck, E. Terrill (ARDL), "Comparative Study of Silica, Carbon Black, and Novel Fillers in Tread Compounds," Paper No. 34 presented at the Fall Meeting of the Rubber Division, ACS, October 11-13, 2011, Cleveland, OH.

S-SBR

Dow Chemical reported the development of a new "novel" functionalized S-SBR that should impart better ice grip as measured by tan delta at -10°C. GEN: S. Thiele, D. Bellgardt, M. Holzieg (Dow Chemical), "Novel, Functionalized SSBR for Silica Containing Tires," Paper No. 91 presented at the Fall Meeting of the Rubber Division, ACS, October 16–18, 2007, Cleveland, OH.

Chemically Modified SBR

Chemically modified SBR polymers are reported to impart better ice skid resistance than unmodified ones. F. Suzuki (Nippon Zeon Co.), "Rubbers for Low Rolling Resistance," *Tire Technology International*, 1997, p. 87.

IBR

Consider the use of isoprene butadiene rubber (IBR) for better ice traction and better wet traction. RP: R. Dailey.

Silica and Coupling Agent

Mroczkowski reported a significant improvement in ice traction when he used a relatively high level of coupling agent with precipitated silica as a partial replacement for carbon black. He also noted a significant improvement in rolling resistance without much loss in tread wear or wet traction for an NR/BR/SBR passenger tread blend. GEN: W. Waddell, L. Evans, "Use of Nonblack Fillers in Tire Compounds," *Rubber Chemistry and Technology*, July-August, 1996, Vol. 69, p. 377.

SBR/BR and NR/BR Blends

Natural rubber has the capability to impart a higher friction on a smooth ice surface. However, depending on the carbon black loading, it varies in its temperature range for ice traction. In addition, a rubber blend at equal proportions of SBR and BR has been reported to impart higher friction on a smooth ice surface. An NR/BR blend can also impart a high friction on a smooth ice surface in a broad temperature range. However, the selection of polymer or polymer blends with the type and amount of carbon black and oil loading has a significant effect on the level of ice friction and the temperature range over which this friction is imparted. GEN: K. Grosch, "The Rolling Resistance, Wear and Traction Properties of Tread Compounds," *Rubber Chemistry and Technology*, July-August, 1996, Vol. 69, p. 495.

Softener

In a tread compound, it was found that using a softener that has a solubility parameter close to the solubility parameter of the base rubber will likely improve the ice traction performance. GEN: A. Ahagon, T. Kobayashi, M. Misawa, "Friction on Ice," *Rubber Chemistry and Technology*, March-April, 1988, p. 14.

6.4 Improving Tire Rolling Resistance

With the government requirements for better gasoline consumption (CAFE) in the past 25 years, this tire performance characteristic has become very important.

The following ***experimental*** *ideas may impart lower rolling resistance to a pneumatic tire. Also, all relevant literature sources, including the ones cited below, should be researched and read.* ***Caution:*** *these general experimental ideas may not work in all specific situations. Changes to improve tire rolling resistance will certainly affect other properties as well, for better or for worse, and this book does not purport to show how these other properties are affected. Also, this book does not purport to address safety and health issues.*

Comparison of Base Elastomers

Back in 1977, J. Hunt, J. Walker, and G. Hall ranked various elastomers regarding their effect on rolling resistance. Their RR index was as follows:

Polymer	Rolling Resistance Index
NR	100
High *cis*-BR	94
Solution SBR	93
Emulsion SBR (OE)	90
IIR	73

Since this ranking, there has been a great deal of patent activity with new solution SBR and BR polymers that would probably give a much higher ranking for rolling resistance to these elastomers and still provide acceptable tire traction. GEN: L. Chang, J. Shackleton, "An Overview of Rolling Resistance," *Elastomerics*, March, 1983, p. 22.

BR

Neodymium-catalyzed high *cis*-BR can have very high linearity and very low vinyl content and provide the highest level of crosslinking compared to other BRs, which usually results in the highest rebound values being imparted to its respective vulcanizates. Rolling resistance is reduced significantly through the use of more high *cis*-BR in the tire tread. GEN: Lim Yew Swee (Lanxess), "Benefits of Butadiene Rubber in Natural Rubber-Based Truck and Sidewall," Presented at the India Rubber Exposition and Conference (IRE 2011), January 19, 2011, Chennai, India.

Solution SBR

Consider basing a compound on solution SBR (S-SBR) instead of emulsion SBR (E-SBR) for better rebound and tire rolling resistance. S-SBR is usually more linear than E-SBR. RT: Chapter 7, "General Purpose Elastomers and Blends," G. Day, p. 149.

Other chain-end modifications of S-SBR during the polymerization process have been considered, such as the unique chemistry involving alkyl pyrrolidinones to improve tire rolling resistance. GEN: S. Thiele, S. Knoll (Styron Deutschland GmbH, Merseburg, Germany), "Novel Functionalized SSBR for Silica and Carbon Black Containing Tires," Presented at the Fall Meeting of the Rubber Division, ACS, October 11, 2011, Cleveland, OH.

S-SBR polymers modified with GPMOS (glycidoxypropyltrimethoxysilane) and DMI (dimethylimidazolidinone) during the polymerization process were found to reduce hysteresis loss for silica-based compounds with TEPST coupling agent in order to reduce rolling resistance of tires. GEN: Akira Saito, Haruo Yamada, Takaaki Matsuda, Nobuaki Kubo, Norifusa Ishimura, "Improvement of Rolling Resistance of Silica Tire Compounds by Modified S-SBR," Paper No. 39 presented at the Spring Meeting of the Rubber Division, ACS, April 29, 2002, Savannah, GA.

Functionalizing a high-vinyl solution SBR with carboxylic acid along the backbone during the polymerization process has been shown to improve filler polymer interactions for lower hysteresis with a silica-loaded compound, thus improving tire rolling resistance. GEN: Thomas Gross, Judy Hannay (Lanxess), "New Solution SBRs to Meet Future Performance Demands," Paper No. 11A presented at a meeting of ITEC, September, 2008.

Tin Coupling

Consider using an anionic polymerized BR that has been terminated in the polymerization process by the polymer supplier with a multifunctional agent such as tin tetrachloride. RT: Chapter 7, "General Purpose Elastomers and Blends," G. Day, p. 146.

Chemically Modified S-SBR

S-SBR can be chain-end modified through the addition of tin tetrachloride, silicon tetrachloride, or carbon dioxide during the polymerization process. One reason for chain-end modification is to couple multiple polymer chains together. During mixing with other compounding ingredients such as carbon blacks, the tin-coupled elastomer can decouple chain ends, then react with carbon black aggregates, thus providing lower hysteresis and better tire rolling resistance from the tread. However, this reaction appears generally not to lower the hysteresis at 0°C (32°F), which means that wet traction is not necessarily lost. Thus the compounder can improve the rolling resistance without loss in wet traction. RT: Chapter 7, "General Purpose Elastomers and Blends," G. Day, p. 150. GEN: V. Quiteria, C. Sierra, J. Fatou, C. Calan, L. Fraga, Presented at ACS Rubber Div. Meeting, Fall, 1995, Paper No. 78.; F. Tsutsumi, M. Sakakibara, N. Oshima, "Structure and Dynamic Properties of Solution SBR Coupled with Tin Compounds," *Rubber Chemistry and Technology*, March-April, 1990, Vol. 63, p. 8.

Use S-SBR polymers that have chain ends modified with 4,4-bis(diethylamino)-benzophenone (EAB). This modification increases polymer reactivity with carbon black, which improves tire tread compound rolling resistance. RT: Chapter 7, "General Purpose Elastomers and Blends," G. Day, p. 150.

Use an SBR containing butadiene in the vinyl or 1,2 configuration and with greatly increasing styrene content at the chain ends (SSCP-901 from Shell) with a blend in a tire tread compound to achieve better rolling resistance and maintain wet traction. GEN: L. Chang, J. Shackleton, "An Overview of Rolling Resistance," *Elastomerics*, March, 1983, p. 23.

SIBR

Consider using styrene-isoprene-butadiene rubber (SIBR) for better rolling resistance in a tire tread compound. RT: Chapter 7, "General Purpose Elastomers and Blends," G. Day, pp. 150, 169.

Consider using SIBR polymers that have been modified with tin coupling or EAB for better rolling resistance from a tread compound. RT: Chapter 7, "General Purpose Elastomers and Blends," G. Day, p. 150.

The lower T_g of SIBR relates to tire rolling resistance and fuel economy. GEN: A. Halasa, B. Gross, W. Hsu (Goodyear Tire and Rubber Company), "Multiple Glass Transition Terpolymers of Isoprene, Butadiene, and Styrene," Paper No. 91 presented at the Fall Meeting of the Rubber Division, ACS, October, 2009, Cleveland, OH.

Natural Rubber

Avoid over-masticating or using too strong a chemical peptizer in breaking down the natural rubber "backbone" in order to maintain better dynamic properties and less heat buildup to meet new, tighter tire rolling resistance targets. GEN: C. Clarke, R. Galle-Gutbrecht, M. Hensel, K. Menting, T. Mergenhagen (Struktol), "A New 'Intelligent' Peptizer Concept, Better Dynamic Properties Plus Improved Processing," Paper No. 33 presented at the Fall Meeting of the Rubber Division, ACS, October 16, 2001, Cleveland, OH.

Rubber Blends

Typically, tread compounds are based on a blend of natural rubber, polybutadiene, and one or more solution SBRs to meet specific rolling resistance goals while maintaining the needed traction. RT: Chapter 12, "Compounding with Carbon Black and Oil," S. Laube, S. Monthey, M-J. Wang, p. 318.

Carbon Black Loading

Many times tire rolling resistance may improve if the tread compound has a lower loading of carbon black. GEN: W. Hess, W. Klamp, "The Effects of Carbon Black and Other Compounding Variables on the Tire Rolling Resistance and Traction," *Rubber Chemistry and Technology*, Vol. 56, p. 390.

Carbon Black Surface Area

To improve rolling resistance targets, typically "lower surface area" fully-reinforcing carbon blacks are used, commonly from the N300 types at relatively low loadings such as 40 to 60 phr. RT: Chapter 12, "Compounding with Carbon Black and Oil," S. Laube, S. Monthey, M-J. Wang, p. 318.

Carbon Black Aggregate Size Distribution

Select a grade of carbon black with a broader aggregate size distribution for a tread to achieve better rolling resistance. GEN: W. Hess, W. Klamp, "The Effects of Carbon Black and Other Compounding Variables on the Tire Rolling Resistance and Traction," *Rubber Chemistry and Technology*, Vol. 56, p. 390.

Consider a carbon black with a broad aggregate size distribution because this creates a situation in which there is more efficient packing with reduced interaggregate contact and resulting lower hysteresis. This should result in improved rolling resistance. GEN: R. Swor, "Utilization of Very High Structure Tread Blacks to Lower the Rolling Resistance of U.S. and European Radial Tires," *Tire Technology International*, 1994.

Consider using the new Sid Richardson SR129 (tread grade) or SR401 (non-tread-grade) carbon blacks from a "special reactor" technology; these are called "low hysteresis" carbon blacks, with wider aggregate size distributions and higher structures to impart lower tire rolling resistance. GEN: Leszek Nikiel, Wesley Wampler, Henry Yang, Tom Carlson (Sid Richardson Carbon and Energy Company), "Improved Carbon Blacks for Low Hysteresis Applications in Rubber," Paper No. 93 presented at the Fall Meeting of the Rubber Division, ACS, October 16, 2007, Cleveland, Ohio.

Very High Structure Carbon Black at Low Concentration

Lower rolling resistance from the tread compound can be achieved by using a *lower* concentration of the so-called "very high structure" carbon black in place of a conventional reinforcing carbon black. These special, very high structure blacks (such as CD-2038 from Columbian) have highly branched shaped aggregates with such high void space within that they are reportedly capable of absorbing a high volume in rubber or oil. GEN: R. Swor, "Utilization of Very High Structure Tread Blacks to Lower the Rolling Resistance of U.S. and European Radial Tires," *Tire Technology International*, 1994.

Nanoblacks

Using nanoblacks with higher surface roughness and higher surface activity than regular carbon blacks will significantly reduce hysteresis [cured tan δ at 60°C (140°F) over a wide range of applied strains or deformations] and impart better rolling resistance to the tread and most likely not harm wet traction. GEN: A. McNeish, "Nanoblacks for Rolling Resistance," Presented at the Fall 2000 ITEC Meeting, Paper No. 23A; B. Freund, F. Forster, "New Filler Concepts for Tire Treads with Optimized Traction, Abrasion and Rolling Resistance," Presented at ACS Rubber Div. Meeting, Spring, 1996, Paper No. 4.

Carbon Black Loading and Cure

Lower rolling resistance might be achieved by reducing carbon black loadings in a tread formulation and compensating for the lower physicals by increasing the accelerator/sulfur loadings. GEN: L. Chang, J. Shackleton, "An Overview of Rolling Resistance," *Elastomerics*, March, 1983, p. 24.

Carbon Black Disproportionate Distribution

Sometimes an uneven polymer phase distribution of carbon black between different elastomer components may impart lower rolling resistance. GEN: W. Hess, W. Klamp, "The Effects of Carbon Black and Other Compounding Variables on the Tire Rolling Resistance and Traction," *Rubber Chemistry and Technology*, Vol. 56, p. 390.

Reportedly, by achieving controlled disproportionate distribution of carbon black between elastomers by adding halobutyl rubber late in a mixing cycle with a polymer such as BR, better rolling resistance and good wear may result. This so-called disproportionate distribution technique is also reported to be used when blending natural rubber and 1,2-polybutadiene with similar results. This uneven carbon black distribution reportedly can also be achieved through the use of high Mooney viscosity, labile crosslinked SBR. GEN: L. Chang, J. Shackleton, "An Overview of Rolling Resistance," *Elastomerics*, March, 1983, p. 23.

Carbon Black Chemical Promotors

Use of a carbon black-rubber coupling agent or "chemical promotor" in a carbon black loaded rubber compound may increase resilience and modulus as well as improve abrasion loss. It may also improve tire rolling resistance and wet grip. In the past, such coupling agents as *N*-(2-methyl-2-nitropropyl)-4-nitrosoaniline, *N*-4-dinitroso-*N*-methylaniline, *p*-nitrosodiphenylamine, and *p*-nitroso-*N*,*N*-dimethylaniline were used. These nitroso compounds are no longer used because they give off nitrosamines, which are suspected carcinogens. Therefore other coupling agents have been tried. A new coupling agent *p*-aminobenzenesulfonyl azide (or amine-BSA) has recently been investigated that may impart these performance advantages to a compound, including improved tire rolling resistance. GEN: L. Gonzalez, A. Rodriguez, J. deBenito, A. Marcos, "A New Carbon Black-Rubber Coupling Agent to Improve Wet Grip and Rolling Resistance of Tires," *Rubber Chemistry and Technology*, May-June, 1996, Vol. 69, p. 266.

Recent work has shown that a chemical "coupler" can improve carbon black based rubber compounds by lowering their hysteresis and imparting lower rolling resistance to a tire. These carbon black "couplers" are a 3-thiopropionic acid and ethylene diamine diamide, both of which have been reported to lower hysteresis of a carbon black loaded rubber compound. GEN: James Burrington (Lubrizol Corp.), "Carbon Black Coupler Technology for Low Hysteresis Tire," Paper No. 108 presented at the Fall Meeting of the Rubber Division, ACS, October 16, 2007, Cleveland, OH.

Consider using a special "surface modified" carbon black with in-chain functionalized solution S-SBR to achieve a significant reduction in compound hysteresis and a lowering of tire rolling resistance. GEN: J. Douglas, S. Crossley, J. Hallett, J. Curtis, D. Hardy, T. Cross, N. Steinhauser, A. Lucassen, H. Kloppenburg (Lanxess and Columbian Chemicals), "The Use of a Surface-Modified Carbon Black with an In-Chain Functionalized Solution SSBR as an Alternative to Higher Cost Green Tire Technology," Paper No. 38 presented at the Fall Meeting of the Rubber Division, ACS, October 11, 2011, Cleveland, OH.

Organosilane and Carbon Black

Using an organosilane with a carbon black can improve a tread compound's rolling resistance. GEN: R. Swor, "Utilization of Very High Structure Tread Blacks to Lower the Rolling Resistance of U.S. and European Radial Tires," *Tire Technology International*, 1994.

"Radical Compounding"

Use "radical compounding" to achieve a reduction in rolling resistance while maintaining or improving tread wear. This radical compounding is simply application of the principles of optimum loading with a higher surface area carbon black. First, the optimum loading of a higher surface area (finer particle size) carbon black is lower than for a lower surface area carbon black. Second, hysteresis varies by the square of the loading but changes linearly with surface area. Thus a lower loading of a carbon black with a very high surface area may result in a reduction in tire rolling resistance and an improvement in tire wear resistance. RT: Chapter 12, "Compounding with Carbon Black and Oil," S. Laube, S. Monthey, M-J. Wang, p. 316.

Silica Loading

Rolling resistance was reportedly significantly reduced when a TESPT-modified silica was substituted for an ISAF carbon black in a tread compound with no significant loss in traction. The use of DPG as a "kicker" with a sulfenamide cure also helps. RT: Chapter 13, "Precipitated Silica and Non-black Fillers," W. Waddell, L. Evans, pp. 340, 342.

S. Ahmad and R. Schaefer reported in a patent a significant improvement in the rolling resistance of a passenger tire when they partially substituted precipitated silica and a silane coupling agent for carbon black in an SBR/BR blend without experiencing a significant loss in wet or dry traction. GEN: W. Waddell, L. Evans, "Use of Nonblack Fillers in Tire Compounds," *Rubber Chemistry and Technology*, July-August, 1996, Vol. 69, p. 377.

Use higher loadings of an effective organosilane coupling agent (such as TESPT) with a silica loading and allow an effective silanization reaction to occur during mixing to establish better rubber-filler interaction and correspondingly reduce filler-filler interaction (which is energy absorbing) to achieve low vulcanizate hysteresis and better tire rolling resistance. GEN: A. McNeish, "Nanoblacks for Rolling Resistance," Presented at ITEC Meeting, Fall, 2000, Paper No. 23A.

Chemically modified solution SBR used with silica reportedly gives a better balance between wet skid resistance and rolling resistance. NS-116® reportedly also gives a better rolling resistance vs. wet traction performance trade-off when used with silica. GEN: F. Suzuki (Nippon Zeon), "Rubbers for Low Rolling Resistance," *Tire Technology International*, 1997, p. 87.

Silica/Silane

Consider using a highly dispersible grade of precipitated, hydrated silica to achieve better reduction in hysteresis and rolling resistance. GEN: S. Daudey, L. Guy (Rhodia), "High Performance Silica Reinforced Elastomers from Standard Technology to Advanced Solutions," Paper No. 37 presented at the Fall Meeting of the Rubber Division, ACS, October 11, 2011, Cleveland, OH.

Also, for silica-based compounds, consider an optimal loading of Nanoprene® BM750H VP RW from Lanxess, a BR-acrylate terpolymer additive, to reduce hysteresis and improve rolling resistance. GEN: C. Flanigan, L. Beyer, D. Klekamp, D. Rohweder (Ford), B. Stuck, E. Terrill (ARDL), "Comparative Study of Silica, Carbon Black and Novel Fillers in Tread Compounds," Paper No. 34 presented at the Fall Meeting of the Rubber Division, ACS, October 11, 2011, Cleveland, OH.

Consider using a special zinc soap such as ZB47 from Struktol in a TESPT-treated silica-filled natural rubber based compound in order to improve the heat buildup (HBU) and blowout (BO) times from Firestone Flexometer testing as well as improve tire rolling resistance. GEN: Kwang-Jea Kim, John Vanderkooi (Struktol), "Effects of Zinc Soaps on TESPT and TESPD-Silica Mixtures in Natural Rubber Compounds," Paper No. 70 presented at the Fall Meeting of the Rubber Division, ACS, October 8, 2002, Pittsburgh, PA.

It has been reported that using "moisture-treated" silica with organosilanes can reduce the cured heat buildup of the silica–organosilane compound and improve rolling resistance of tire compounds. GEN: Kwang-Jea Kim, John VanderKooi (Struktol), "Moisture Level Effects on Hydrolysis Reaction in TESPD/Silica/CB/S-SBR Compound," Paper No. 57 presented at the Fall Meeting of the Rubber Division, ACS, October 5, 2004, Columbus, OH.

Substituting the new coupling agent AEO-MPES (alkylethoxylate-substituted mercaptopropylethoxysilane, Degussa's VP Si 363) in place of TESPT can reduce hysteresis by as much as 13% in a silica-loaded compound for an improvement in tire rolling resistance. GEN: O. Klockmann, A. Blume, A. Hasse, "Fuel Efficient Silica Tread Compounds with a New Mercaptosilane: A Practical Way to Improve Its Processing," Paper No. 87 presented at the Fall Meeting of the Rubber Division, ACS, October 16, 2007, Cleveland, OH.

ATH Filler

Alumina trihydrate (ATH) was used as a partial substitute for reinforcing carbon black in a tread formulation to achieve an improvement in rolling resistance and wet traction with only a small reported loss in tread wear. GEN: H. Mouri, K. Akutagawa, "Improved Tire Wet Traction Through the Use of Mineral Fillers," *Rubber Chemistry and Technology*, November–December, 1999, Vol. 72, p. 960.

Sepiolite Filler

Sepiolite, a mineral filler that is a form of magnesium silicate from Spain, reportedly can be used as a partial substitute for carbon black (up to 30%) in tire treads to reduce compound costs. It can reportedly improve wet traction.

If used with silane coupling agents, it may also possibly improve tire rolling resistance. GEN: L. Hernandez, L. Rueda, C. Anton, "Magnesium Silicate Filler in Rubber Tread Compounds," *Rubber Chemistry and Technology,* September-October, 1987, Vol. 60, p. 606.

Nanofillers

Consider experimenting with the new carbon nanotubes (CNT), organically modified layered silicates (OC), graphite with high surface area (G), or chemically reduced graphite oxide (CRGO) to achieve lower hysteresis in a rubber compound and better tire rolling resistance. GEN: M. Galimbeth, M. Coombs, V. Cipolletti, L. Giannini, L. Conzatti, T. Ricco, M. Mauro, G. Guerra (Politecnico di Milano, Department of Chemistry, and Pirelli Tyre Study), "Nano and Nano-structured Fillers and Their Synergistic Behavior in Rubber Composites Such as Tires," Paper No. 35 presented at the Fall Meeting of the Rubber Division, ACS, October 11, 2011, Cleveland, OH.

Reinforcing Resins

Use reinforcing resins to make a hard "subtread" compound to improve the rolling resistance of a tire. Use a reinforcing novolak phenol-formaldehyde resin with a methylene donor such as hexamethylenetetramine (HMT) or hexamethoxymethylomelamine (HMMM). These two ingredients *in situ* react with each other during the normal vulcanization of the rubber compound. The use of this reinforcing resin and a methylene donor in a rubber compound can significantly increase the hardness of the vulcanizate. RT: Chapter 18, "Tackifying, Curing, and Reinforcing Resins," B. Stuck, p. 440

Processing Oil

With the 2010 banning of aromatic oils for use in tires in Europe, SBR-aromatic oil masterbatches have been eliminated for this use. The new "environmentally friendly" oil substitutes (such as medium extracted solvate or MES) used to make these new oil masterbatches have lower glass transition temperature (T_g) values than the discontinued aromatic oils. Therefore, some of these new substitutes should impart lower hysteresis to the cured compound and better tire rolling resistance. GEN: Rudiger Engehausen (Bayer AG), "Overview of New Developments in BR and SBR and Their Influence on Tire-Related Properties," Paper No. 37 presented at the Spring Meeting of the Rubber Division, ACS, April 29, 2002, Savannah, GA.

Tire Design

Tires of radial construction have better rolling resistance than tires of a bias construction because the radial carcass is more flexible. As the radial carcass does not have large interply strains, there is less energy dissipation to lead to poorer rolling resistance. GEN: W. Hess, W. Klamp, "The Effects of Carbon Black and Other Compounding Variables on the Tire Rolling Resistance and Traction," *Rubber Chemistry and Technology*, Vol. 56, p. 390.

Reduce the gage thickness of the tread to reduce rolling resistance. GEN: L. Chang, J. Shackleton, "An Overview of Rolling Resistance," *Elastomerics*, March, 1983, p. 20; W. Hess, W. Klamp, "The Effects of Carbon Black and Other Compounding Variables on the Tire Rolling Resistance and Traction," *Rubber Chemistry and Technology*, Vol. 56, p. 390.

Tire Load

Tire rolling resistance improves when tire load is decreased. GEN: W. Hess, W. Klamp, "The Effects of Carbon Black and Other Compounding Variables on the Tire Rolling Resistance and Traction," *Rubber Chemistry and Technology*, Vol. 56, p. 390.

Tire Inflation Pressure

Design a tire to operate with a higher air pressure. GEN: L. Chang, J. Shackleton, "An Overview of Rolling Resistance," *Elastomerics*, March, 1983, p. 20; W. Hess, W. Klamp, "The Effects of Carbon Black and Other Compounding Variables on the Tire Rolling Resistance and Traction," *Rubber Chemistry and Technology*, Vol. 56, p. 390.

Vehicle Speed

Tire rolling resistance improves when vehicle speeds are decreased. GEN: W. Hess, W. Klamp, "The Effects of Carbon Black and Other Compounding Variables on the Tire Rolling Resistance and Traction," *Rubber Chemistry and Technology*, Vol. 56, p. 390.

Tire Running Temperature

A tire's running temperature has an effect on imparted rolling resistance. A higher running temperature may actually impart less rolling resistance. GEN: L. Chang, J. Shackleton, "An Overview of Rolling Resistance," *Elastomerics*, March, 1983, p. 18.

Appendix 1: Blends

A.1 Introduction

Many different types of elastomers are used today in commercial rubber compounds. Many of these commercial compounds are based not just on one polymer, but on a blend of two or more different elastomers. The reasons for formulating with two or more different elastomers are to (1) reduce compound costs, (2) improve processability, and/or (3) achieve a better balance in cured physical properties.

Cost: The prices of different raw elastomers can vary greatly. Sometimes a higher priced specialty elastomer can be blended (or extended) with a less expensive general purpose rubber to reduce the cost of the compound without taking too severe a loss on cured physical properties. However, this trade-off may not always be technically feasible.

Processability: Some elastomers with very high stereospecificity or very narrow molecular weight distributions may not process well in the factory when used by themselves in compounding. Sometimes when these types of elastomers are blended with other elastomers, the processing characteristics of the compound will improve greatly.

Physical Properties: Often more than one elastomer is chosen for a compound to achieve a better balance in cured physical properties, which might not be possible if only one elastomer were used in the formulation. This is quite common.

Blending Limitations

As one remembers from basic chemistry, "likes dissolve likes." This rule can partially be applied to rubber compounding. Elastomers with very similar chemical structures tend to be more compatible with each other or even miscible. On the other hand, if raw elastomers' chemical structures result in differences in imparted solubility parameters, polarity, hydrogen bonding, and so forth, then problems can result in blending these elastomers. Chemical differences between

RT: *Rubber Technology, Compounding and Testing for Performance*. 2nd ed. Edited by J. S. Dick, Hanser Publications, 2009 ■ GEN: General References from the Literature ■ RP: Review Panel (Advisors for this book) ■ IMPORTANT: The information contained herein is experimental in nature and is meant for individuals with advanced scientific training and rubber compounding experience. All safety precautions should be taken when replicating or applying this information. The author makes no representation or warranty, either expressed or implied.

polymers therefore can sometimes prevent them from blending together well. Sometimes through the use of different mixing techniques or special additives such as certain process aids, homogenizers, or compatibilizers, a better blend can be achieved. However, sometimes the thermodynamic barriers may just be too great to overcome. Even when a compound is based on two different elastomers that are close in chemical structure and have very similar solubility parameters, they still may not be completely miscible. They will instead form microscopic continuous and discontinuous domains from the mixing process. This happens because commercial elastomers consist of macromolecules that are relatively large.

Rubber compounds based on two different elastomer types are usually two-phase systems with continuous and discontinuous domains. Insoluble compounding ingredients, such as carbon blacks, have different affinities for each phase. Therefore, the type of base elastomers selected for a new compound will result in a particular distribution of carbon black between these phases. The nature of this carbon black distribution can greatly affect the physical properties of the compound. However, through phase mixing techniques, this distribution can be altered depending on what particular cured physical property targets the compounder is trying to meet.

Another problem that is commonly encountered when designing a rubber compound to be based on different elastomers is covulcanization. Different types of elastomers have different degrees of unsaturation and may normally require a different curative package. Also, rubber curatives may have different degrees of solubility in different elastomer domains (whether continuous or discontinuous) that result from the blending of these different elastomers. The challenge is to select the appropriate cure system that will covulcanize both elastomers in a blend in order to achieve acceptable cured physical properties. If too much curative migrates out of one elastomer domain into another elastomer domain where it is more soluble, then the elastomer domain having this loss in curative concentration will be severely undercured, causing a very low crosslink density in that domain and resulting in poor cured physical properties for the compound. In these cases, a kind of negative synergism results where the resulting cured physical properties of the blend are much worse than the cured physical properties of the rubber compound when 100 parts of either one or the other elastomer is used. Therefore, it is important that the selected cure system work well with both elastomers in the blend.

The following gives examples of how many of these different blends are used in various applications in the rubber industry. This overview covers both two-way blends and triblends. References are included that provide the literature source(s) for each application.

Blend Matrix Chart Referencing Section Numbers for Applications and Literature

	NR	IR	SBR	BR	CIIR	BIIR	BIMS	IIR	CR	NBR	XNBR	PVC	CSM	EPDM	Q	LDPE	EVA	ECO	T
NR	-																		
IR	A1.5	-																	
SBR	A1.1	A1.2	-																
BR	A1.4		A1.3	-															
CIIR	A1.6				-														
BIIR	A1.7					-													
BIMS	A1.8						-												
IIR	A1.9							-											
CR			A1.25	A1.26				A1.10	-										
NBR	A1.16	A1.14	A1.11	A1.15						-									
XNBR											-								
PVC										A1.12	A1.13	-							
CSM				A1.27									-						
EPDM	A1.20	A1.19	A1.17						A1.18				A1.21	-					
Q														A1.22	-				
LDPE														A1.23		-			
EVA														A1.24			-		
ECO									A1.28									-	
T										A1.29									-

RT: *Rubber Technology, Compounding and Testing for Performance*. 2nd ed. Edited by J. S. Dick, Hanser Publications, 2009 ■ GEN: General References from the Literature ■ RP: Review Panel (Advisors for this book) ■ IMPORTANT: The information contained herein is experimental in nature and is meant for individuals with advanced scientific training and rubber compounding experience. All safety precautions should be taken when replicating or applying this information. The author makes no representation or warranty, either expressed or implied.

A.1.1 SBR/NR Blends

Tires

Light Truck Treads, P. Ciullo, N. Hewitt, *The Rubber Formulary*, Noyes, Norwich, NY, 1999.

Tire Chaffer Rim Flange, P. Ciullo, N. Hewitt, *The Rubber Formulary*, Noyes, Norwich, NY, 1999, p. 77.

Bead Filler, L. Goettler, J. Sezna, P. DiMauro, "Short Fiber Reinforcement of Extruded Rubber Profiles," *Rubber World*, October, 1982, p. 33.

Textile Adhesion Compound, *Hi-Sil Silic Pigments Formulary*, PPG Industries.

Tire Adhesion Compound, Technical Symposiums, 1977–1978, Akron Rubber Group, Inc.

Carcass Coat, N. Hewitt, "Developing New Cord Adhesion Test Methods," *Rubber & Plastics News*, October 7, 1985; A. Werner, "Emulsion Polybutadiene for Tire Carcass Applications," *Rubber India*, April, 1971, p. 13; A. Barbour, "New Clays Improve Compound Properties, *Rubber & Plastics News*, August 24, 1987.

Motorcycle Tire Tread, R. Pandit, "Rubber Compounds for Cycle Tire Manufactures," *Rubber India*, March, 1970, p. 43.

Tire Cushion and Repair Gum, R. Babbit, *The Vanderbilt Rubber Handbook*, R. T. Vanderbilt Co., Norwalk, CT, 1978.

Retread for Truck Tires, *Synaprene Formulary*, Synthetics & Chemicals Ltd., Bombay, India, 1967.

Earthmover Retread Rubber, R. Babbit, *The Vanderbilt Rubber Handbook*, R. T. Vanderbilt Co., Norwalk, CT, 1978.

Belts

Conveyor Belt Cover, "Technical Notes," *Natural Rubber Technology*, 1979, Vol. 10, Part 1, 1979, p. 21; R. Babbit, *The Vanderbilt Rubber Handbook*, R. T. Vanderbilt Co., Norwalk, CT, 1978; "Conveyor Belt Formulation," *Natural Rubber Technology*, 1971, No. 4.

Conveyor Belt Friction Compound, R. Babbit, *The Vanderbilt Rubber Handbook*, R. T. Vanderbilt Co., Norwalk, CT, 1978; "Conveyor Belt Formulation," *Natural Rubber Technology*, 1971, No. 4.

V-Belt Non-oil Resistance, R. Babbit, *The Vanderbilt Rubber Handbook*, R. T. Vanderbilt Co., Norwalk, CT, 1978.

V-Belt Non-oil Resisting Cushion, R. Babbit, *The Vanderbilt Rubber Handbook*, R. T. Vanderbilt Co., Norwalk, CT, 1978.

V-Belt Base, H. Railsback, N. Stumpe, Jr., "Medium Vinyl Polybutadiene/SBR Blends," *Rubber India*, November 1976, p. 14.

Hose

Welding Hose Cover, *Synaprene Formulary*, Synthetics & Chemicals Ltd., Bombay, India, 1967.

Welding Hose Tube, *Synaprene Formulary*, Synthetics & Chemicals Ltd., Bombay, India, 1967.

Braided Water Hose, *Synaprene Formulary*, Synthetics & Chemicals Ltd., Bombay, India, 1967.

High-Pressure Braided Spray Hose Tube, *Synaprene Formulary*, Synthetics & Chemicals Ltd., Bombay, India, 1967.

Vacuum Brake Hose, *Synaprene Formulary*, Synthetics & Chemicals Ltd., Bombay, India, 1967.

Boiler Wash-Out Hose, *Synaprene Formulary*, Synthetics & Chemicals Ltd., Bombay, India, 1967.

Lab Tubing, *Synaprene Formulary*, Synthetics & Chemicals Ltd., Bombay, India, 1967.

Shoes

Footwear, Knee Boot Upper Compound, *Synaprene Formulary*, Synthetics & Chemicals Ltd., Bombay, India, 1967.

Footwear, Microcellular Compound, *Synaprene Formulary*, Synthetics & Chemicals Ltd., Bombay, India, 1967.

Footwear, Injection Molded Outsole, R. Babbit, *The Vanderbilt Rubber Handbook*, R. T. Vanderbilt Co., Norwalk, CT, 1978.

Footwear, Upper for Oxford Rainwear, *Synaprene Formulary*, Synthetics & Chemicals Ltd., Bombay, India, 1967.

Footwear, Calendered Outsole, R. Babbit, *The Vanderbilt Rubber Handbook*, R. T. Vanderbilt Co., Norwalk, CT, 1978.

Footwear, Boot Upper, R. Babbit, *The Vanderbilt Rubber Handbook*, R. T. Vanderbilt Co., Norwalk, CT, 1978.

Footwear, Calendered, R. Babbit, *The Vanderbilt Rubber Handbook*, R. T. Vanderbilt Co., Norwalk, CT, 1978.

Sole Compound (High-Styrene Resin), P. Ciullo, N. Hewitt, *The Rubber Formulary*, Noyes, Norwich, NY, 1999, p. 144.

Footwear, Heavy-Duty Black Soling Compound (Injection Molding, High-Styrene Resin), "Waffle Sponge Carpet Underlay: Formulation and Method of Manufacture," *Natural Rubber Technology*, Vol. 7, Part 3, 1976, p. 66.

Seals and Gasket

Refrigerator Gasket, *Synaprene Formulary*, Synthetics & Chemicals Ltd., Bombay, India, 1967.

Gaskets and Rubber Fasteners, *Synaprene Formulary*, Synthetics & Chemicals Ltd., Bombay, India, 1967.

Trunk Lid Seals, *Synaprene Formulary*, Synthetics & Chemicals Ltd., Bombay, India, 1967.

Rollers

Soft Rubber Roller Compound, *Synaprene Formulary*, Synthetics & Chemicals Ltd., Bombay, India, 1967.

Roll Compound, P. Ciullo, N. Hewitt, *The Rubber Formulary*, Noyes, Norwich, NY, 1999, p. 137.

Motor Mounts and Bushings

Shock Absorber Bushing, *Synaprene Formulary*, Synthetics & Chemicals Ltd., Bombay, India, 1967.

Automotive Engine Mount, *Synaprene Formulary*, Synthetics & Chemicals Ltd., Bombay, India, 1967.

45 Duro Motor Mount, P. Ciullo, N. Hewitt, *The Rubber Formulary*, Noyes, Norwich, NY, 1999, p. 102.

Tank Lining

Tank Lining Repair Compound, "Questions and Answers," *Rubber India*, April, 1985, p. 35.

Ebonite Lining, *Synaprene Formulary*, Synthetics & Chemicals Ltd., Bombay, India, 1967.

Athletic Goods

Play Ball, *Synaprene Formulary*, Synthetics & Chemicals Ltd., Bombay, India, 1967.

Football Bladder, *Synaprene Formulary*, Synthetics & Chemicals Ltd., Bombay, India, 1967.

Proofing Compound for DT Cloth, *Synaprene Formulary*, Synthetics & Chemicals Ltd., Bombay, India, 1967.

Miscellaneous Applications

Cable Compound, *Synaprene Formulary*, Synthetics & Chemicals Ltd., Bombay, India, 1967.

Feed Pipe for Locomotives, *Synaprene Formulary*, Synthetics & Chemicals Ltd., Bombay, India, 1967.

Hot Water Bottles, *Synaprene Formulary*, Synthetics & Chemicals Ltd., Bombay, India, 1967.

A.1.2 SBR/IR Blends

Shoes

Translucent Soling, *Hi-Sil Silica Pigments Formulary*, PPG Industries.

Footwear, Quality Heel, R. Babbit, *The Vanderbilt Rubber Handbook*, R. T. Vanderbilt Co., Norwalk, CT, 1978.

Belts

Conveyor Belt Compound, "New B. F. G. (Thiocarbamyl Sulfenamide) Accelerator," *Rubber World*, March, 1974, p. 37.

A.1.3 SBR/BR Blends

Tires

Wet Traction Treads, P. Ciullo, N. Hewitt, *The Rubber Formulary*, Noyes, Norwich, NY, 1999, p. 134.

Passenger Treads, P. Ciullo, N. Hewitt, *The Rubber Formulary*, Noyes, Norwich, NY, 1999, pp. 125, 126, 124.

High Performance Treads, P. Ciullo, N. Hewitt, *The Rubber Formulary*, Noyes, Norwich, NY, 1999, p. 129.

Bias Sidewall, D. Miller, R. Bessent, J. Kuczkowski, “Long Term Antiozonant Protection of Sidewalls,” *Rubber World*, October, 1985, p. 31.

Camelback, High Quality, *Synaprene Formulary*, Synthetics & Chemicals Ltd., Bombay, India, 1967.

Retread Rubber, Precured, R. Babbit, *The Vanderbilt Rubber Handbook*, R. T. Vanderbilt Co., Norwalk, CT, 1978.

Heavy Duty Retread, R. Babbit, *The Vanderbilt Rubber Handbook*, R. T. Vanderbilt Co., Norwalk, CT, 1978.

Passenger Treads, ASTM E1136, A Radial Standard Reference Test Tire, 1988; A. Patel, J. Byers, “Influence of Tread Grade Carbon Blacks, at Optimum Loadings, on Compound Properties,” *Elastomerics*, August, 1980, p. 17; Technical Symposiums, Akron Rubber Group Inc., 1975–1976; C. Stacy, P. Johnson, G. Kraus, “Effect of Carbon Black Structure Aggregate Size Distribution on Properties of Reinforced Rubber,” *Rubber Chemistry and Technology*, Vol. 48, p. 538; B. Boonstra, “Resistivity of Unvulcanized Compounds of Rubber and Carbon Black,” *Rubber Chemistry and Technology*, Vol. 50, p. 194; E. Dizon, L. Papazian, “The Processing of Filler-Reinforced Rubber,” *Rubber Chemistry and Technology*, Vol. 50, p. 765; A. Veith, V. Chirico, “A Quantitative Study of the Carbon Black Reinforcement System for Tire Tread Compounds, *Rubber Chemistry and Technology*, Vol. 52, p. 748; W. Hess, W. Klamp, “The Effects of Carbon Black and Other Compounding Variables on Tire Rolling Resistance and Traction,” *Rubber Chemistry and Technology*, Vol. 56, p. 390; G. Day, S. Futamura, “Comparing Polymer Effects on Tread Compounds,” *Rubber and Plastics News*, October 6, 1986, p. 74; B. Shank, “Compounding for Quality and Efficiency,” *Elastomerics*, July, 1987, p. 17; A. Werner, “Emulsion BR Improves, Modifies Specialty, General-Use Elastomers,” *Rubber & Plastics News*, December 17, 1984, p. 22; B. Chiago, R. Hallman, “Polymer Blends Produce Quality Tread Rubber,” *Rubber & Plastics News*, February 20, 1978, p. 18; J. Byers, A. Patel, “Carbon Black’s Influence on Treadwear and Hysteresis,” *Rubber World*, June, 1983, p. 21; B. Kastein, “Peptizers and Other Processing Aids,” *Rubber World*, May, 1982, p. 38; K. Moore, “A Practical System for Elastomer Substitution,” *Rubber & Plastics News*, May 15, 1989, p. 16; M. Deviney, L. Whittington, B. Corman, “Migration of Oil in Elastomers,” *Rubber World*, January, 1972.

Heavy Duty Truck Tread, D. Barnard, C. Baker, I. Wallace, “Natural Rubber Compounds for Truck Tires,” *Rubber Chemistry and Technology*, Vol. 58, p. 740.

Belts

Conveyor Belt Compound, A. Werner, "Emulsion BR Improves, Modifies Specialty, General-Use Elastomers," *Rubber & Plastics News*, December 17, 1984.

Belt Friction, Skim and Coat Stock, P. Ciullo, N. Hewitt, *The Rubber Formulary*, Noyes, Norwich, NY, 1999, p. 156.

Conveyor Belt Compound, A. Werner, "Emulsion BR Improves, Modifies Specialty, General-Use Elastomers," *Rubber & Plastics News*, December 17, 1984.

Hose

General Purpose Hose Tube and Cover, A. Werner, "Emulsion PB Blends for Mechanicals," *Rubber India*, December 1970, p. 13.

Gaskets

Pipe Gasket Compound, A. Werner, "Emulsion BR Improves, Modifies Specialty, General-Use Elastomers," *Rubber & Plastics News*, December 17, 1984.

Shoes

Sole Compound for Injection Molding, R. Hallman, M. Cavicchia, P. Stricharczuk, "Elastomers for Injection Moulding," *Rubber India*, 1973, p. 12.

Firm Cellular Soling, Footwear, P. Ciullo, N. Hewitt, *The Rubber Formulary*, Noyes, Norwich, NY, 1999, p. 749.

Footwear Crepe (High-Styrene Resin), R. Babbit, *The Vanderbilt Rubber Handbook*, R. T. Vanderbilt Co., Norwalk, CT, 1978.

A.1.4 NR/BR Blend

Tires

Low Rolling Resistance Truck Tread, P. Ciullo, N. Hewitt, *The Rubber Formulary*, Noyes, Norwich, NY, p. 108.

Tire Undertread, Technical Symposiums, Akron Rubber Group, Inc., 1983–1984.

Tire Toe Strip, P. Ciullo, N. Hewitt, *The Rubber Formulary*, Noyes, Norwich, NY, p. 136.

Black Sidewalls, W. Fischer, D. Young, "Contributions of Innerliners and Sidewalls to Tubeless Radial Ply Tire Performance," *Rubber Industry*, August,

1975, p. 140; P. Ciullo, N. Hewitt, *The Rubber Formulary*, Noyes, Norwich, NY, p. 75; R. Babbit, *The Vanderbilt Rubber Handbook*, R. T. Vanderbilt Co., Norwalk, CT, 1978; *Compounders' Pocket Book*, Monsanto Corp., Akron, OH, 1981; L. Walker, J. Luecken, "Antidegradants for Ozone and Fatigue Resistance: Laboratory and Tire Tests," *Elastomerics*, May, 1980, p. 36; D. Kong, J. White, "Inflation Characteristics of Unvulcanized Gum and Compounded Rubber Sheets," *Rubber Chemistry and Technology*, Vol. 59, p. 315; L. Walker, W. Helt, "High-Temperature Curing of Radial Passenger Tires," *Rubber Chemistry and Technology*, Vol. 59, p. 285.

Steel Cord Adhesion Compound, H. Ehrend, K. Morche, "Polymer-Bound Rubber Chemicals: Some Aspects of Optimized Polymer Binders," *Rubber World*, October, 1984, p. 24.

Truck Body Plies, Bias, R. Babbit, *The Vanderbilt Rubber Handbook*, R. T. Vanderbilt Co., Norwalk, CT, 1978.

Passenger Body Plies, Outer Ply, R. Babbit, *The Vanderbilt Rubber Handbook*, R. T. Vanderbilt Co., Norwalk, CT, 1978.

Motorcycle Tire Compound, S. Malik, "Rubber Compounding Natural & Polybutadiene Rubber: Technique & Technology," *Rubber India*, April, 1985, p. 23.

Retread for Winter Car Tires, "Winter Car Tires: Retread Formulation for Good Grip and Wear Resistance," *Natural Rubber Technology*, Vol. 9, Part 2, 1978, p. 41.

Retread for Light Truck Tires, "Winter Car Tires, Tire Retread Formulations, *Natural Rubber Technology*, Rubber Developments Supplement, 1971, No. 11.

Winter Treads, Derham, R. Newell, P. Swift, "The Use of Silica for Improving Tread Grip in Winter Tires," *Natural Rubber Technology*, Vol. 19, No. 1, 1988, p. 1; *Natural Rubber Technology*, 1970, No. 9, p. 2.

Treads, P. Hamed, L. Walker, "Comparative Properties of Natural Rubber and *cis*-1,4-Polyisoprene," *Rubber World*, July, 1971, p. 51.

OTR and Truck Tread, *Hi-Sil Silica Pigments Formulary*, PPG Industries.

Truck Treads, Elastomeric Materials, *Desk-Top Data Bank*, Cordura Publications, Inc., 1977; W. Hess, P. Vegvari, R. Swor, "Carbon Black in NR/BR Blends for Truck Tires," *Rubber Chemistry and Technology*, Vol. 58, p. 350; P. Vegvari, W. Hess, V. Chirico, "Measurement of Carbon Black Dispersion in Rubber by Surface Analysis," *Rubber Chemistry and Technology*, Vol. 51, p. 817; J. Cardina, "The Determination of Rubber in Atmospheric Dusts," *Rubber Chemistry and Technology*, Vol. 46, p. 232; C. Baker, I. Gelling, I. Wallace, "Advances in Natural Rubber for Tires: Compounding for Improved Wear," *Elastomerics*, July, 1989, p. 20; D. Barnard, C. Baker, I. Wallace, "NR Compounds for Truck Tires," *Natural Rubber Technology*, Vol. 16, Part 1,

1985, Publication 1054; S. Malik, "Rubber Compounding Natural & Polybutadiene Rubber: Technique and Technology," *Rubber India*, April, 1985, p. 23; A. Patel, J. Byers, "Influence of Tread Grade Carbon Blacks, at Optimum Loadings, on Compound Properties," *Elastomerics*, August, 1980, p. 17; T. Kempermann, "The Relationship Between Heat Buildup and the Chemistry of Crosslinking Systems," *Rubber Chemistry and Technology*, Vol. 55, p. 391.

Belts

Conveyor Belt Cover Stock, R. Babbit, *The Vanderbilt Rubber Handbook*, R. T. Vanderbilt Co., Norwalk, CT, 1978.

P. Ciullo, N. Hewitt, *The Rubber Formulary*, Noyes, Norwich, NY, p. 91.

Shoes

Soling, 80 Duro, P. Ciullo, N. Hewitt, *The Rubber Formulary*, Noyes, Norwich, NY, p. 117.

A.1.5 NR/IR Blends

Shoes

Heels, *Compounders' Pocket Book*, Monsanto Corp., Akron, OH, 1981.

Footwear, Outsole, R. Babbit, *The Vanderbilt Rubber Handbook*, R. T. Vanderbilt Co., Norwalk, CT, 1978.

Tires

Skim (Belt) Compound, L. Walker, W. Helt, "High-Temperature Curing of Radial Passenger Tires," *Rubber Chemistry and Technology*, Vol. 59, p. 285.

Radial Sidewall, D. Miller, R. Dessent, J. Kuczkowski, "Long Term Antiozonant Protection of Sidewalls," *Rubber World*, October, 1985, p. 31.

A.1.6 NR/CIIR Blends

Tires

Innerliner, W. Fischer, D. Young, "Contributions of Innerliners and Sidewalls to Tubeless Radial Ply Tire Performance," *Rubber Industry*, August, 1975,

p. 140; P. Ciullo, N. Hewitt, *The Rubber Formulary*, Noyes, Norwich, NY, pp. 193, 195; A. Werner, "SBR Polymers Help Users in Solving Rubber Compounding Problems," *Rubber & Plastics News*, August 16, 1982, p. 20; F. O'Connor, J. Slinger, "Processing Aids: The All-Inclusive Category," *Rubber World*, October, 1982, p. 19; H. Barager, "High Performance Innerliners Through the Use of Neoprene Blends," *Rubber World*, March, 1984, p. 30.

Tire Cover Strip, P. Ciullo, N. Hewitt, *The Rubber Formulary*, Noyes, Norwich, NY, p. 76.

Large Tire Inner Tubes, "Tire Inner Tubes: Formulations Suitable for Large Tires," *Natural Rubber Technology*, Vol. 13, Part 4, 1982, p. 87.

White Sidewall, "Tire Inner Tubes: Formulations Suitable for Large Tires," *Natural Rubber Technology*, Vol. 13, Part 4, 1982, p. 90.

A.1.7 NR/BIIR Blends

Sports Apparel

Diving Suit Proofing Compound, P. Ciullo, N. Hewitt, *The Rubber Formulary*, Noyes, Norwich, NY, p. 104.

Squash Ball, 45 Duro, P. Ciullo, N. Hewitt, *The Rubber Formulary*, Noyes, Norwich, NY, p. 199.

Tires

Innerliner, R. Babbit, *The Vanderbilt Rubber Handbook*, R. T. Vanderbilt Co., Norwalk, CT, 1978.

A.1.8 NR/BIMS Blends

Tires

White Sidewall, P. Ciullo, N. Hewitt, *The Rubber Formulary*, Noyes, Norwich, NY, p. 708.

Motor Mounts

Heat-Resistant Engine Mount, P. Ciullo, N. Hewitt, *The Rubber Formulary*, Noyes, Norwich, NY, p. 710.

A.1.9 NR/IIR Blends

Sports Apparel

Squash Ball, *Compounders' Pocket Book*, Monsanto Corp., Akron, OH, 1981.

A.1.10 CR/IIR Blends

Hose

Hose Jacket Compound, J. Horvath, "High Performance Nitrile Rubber Automotive Fuel Hose," *Rubber Chemistry and Technology*, Vol. 52, p. 883.

A.1.11 NBR/SBR Blends

Hose

Hydraulic Hose Tube, J. Dunn, R. Vara, "Oil Resistant Elastomers for Industrial Hose Applications," *Rubber India*, November, 1984, p. 11.

General Purpose Hydraulic Service, P. Ciullo, N. Hewitt, *The Rubber Formulary*, Noyes, Norwich, NY, p. 390.

High-Pressure Hose, P. Ciullo, N. Hewitt, *The Rubber Formulary*, Noyes, Norwich, NY, p. 405.

Belts

Belt Cover, G. Blackshaw, R. Burkey, H. Schwarz, D. Kershaw, R. Melley, "Designing High Performance Compounds for Severe Mining Service Conditions," *Rubber World*, p. 18.

Conveyor Belt Cover, G. Blackshaw, R. Burkey, H. Schwaz, D. Kershaw, R. Melley, "Designing High Performance Compounds for Severe Mining Service Conditions," *Rubber World*, April, 1985, p. 18.

Rollers

Idler Roll Compound, G. Blackshaw, R. Burkey, H. Schwarz, D. Kershaw, R. Melley, "Designing High Performance Compounds for Severe Mining Service Conditions," *Rubber World*, April, 1985, p. 18.

Gaskets

Gasket Compound, P. Ciullo, N. Hewitt, *The Rubber Formulary*, Noyes, Norwich, NY, p. 430.

A.1.12 NBR/PVC Blends

Rollers

Roll Cover Compound, 60 Shore A, R. School, "NBR/PVC Blends Combine Polymers' Properties," *Rubber & Plastics News*, December 16, 1985, p. 14.

De-husking Rolls, "De-husking Rolls," *Rubber India*, March, 1989, p. 33.

Soft Roll Cover Compound, R. School, "NBR/PVC Blends Combine Polymers' Properties," *Rubber & Plastics News*, December 16, 1985, p. 14.

Cable

Molded Electrical Connectors, R. School, "NBR/PVC Blends Combine Polymers' Properties," *Rubber & Plastics News*, December 16, 1985, p. 14; G. Blackshaw, R. Burkey, H. Schwaz, D. Kershaw, R. Melley, "Designing High Performance Compounds for Severe Mining Service Conditions," *Rubber World*, April, 1985, p. 18; *Rubber World*, April, 1985, p. 19.

Cable Jacket, H. Schwarz, "Carboxylated Nitrile/PVC Fluxed Blends for Premium Quality Products," *Elastomerics*, November, 1980, p. 17.

Sponge

Closed Cell Sponge Insulation Tubing, P. Ciullo, N. Hewitt, *The Rubber Formulary*, Noyes, Norwich, NY, pp. 568, 570.

Vinyl Closed Cell Sponge, Flotation, P. Ciullo, N. Hewitt, *The Rubber Formulary*, Noyes, Norwich, NY, p. 574.

Vinyl Closed Cell Sheet Sponge, P. Ciullo, N. Hewitt, *The Rubber Formulary*, Noyes, Norwich, NY, p. 572.

Miscellaneous Products

Spinning Apron Compound, H. Schwarz, "NBR/PVC Blends Combine Polymers' Properties," *Rubber & Plastics News*, December 16, 1985, p. 14.

Protective Clothing Nonblack Compound, R. School, "NBR/PVC Blends Combine Polymers' Properties," *Rubber & Plastics News*, December 16, 1985, p. 14.

Auto Fuel Cover, P. Ciullo, N. Hewitt, *The Rubber Formulary*, Noyes, Norwich, NY, p. 365.

A.1.13 XNBR/PVC Blend

Hose

Hose Cover Stock, G. Blackshaw, R. Burkey, H. Schwaz, D. Kershaw, R. Melley, "Designing High Performance Compounds for Severe Mining Service Conditions," *Rubber World*, April, 1985, p. 18; R. Weir, "Carboxylated Nitrile Rubber: A High Performance Elastomer," *Rubber India*, February, 1980, p. 13.

Oil Suction and Discharge Hose Cover, J. Dunn, R. Vara, "Oil Resistant Elastomers for Industrial Hose Applications, *Rubber India*, November, 1984, p. 11.

Cable

Heavy Duty Cable Jacket, G. Blackshaw, R. Burkey, H. Schwaz, D. Kershaw, R. Melley, "Designing High Performance Compounds for Severe Mining Service Conditions," *Rubber World*, April, 1985, p. 18; H. Schwarz, "Light-Colored Extra-Heavy Duty Cable Jackets Based on Carboxylated Nitrile/PVC Blends," *Rubber World*, November, 1982, p. 22.

Lining

Lining Compound, G. Blackshaw, R. Burkey, H. Schwaz, D. Kershaw, R. Melley, "Designing High Performance Compounds for Severe Mining Service Conditions," *Rubber World*, April, 1985, p. 18; *Rubber World*, April, 1985, p. 19.

Rollers

Roll Covers, R. Weir, "Carboxylated Nitrile Rubber: A High Performance Elastomer," *Rubber India*, February, 1980, p. 13.

Belts

Belt Cover, H. Schwarz, "Carboxylated Nitrile/PVC Fluxed Blends for Premium Quality Products," *Elastomerics*, November, 1980, p. 17.

A.1.14 NBR/IR Blends

Rollers

90 Duro Nonblack Rolls, General Purpose, P. Ciullo, N. Hewitt, *The Rubber Formulary*, Noyes, Norwich, NY, p. 486.

A.1.15 NBR/BR Blends

Shoes

Soling, 63 Duro, P. Ciullo, N. Hewitt, *The Rubber Formulary*, Noyes, Norwich, NY, p. 501.

Belts

High Abrasion Resistant Conveyor Belt, P. Ciullo, N. Hewitt, *The Rubber Formulary*, Noyes, Norwich, NY, p. 533.

A.1.16 NBR/NR Blends

Dampers

High Damping, Low Creep Rubber Compound, *Natural Rubber Technology*, Vol. 7, Part 3, 1976; J. S. Dick, H. Pawlowski, "Alternate Instrumental Methods of Measuring Scorch and Cure Characteristics," *Polymer Testing*, Vol. 14, 1995, pp. 45-84.

A.1.17 EPDM/SBR Blends

Gaskets

Auto Window Gaskets, P. Ciullo, N. Hewitt, *The Rubber Formulary*, Noyes, Norwich, NY, p. 270.

Sponge

Press Molded Closed Cell Sheet Sponge, P. Ciullo, N. Hewitt, *The Rubber Formulary*, Noyes, Norwich, NY, p. 181.

Hose

Hose Stock, *Elastomeric Materials, Desk-Top Data Bank*, The International Plastics Selector, Inc., Cordura, San Diego, CA, 1977.

A.1.18 EPDM/CR Blends

Gaskets

Auto Window Gaskets, P. Ciullo, N. Hewitt, *The Rubber Formulary*, Noyes, Norwich, NY, p. 270.

A.1.19 EPDM/IR Blends

Wipers

Wiper Blades, P. Ciullo, N. Hewitt, *The Rubber Formulary*, Noyes, Norwich, NY, p. 85.

A.1.20 EPDM/NR Blends

Sheet

Duro 40 for Tear Strength, P. Ciullo, N. Hewitt, *The Rubber Formulary*, Noyes, Norwich, NY, p. 101.

Miscellaneous

70 Duro Exercise Band, P. Ciullo, N. Hewitt, *The Rubber Formulary*, Noyes, Norwich, NY, p. 114.

A.1.21 EPDM/CSM Blends

Belts

Transmission Belt, P. Ciullo, N. Hewitt, *The Rubber Formulary*, Noyes, Norwich, NY, p. 269.

Conveyor Belt Cover Stock, R. Babbit, *The Vanderbilt Rubber Handbook*, R. T. Vanderbilt Co., Norwalk, CT, 1978.

Hose

Hose Jacket Compound, *Rubber Chemistry and Technology*, Vol. 62, p. 883, Gdy, 3/79.

A.1.22 EPDM/Silicone Compatibilized

Rollers

Heat Resistant Roll Compound, T. Jablonowski, "Compounding Requirements and Techniques for Rubber Covered Rolls," *Rubber World*, July, 1989, p. 25.

A.1.23 EPDM/LDPE Blend

Cable

Cable Compound, U. Vaidya, "Flame-Retardant EPDM Integral Wire and Cable Insulation/Cover," *Elastomerics*, April, 1979, p. 22.

A.1.24 EPDM/EVA Blend

Miscellaneous

Duro 92 EVA Compound, P. Ciullo, N. Hewitt, *The Rubber Formulary*, Noyes, Norwich, NY, p. 722.

Duro 89 and 85 EVA Compound for Transparency, P. Ciullo, N. Hewitt, *The Rubber Formulary*, Noyes, Norwich, NY, p. 720.

A.1.25 CR/SBR Blends

Hose

Air Conditioner Hose, P. Ciullo, N. Hewitt, *The Rubber Formulary*, Noyes, Norwich, NY, p. 208.

Oil Suction and Delivery Hose, P. Ciullo, N. Hewitt, *The Rubber Formulary*, Noyes, Norwich, NY, p. 156.

Sponge

Cellular Shoe Soling, P. Ciullo, N. Hewitt, *The Rubber Formulary*, Noyes, Norwich, NY, p. 240.

Press Molded Closed Cell Sheet Sponge, Oil Resistant, Medium Swell, P. Ciullo, N. Hewitt, *The Rubber Formulary*, Noyes, Norwich, NY, p. 239.

Shoe

Sole Compound, P. Ciullo, N. Hewitt, *The Rubber Formulary*, Noyes, Norwich, NY, p. 140.

A.1.26 CR/BR Blends

Hose

Hose Compound, P. Ciullo, N. Hewitt, *The Rubber Formulary*, Noyes, Norwich, NY, pp. 209, 210.

Oil Suction and Discharge Hose, J. Dunn, R. Vara, "Oil Resistant Elastomers for Industrial Hose Applications," *Rubber India*, November, 1984, p. 11; J. Dunn, R. Vara, "Oil Resistant Elastomers for Hose Applications," *Rubber Chemistry and Technology*, Vol. 56, p. 557.

Hose Cover, A. Werner, "Emulsion PB Blends for Mechanicals," *Rubber India*, December 1970, p. 13; A. Werner, "Emulsion BR Improves, Modifies Specialty, General-Use Elastomers," *Rubber & Plastics News*, December 17, 1984.

Miscellaneous Products

Neoprene Extrusion Compound, A. Werner, "Emulsion BR Improves, Modifies Specialty, General-Use Elastomers," *Rubber & Plastics News*, December 17, 1984.

A.1.27 CSM/BR Blends

Miscellaneous Products

General Purpose Cover, P. Ciullo, N. Hewitt, *The Rubber Formulary*, Noyes, Norwich, NY, p. 587.

A.1.28 CR/Hydrin Blends

Miscellaneous Products

Semiconductive Direct Bonding, P. Ciullo, N. Hewitt, *The Rubber Formulary*, Noyes, Norwich, NY, p. 232.

A.1.29 NBR/FA Polysulfide Rubber

Printing

Printing Blankets, S. Flanders, "Polysulfide Rubbers," *Rubber India*, August, 1985, p. 19.

A.1.30 Triblends

A.1.30.1 NR/SBR/BR Blends

Tires

Low Rolling Resistance Tread, P. Ciullo, N. Hewitt, *The Rubber Formulary*, Noyes, Norwich, NY, p. 131.

High Performance Tread, P. Ciullo, N. Hewitt, *The Rubber Formulary*, Noyes, Norwich, NY, p. 132.

Tire Carcass, 83, P. Hamed, L. Walker, "Comparative Properties of NR and *cis*-1,4-Polyisoprene," *Rubber World*, July, 1971, p. 51; H. Barager, "High Performance Innerliners Through the Use of Neoprene Blends," *Rubber World*, March, 1984, p. 30; P. Ciullo, N. Hewitt, *The Rubber Formulary*, Noyes, Norwich, NY, p. 74; R. Babbit, *The Vanderbilt Rubber Handbook*, R. T. Vanderbilt Co., Norwalk, CT, 1978; M. Deviney, L. Whittington, C. Good, R. Sperley, "Mobility of Extender Oil in Tires under Service Conditions," *Rubber Chemistry and Technology*, Vol. 47, p. 127; M. Deviney, L. Whittington, B. Corman, "Migration of Oil in Elastomers," *Rubber World*, January, 1972, p. 31; E. Buckler, C. Briggs, J. Henderson, E. Lasis, "New Findings on General Purpose Synthetic Rubbers," *Elastomerics*, December, 1977, p. 32.

Tank Tracks

Rubber Tank Tracks, C. Pergantis, T. Murray, J. Mead, R. Shuford, A. Alesi, "Field Observations on Rubber Tank Tracks," *Rubber World*, April, 1989, p. 31.

Belts

Conveyor Belt Cover, P. Ciullo, N. Hewitt, *The Rubber Formulary*, Noyes, Norwich, NY, p. 155.

Shoes

60 Duro Footwear, P. Ciullo, N. Hewitt, *The Rubber Formulary*, Noyes, Norwich, NY, p. 146.

85 Duro Sole, P. Ciullo, N. Hewitt, *The Rubber Formulary*, Noyes, Norwich, NY, p. 154.

65 Duro Footwear, P. Ciullo, N. Hewitt, *The Rubber Formulary*, Noyes, Norwich, NY, p. 149.

A.1.30.2 BR/SBR/IR Blends

Shoes

62 Duro Footwear, P. Ciullo, N. Hewitt, *The Rubber Formulary*, Noyes, Norwich, NY, p. 147.

High Quality Heel, R. Babbit, *The Vanderbilt Rubber Handbook*, R. T. Vanderbilt Co., Norwalk, CT, 1978.

A.1.30.3 NR/IR/SBR Blends

Shoes

Boot Lining Gum, R. Babbit, *The Vanderbilt Rubber Handbook*, R. T. Vanderbilt Co., Norwalk, CT, 1978.

Tires

Apex Strip, L. Walker, W. Helt, "High-Temperature Curing of Radial Passenger Tires," *Rubber Chemistry and Technology*, May-June, Vol. 59, 1986, p. 285

Body Skim, L. Walker, W. Helt, "High-Temperature Curing of Radial Passenger Tires," *Rubber Chemistry and Technology*, Vol. 59, 1986, p. 285.

A.1.30.4 NR/EPDM/BIIR Blends

Tires

White Sidewalls, R. Babbit, *The Vanderbilt Rubber Handbook*, R. T. Vanderbilt Co., Norwalk, CT, 1978; *Hi-Sil Silica Pigments Formulary*, PPG Industries; "Tire Sidewalls: Whitewall Formulations," *Natural Rubber Technology*, Vol. 13, Part 4, 1982, p. 90.

A.1.30.5 NR/EPDM/CIIR Blends

Tires

White Sidewalls, B. Shank, "Compounding for Quality and Efficiency," *Elastomerics*, July, 1987, p. 17; A. Barbour, A. Rice, "New Clays Improve Compound Properties," *Rubber & Plastics News*, August 24, 1987; R. Babbit, *The Vanderbilt Rubber Handbook*, R. T. Vanderbilt Co., Norwalk, CT, 1978.

A.1.30.6 NR/CR/CSM Blends

Tires

White Sidewalls, "New Titanium Dioxide Extender Replaces up to 25% with No Elasticity Loss in Initial Tests," *Elastomerics*, August, 1988, p. 14; A. Sircar, T. Lamond, "ID of Elastomers in Tire Sections by Total Thermal Analysis," *Rubber Chemistry and Technology*, Vol. 48, p. 640.

A.1.30.7 NBR/PVC/SBR Blends

Hose

Pump Hose, *Synaprene Formulary*, Synthetics & Chemicals Ltd., Bombay, India, 1967.

Cord

Low-Cost Heater Cord Jacket, R. School, "NBR/PVC Blends Combine Polymers' Properties," *Rubber & Plastics News*, December 16, 1985, p. 14.

A.1.30.8 XNBR/PVC/NBR Blends

Shoes

Footwear Compound, R. Weir, "Carboxyated Nitrile Rubber: A High Performance Elastomer," *Rubber India*, February, 1980, p. 13.

A.1.30.9 XNBR/PVC/BR Blends

Shoe

Highly Abrasion Resistant, Direct Molded Soling. 3

A.1.30.10 XNBR/NBR/BR Blends

Shoe

Duro 60 Soling, P. Ciullo, N. Hewitt, *The Rubber Formulary*, Noyes, Norwich, NY, p. 500.

A.1.30.11 NBR/SBR/BR Blends

Miscellaneous

Oil Resistant Compound, 70 Duro, P. Ciullo, N. Hewitt, *The Rubber Formulary*, Noyes, Norwich, NY, p. 174.

A.1.30.12 NBR/BR/NR Blends

Shoes

70 Duro Sole, P. Ciullo, N. Hewitt, *The Rubber Formulary*, Noyes, Norwich, NY, p. 151.

A.1.30.13 NR/EPDM/SBR Blends

Tires

White Sidewalls, "Epsyn55," *Rubber World*, November, 1973; "White Sidewalls," *Rubber World*, November, 1973.

Miscellaneous

Handle Grips, P. Ciullo, N. Hewitt, *The Rubber Formulary*, Noyes, Norwich, NY, p. 103.

A.1.30.14 EPDM/SBR/CR Blends

Sponge

Closed Cell Sheet Sponge, Press Molded, P. Ciullo, N. Hewitt, *The Rubber Formulary*, Noyes, Norwich, NY, p. 182.

A.1.31 Four-Way Blends

A.1.31.1 NR/EPDM/SBR/CIIR Blends

Tires

White Sidewalls, A. Sircar, T. Lamond, "ID of Elastomers in Tire Sections by Total Thermal Analysis, 2. White Sidewall Compounds of EPDM and Blends," *Rubber Chemistry and Technology*, Vol. 48, p. 631.

Appendix 2: Cure Systems

A.2 Introduction

Thousands of different curative combinations are used in the rubber industry to cure a wide variety of different elastomers as well as elastomer blends. In this appendix we discuss some typical curative combinations as a starting point in the development of a new formulation. Certainly these example cure systems can be adjusted in a "fine-tuning" process to improve or meet certain target compound properties. *Most of the cure systems discussed below use zinc oxide and stearic acid as activators* even though they may not be mentioned or noted. However, ZnO is a vulcanizing agent with many halogenated elastomers.

A.2.1 General Purpose Elastomer Cure Packages

Many compounds based on general purpose elastomers (NR, SBR, BR) and their blends use cure systems based on sulfenamide accelerators. The exact selection of a cure package is usually based on the following required curing characteristics:

- Processing safety
- Cure rate index
- Activity (ultimate modulus buildup per accelerator loading)

GEN: *Compounders' Pocket Book*, Monsanto, 1981, p. 83.

Processing or scorch safety can be controlled by the selection of the primary accelerator used in the cure system. By far, sulfenamide accelerators are the most commonly selected primary accelerator for general purpose elastomer-based compounds because, as a class, they give the best processing protection (scorch protection). Next, by accelerator class, the thiazoles (MBT and especially MBTS) give intermediate scorch safety while cure systems based on thiurams, dithiocarbamates, peroxides, or even guanidines are far more scorchy (the least scorch safety). GEN: R. N. Datta, *Rubber Curing Systems*, RAPRA Technology, Ltd., Report 144, 2001, p. 6.

For the most commonly used primary accelerators (the sulfenamides), processing safety (scorch safety time) is longer with the secondary amine based sulfenamide cures but shorter with the primary amine sulfenamide based cures. Therefore, the scorch safety time of a general purpose rubber based compound is mainly determined based on which sulfenamide accelerator is selected.

CBS (short scorch safety) < TBBS < MBS < DCBS (long scorch safety)

Cure rate is also determined from the selection of the primary accelerator for curing general purpose rubber.

As shown below, the selection of a given class of accelerator is a great determiner of cure rate.

Again, the most common systems for curing general purpose elastomers involve the sulfenamide accelerators because of their scorch safety. The proper selection of the specific sulfenamide will have a large impact on the cure rate of the compound. The following is the general ordinal relationship among some of the sulfenamide accelerators:

DCBS (slow cure rate) < MBS < CBS < TBBS (fast cure rate)

Class	Vulcanization Rate	Specific Examples
Aldehyde-amine	Slow	
Guanidines	Medium	DPG, DOTG
Thiazoles	Semifast	MBT, MBTS
Sulfenamides	Fast, delayed action	CBS, TBBS, MBS, DCBS
Sulfenimides	Fast, delayed action	TBSI
Dithiophosphates	Fast	ZBPD
Thiurams	Very fast	TMTD, TMTM, TETD, TBZTD
Dithiocarbamates	Very fast	ZnDMC, ZnDBC

RT: B. H. To, Chapter 15, "Sulfur Cure Systems," p. 383

One very effective way for a compounder to increase the cure rate of a sulfenamide cure system is to add a "secondary" accelerator (or "kicker"). Thiurams, dithiocarbamates, and guanidines are used at low part loadings with sulfenamide accelerators to increase the cure rate. Sometimes, one of the trade-offs in using one of these secondary accelerators is a loss in scorch safety time. Often this loss can be minimized by selecting the optimum combination of primary sulfenamide accelerator with the proper selection of one of the secondary accelerators at the optimal part loading. However, it is also possible to achieve a better trade-off between cure rate and scorch safety by adding CTP prevulcanization inhibitor to the primary and secondary accelerator combination. RT: Chapter 15, "Sulfur Cure Systems," B. H. To, p. 387.

Cured modulus depends on ultimate crosslink density formed during cure and is a very important consideration. This compound property can be adjusted by changing either the sulfur concentration, the accelerator concentration, or a combination of both. Increasing either the part loading of the sulfur or accelerator will cause the cured modulus and ultimate crosslink density to rise (and vice versa). These adjustments can affect scorch safety and cure rate properties as well. GEN: *Compounders' Pocket Book*, Monsanto, 1981, p. 88.

The ratio of sulfur to accelerator is very important in determining the type of sulfur crosslinks. A higher ratio of sulfur to accelerator may result in a higher preponderance of polysulfide crosslinks, which may impart better flex fatigue properties but poorer aging and set properties (these are called "conventional" cure systems). On the other hand, a lower ratio of sulfur to accelerator (or even no sulfur at all, using a sulfur donor such as TMTD or DTDM instead) may result in a vulcanizate with better age resistance properties but perhaps poorer flex fatigue because there will be a greater preponderance of mono- and disulfidic crosslinks and fewer polysulfidic crosslinks. These are called "efficient vulcanization" (or EV) cures. RT: Chapter 15, "Sulfur Cure Systems," B. H. To, p. 387. Common examples of these conventional, semi-EV and EV cures are shown below.

For Natural Rubber-Based Compounds

Conventional cure		EV cure (1)	
Sulfur	2.5 phr	Sulfur	0.5 phr
Sulfenamide accelerator	0.6	Sulfenamide accelerator	5.0
Semi-EV cure (1)		**EV cure (2)**	
Sulfur	1.5 phr	Sulfenamide accelerator	1.0 phr
Sulfenamide accelerator	1.5	DTDM	1.0
		TMTD	1.0
Semi-EV cure (2)			
Sulfur	1.5 phr		
Sulfenamide accelerator	0.5		
DTDM	0.6		

For SBR-Based Compounds

Conventional cure		EV cure	
Sulfur	2.0 phr	Sulfenamide accelerator	1.0 phr
Sulfenamide accelerator	1.0	DTDM	2.0
		TMTD	0.4
Semi-EV cure			
Sulfur	1.0 phr		
Sulfenamide accelerator	3.0		

GEN: *Compounders' Pocket Book*, Monsanto, 1981, pp. 116–118

A.2.2 EPDM Cure Packages

The following are examples of some commonly used cure systems for EPDM-based compounds. Each cure system has its unique advantages and disadvantages. RT: B. H. To, Chapter 16, "Cures for Specialty Elastomers," p. 395.

Cure System 1: Low Cost

- Advantage: Low cost
- Disadvantage: Tendency to bloom

Sulfur	1.5 phr
TMTD	1.5
MBT	0.5

Cure System 2: Triple 8

- Advantage: Very fast cure, excellent physical properties, very low or no bloom
- Disadvantage: Relatively expensive, scorchy

Sulfur	2.0 phr
MBT	1.5
TeDEC	0.8
DPTT	0.8
TMTD	0.8

Cure System 3: Low Set

- Advantage: Good heat aging resistance and excellent compression set properties
- Disadvantage: High cost, tendency to bloom

Sulfur	0.5 phr
ZBDC	3.0
ZMDC	3.0
DTDM	2.0
TMTD	3.0

Cure System 4: General Purpose

- Advantage: A general purpose low or nonblooming cure
- Disadvantage: Moderate cost

Sulfur	2.0 phr
MBTS	1.5
ZBDC	2.5
TMTD	0.8

Cure System 5: 2121 System

- Advantage: Results from a DOE to give optimal balance for good compression set resistance, good heat aging characteristics, and good physical properties.
- Disadvantage: High cost

ZBPD	2.0 phr
TMTD	1.0
TBBS	2.0
Sulfur	1.0

A.2.3 NBR Cure Packages

As a class of elastomers, nitrile rubber is commonly used when very good resistance from oil attack is needed. NBR cure systems are somewhat analogous to those used for general purpose elastomers such as SBR except that a special magnesium carbonate treated grade of sulfur is commonly used in place of conventional "rubber makers" sulfur (for better dispersion). Commonly, cure systems based on 1.5 phr of treated sulfur are used with different accelerators to achieve different cure rates and states of cure as shown in the examples below. RT: B. H. To, Chapter 16, "Cures for Specialty Elastomers," p. 397; GEN: R. N. Datta, "Rubber Curing Systems," RAPRA, Report 144, p. 19.

A Common Thiuram Cure

MC-treated sulfur	1.5 phr
TMTM	0.4

A Common Thiazole Cure

MC-treated sulfur	1.5 phr
MBTS	1.5

A Common Sulfenamide Cure

MC-treated sulfur	1.5 phr
TBBS	1.2
TMTD	0.1

By using more sulfur donors, such as TMTD or DTDM, partially or completely in place of sulfur, one can improve the NBR compound's aging and compression set properties as illustrated below. RT: B. H. To, Chapter 16, "Cures for Specialty Elastomers," Tables 16.7 and 16.8.

A Common Low-Sulfur Cure

MC-treated sulfur	0.3 phr
TBBS	1.0
TMTD	1.0

A Common "Sulfurless" Cure

TBBS	1.0 phr
TMTD	1.0
DTDM	1.0

A Common "Sulfurless" Cure with Higher Cured Hardness

TBBS	3.0 phr
TMTD	3.0
DTDM	1.0

A Common "Sulfurless" Cure with Still Higher Cured Hardness

TBBS	2.0 phr
TMTD	2.0
DTDM	2.0

Also, for these low (or no)-sulfur cure systems, scorch safety and modulus can be controlled as illustrated below. RT: B. H. To, Chapter 16, "Cures for Specialty Elastomers," p. 404.

Effect of Cure Systems on Scorch and Modulus

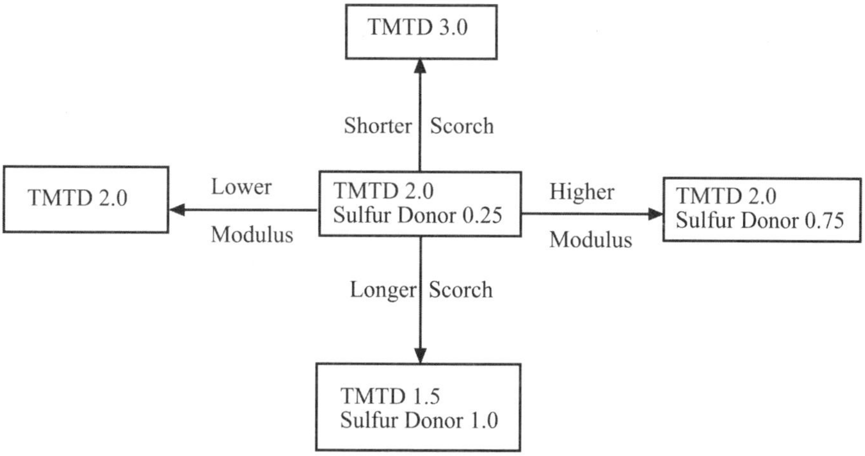

In addition, replacing TMTD concentrations with sulfur donor or TBBS and sulfur donor will lengthen process safety time for the nitrile formulation. Adjustments for state of cure can also be made as illustrated below.

Effect of Cure Systems on Processing Safety and State of Cure

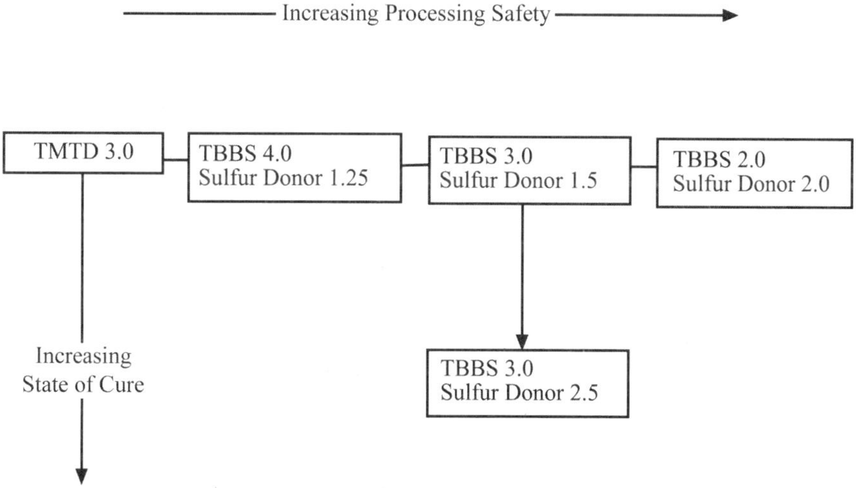

A.2.4 Polychloroprene Cure Packages

It is very common practice to use 5 phr of zinc oxide and 4 phr of magnesium oxide for the majority of polychloroprene (CR) compounds. In polychloroprene curing, the zinc oxide is actually the vulcanizing agent, not simply the activator, while magnesium oxide is used for control of scorch. Usually the magnesium oxide grade used for CR has a high surface area. GEN: R. Murray, D. Thompson, *The Neoprenes*, E. I. DuPont de Nemours, Wilmington, DE, 1963, p. 21.

Historically, ethylene thiourea (ETU) was the accelerator of choice for curing mercaptan-modified CRs such as neoprene W-types. However, during the last 20 years, use of ETU has declined because this accelerator was found to be carcinogenic, posing a potential health hazard. Some cure systems that have been used in place of ETU for compounds based on neoprene W-type polymers are also listed below. RT: B. H. To, Chapter 16, "Cures for Specialty Elastomers," p. 406.

Historic Thiourea Cure

ETU	0.5 phr

Thiuram/Guanidine Cure (good scorch safety)

TMTM	1.0 phr
DOTG	1.0
Sulfur	0.5

Al (Thiocarbanilide) Cure (good compression set)

A-1	0.7 phr

Surprisingly, CTP (prevulcanization inhibitor) can actually work with some non-ETU cure systems provided one can accept a loss in compression set and modulus.

A1 with CTP (improved scorch safety)

A-1	0.7 phr
CTP	0.2

Zinc-*O*,*O*-di-n-butylphosphorodithioate (ZBPD) can also be used in an A-1 cure to provide a very fast cure, but it is scorchy.

A1/ZBPD Cure (fast cure rate/scorchy)

A-1	0.7 phr
ZBPD	0.5

RT: *Rubber Technology, Compounding and Testing for Performance*. 2nd ed. Edited by J. S. Dick, Hanser Publications, 2009 ■ GEN: General References from the Literature ■ RP: Review Panel (Advisors for this book) ■ IMPORTANT: The information contained herein is experimental in nature and is meant for individuals with advanced scientific training and rubber compounding experience. All safety precautions should be taken when replicating or applying this information. The author makes no representation or warranty, either expressed or implied.

A.2.4.1 General Purpose Thiourea Cure Systems for Carbon Black Loaded CR Compounds

Even though a number of health issues are associated with the use of thiourea accelerators, these compounds are still being used as long as proper safety precautions are followed. Some of these general purpose cure systems are given below. GEN: Technical Information, DuPont Dow Elastomers, Rev. 2, January 2002, p. 3.

Thiourea Systems – General Purpose (very good bin stability)

ETU	0.5–0.75 phr
TMTD	0.5–1.0

Thiourea Systems – General Purpose (very good scorch safety, excellent set resistance)

ME_3TU	0.75–1.5 phr
Epoxy resin	1.0–2.0

A.2.4.2 Special Purpose Thiourea Cure Systems for Carbon Black Loaded CR Compounds

Thiourea cure systems are also used sometimes to achieve special properties.

Thiourea Systems – Special Purpose (good heat resistance)

ETU	0.5–1.0 phr

Thiourea Systems – Special Purpose (for very fast high-temperature cures or for low-temperature cures)

ETU	2.0 phr
Vanax® PML	1.0

A.2.4.3 General Purpose Non-thiourea Cure Systems for Carbon Black Loaded CR Compounds

Because of health and safety issues, thiourea cure systems are generally avoided. The following are some non-thiourea cure systems for CR general purpose applications.

Non-thiourea – General Purpose Cure System (good compromise in properties)

HMT	0.75–1.0 phr
TMTD	1.0–1.5
PEG	0–2.0

Non-thiourea – General Purpose Cure System (good compromise in properties, No. 2)

HMT	0.75–1.0 phr
TETD	0.75–1.0
DOTG	0.5–1.0

Non-thiourea – General Purpose Cure System (good compromise, commonly used with CR/SBR blends)

DOTG	1.5 phr
TMTM	1.0
Sulfur	0.25–0.30

Non-thiourea – General Purpose Cure System (good process safety and physical property retention)

Vanax® PML	0.75–1.5 phr
Calcium oxide	4–8

Non-thiourea – General Purpose Cure System (a good alternative to ETU)

CPA	0.75–2.0 phr

Non-thiourea – General Purpose Cure System (good delayed action, but cure properties somewhat sensitive to metal oxides and compounding techniques)

CRV	0.75–2.0 phr

A.2.4.4 Special Purpose Non-thiourea Cure Systems for Carbon Black Loaded CR Compounds

Various non-thiourea cure systems are available for special purpose applications as shown below.

RT: *Rubber Technology, Compounding and Testing for Performance*. 2nd ed. Edited by J. S. Dick, Hanser Publications, 2009 ■ GEN: General References from the Literature ■ RP: Review Panel (Advisors for this book) ■ IMPORTANT: The information contained herein is experimental in nature and is meant for individuals with advanced scientific training and rubber compounding experience. All safety precautions should be taken when replicating or applying this information. The author makes no representation or warranty, either expressed or implied.

Non-thiourea – Special Purpose Cure System (very good process safety)

DOTG	0.5–1.0 phr
TMTM	0.5–1.0
Sulfur	1.0–1.5

Non-thiourea – Special Purpose Cure System (better scorch safety than ETU cure)

m-PBM (HVA-2)	1 phr
MBTS	0.5–1.0
Sulfur	0.25

Non-thiourea – Special Purpose Cure System (better scorch safety)

m-PBM (HVA-2)	1 phr
TMTD	0.5–1.0
Sulfur	0.25

A.2.5 Chlorobutyl Rubber Cure Packages

A significant number of different types of cure systems are commonly used to vulcanize chlorobutyl compounds.

Zinc oxide has a special role in chlorobutyl rubber, functioning as a vulcanizing agent. The use of 3 to 5 phr of zinc oxide alone can cure a chlorobutyl compound (usually 5 phr). The use of ZnO above 5 phr does not really help cure rate or state of cure; however, it might improve heat resistance.

Unlike vulcanization chemistry associated with general purpose elastomers such as natural rubber or SBR, in the case of vulcanization chemistry associated with chlorobutyl, some acidic additives will accelerate the cure while some alkaline additives may retard it. For example, the scorch and cure rate of a zinc oxide cure of a CIIR compound is affected by the presence of stearic acid. A higher concentration of stearic acid will increase the compound's cure rate and reduce its scorch safety. Other acidic compounding ingredients, such as wood rosin, rosin esters, and so forth have similar effects. On the other hand, magnesium oxide can retard the cure of CIIR compounds. GEN: *Chlorobutyl Rubber Compounding Applications*, Exxon Chemical, 1976, pp. 7–10.

Thiuram cure systems are commonly used to cure chlorobutyl rubber compounds. Thiurams such as TMTD reportedly react with the unsaturated carbon sites to yield carbon–sulfur crosslinks. The scorch safety and physical properties of the cured compounds can be greatly improved through the use of MBTS and/or magnesium oxide. Commonly, the grade of MgO used in curing CIIR is significantly lower in surface area than what is normally used in CR cures (discussed earlier). Too much MgO can reduce the cured modulus and hurt compression set and flexing properties. But higher MgO loadings may improve the heat resistance for a CIIR compound. Also, sometimes MBT may be used in place of MBTS; however, MBT imparts less scorch protection than MBTS. GEN: *Chlorobutyl Rubber Compounding Applications*, Exxon Chemical, 1976, p. 11.

Dithiocarbamate cure systems built on accelerators such as ZnDEC or ZnDMC are also used to cure chlorobutyl compounds. One advantage of these cure systems is low compression set; however, they usually have only fair scorch safety.

Alkyl phenol disulfide cures are commonly used to help achieve better adhesion between a CIIR and a general purpose elastomer-based compound (such as NR, etc.). This cure system is commonly used in blends of CIIR and NR.

The following are some typical cure systems that can be used to cure chlorobutyl rubber compounds. GEN: *Chlorobutyl Rubber Compounding Applications*, Exxon Chemical, 1976, p. 18.

Zinc Oxide Cure System

Zinc Oxide Cure
(very good heat resistance, low cured modulus, slow cure rate)

Zinc oxide	5.0 phr

Thiuram Cure Systems

Thiuram/Thiazole Cure System
(good heat resistance, tear resistance, general purpose)

Zinc oxide	5.0 phr
Magnesium oxide	0.5
TMTD	1.0
MBTS	2.0

Thiuram/Thiazole Cure System
(only for silica-loaded compounds for steam resistance)

Zinc oxide	5.0 phr
Magnesium oxide	2.0
TMTD	1.0
MBTS	1.0

Alkyl Phenol Disulfide Cure Systems

Alkyl Phenol Disulfide Cure System
(for blends with NR, better adhesion to other general purpose based compounds)

Zinc oxide	5.0 phr
Magnesium oxide	0.25
Vultac 5	1.0
MBTS	0.75 phr

Sulfur-Guanidine Cure Systems

Sulfur-Guanidine Cure System
(for NR blends; may also give good adhesion)

Zinc oxide	5.0 phr
DOTG	1.5
Sulfur	2.0

Resin Cure Systems

Methylol-Terminated Phenolic Resin Cure System
(good heat and ozone resistance)

Zinc oxide	5.0 phr
SP 1055	4.0
MBTS	2.0

A.2.6 Butyl Rubber Cure Packages

Butyl rubber (IIR) has a backbone with a much lower degree of unsaturation than do other general purpose elastomers such as NR or SBR. As a result, many IIR cures contain lower concentrations of sulfur and more active accelerators such as thiurams in order to achieve a good balance for rate of cure and cured physical properties. Commonly, thiuram accelerators and thiazole accelerators are used to give good heat resistance with low compression set. Also, some newer cure systems have now been developed in order to avoid the use of thiuram accelerators, which are *N*-nitrosamine generators that pose a potential health hazard to factory workers. Finally, there are resin cures available for achieving superior heat resistance for a butyl compound.

The following are some examples of butyl rubber cure systems. RT: B. H. To, Chapter 16, "Cures for Specialty Elastomers," pp. 402–410.

Thiuram Cure Systems

Conventional Thiuram Cure System

TMTD	1.0 phr
MBT	0.5
Sulfur	2.0

Semi-EV Thiuram Cure System

TMTD	1.0 phr
Sulfur	0.5
DTDM	1.2
TBBS	0.5

Non-thiuram Cure Systems

ZBPD/TBBS/Sulfur Cure System (approaches a semi-EV performance)

Sulfur	1.25 phr
MBT	1.0
ZBPD	1.25
TBBS	1.25

ZBPD/TBBS/Sulfur Cure System (higher modulus)

Sulfur	2.0 phr
MBT	1.0
ZBPD	1.25
TBBS	1.25

ZBPD/TBBS/Sulfur Cure System (short cure time)

Sulfur	1.0 phr
ZBPD	5.0
TBBS	1.0

Methylol-Terminated Phenol Formaldehyde Cures

GEN: R. Babbit, *The Vanderbilt Rubber Handbook*, Vanderbilt Co., 1978.

Resin Cure (poor scorch safety, but very fast cure rate)

Methylol-terminated PF resin	12 phr
$SnCl_2$	2

Resin Cure (some scorch safety, but slow cure rate)

Methylol-terminated PF resin	12 phr
Halogenated polymer	5

Review Panel Members' Biographies

L. Roger Dailey

Roger Dailey has practiced compounding for 25 years, starting as a plant chemist and rising quickly to the Sr. Dev. Chemist level. Moving from the cauldron of reality, he spent several years as a Sr. Development Engineer prior to becoming Team Leader for Joint Venture Technology. He has conducted problem-solving projects all over the globe for tire and nontire applications, ranging from medical devices to simple gaskets to complex engineered products. Roger is Six Sigma Black Belt trained and regularly provides compounding talks at various venues, including rubber classes held at the University of Akron, Ohio.

Roger recently has been Team Leader for new product implementation and participates in Goodyear Chemicals new service group Solgenesis as a Process Optimization Specialist. Roger has contributed to many trade secrets and patents through his varied career. While Roger enjoys his global escapades, he will tell you he enjoys his family the most, especially watching his girls grow in their various activities.

Roger_dailey@goodyear.com

R. J. Del Vecchio

After a lengthy career in the rubber industry, working for companies such as B. F. Goodrich, United Technologies, Lord, and Barry Controls, Del founded a full-time consulting service in late 1990. Much of his work is based on his long history of experience with engineering applications of rubber, as well as the normal technical skills involved in compounding and processing a wide variety of elastomers. In addition, he has become an enthusiastic user of, and instructor in, statistical methods, especially as they apply to the rubber industry.

He has published numerous technical papers at ACS Rubber Division meetings, has appeared in the trade literature both here and overseas, and has authored several chapters in technical books, on topics such as rubber testing, SPC for the mixing operation, and basic rubber technology. His consulting practice is based primarily in the molded goods field, including materials development and process optimization, but he also provides services such as statistical analyses, test development, failure analysis, TQM training, and acting as an expert witness.

His book on the practical application of *Design of Experiments* in the rubber and plastics industries, sponsored by SPE, was released by Hanser Press in 1997. He has since published *A Beginner's Guide to Rubber Technology* (October 2001) and edited *Fundamentals of Rubber Technology* for the Rubber Division, ACS (March 2003).

techconsultserv@juno.com

Ronald W. Fuest

Ronald Fuest earned a Ph.D. in physical organic chemistry from Rutgers University, NJ.

He worked in the rubber industry in various positions for more than 40 years, holds 10 patents, and published numerous papers in different publications.

Ronald made major contributions to the field by introducing new high-performance polyurethane products, developing new applications for castable polyurethanes, and assisting customers worldwide in processing urethane prepolymers into final molded parts. Over that period, the Adiprene/Vibrathane business grew by more than ten-fold.

Ronald led the introduction of high-resilience MDI prepolymers into the skateboard and roller skate markets, two of the largest applications for urethane prepolymers.

He also introduced mechanical engineering as a technical service discipline to assist polyurethane users in making the most cost-effective use of high-performance polyurethane elastomers. Ronald is recognized worldwide as a leading expert in this area.

ron@fuest.net

James R. Halladay

James Halladay received his B. S. in physics from the University of Kentucky in 1976 and joined Parker Seal Company in Lexington, KY, as a rubber chemist. In 1981, Jim joined Lord Corp. as Materials and Process Engineering Manager. He is currently a Senior Staff Engineer and specializes in formulating elastomers for dynamic applications. His responsibilities include compound and process development in both organic rubber and silicone elastomers for applications involving vibration isolation and control. Jim has authored or coauthored numerous papers and holds six US patents.

james_halladay@lord.com

Frederick Ignatz-Hoover

Dr. Ignatz-Hoover was awarded his Ph.D. in polymer science from the University of Akron. His Ph.D. research in the area of anionic polymer synthesis was under the direction of Professor Quirk. Dr. Ignatz-Hoover's industrial career in rubber chemistry began in 1986 with Monsanto, Rubber Chemicals. Joining the new Products Group, he worked with Aubert Coran, studying accelerator and antidegradant chemistry as well as carbon black dispersion. He has recently worked in the area of molecular modeling, applying semiempirical quantum mechanical techniques and QSPR/QSAR techniques to various aspects of polymer chemistry and vulcanization chemistry. The modeling work has provided additional insight into the mechanism of zinc-mediated accelerated sulfur vulcanization and the free radical reactivity in antidegradant chemistry.

From his work at Monsanto and now Flexsys America L.P., Dr. Ignatz-Hoover has been a regular contributor to the ACS Rubber Division in contributed papers, educational symposia, and publications. He has served as editor of the journal *Rubber Chemistry and Technology*. He is a member of the American Chemical Society and the Rubber Division, ACS. He is an inaugural member of the Advisory Board at the Florida Center for Heterocyclic Compounds at the University of Florida and the Industrial Advisory Board for the Center for Advanced Polymer Composite Engineering (CAPSE) at Ohio State University, and he is on the industrial advisory board of the Dept. of Chemistry, the University of Akron.

Ignatz-hoover@flexsys.com

Mark A. Lawrence

Mark A. Lawrence is the Technical Director for Rubber Chemicals and Resins at INDSPEC Chemical, a subsidiary of Occidental Chemical Corporation. In 1984, he received his B.S. in chemical engineering from the Pennsylvania State University. Since then he has held various compounding and processing positions at both Uniroyal/Goodrich and Bridgestone/Firestone Tire Companies. He has been working to develop new bonding resins for both steel cord and fabric adhesion since he joined INDSPEC in 1996. He can be reached by e-mail at *Mark_Lawrence@oxy.com* or by phone at 412-826-3675.

John M. Long

Employment:

B. F. Goodrich/Uniroyal Goodrich	1965–1994
Materials Development	1965–1966
Advanced and International Tire Development	1966–1969
Factory Service Chemist, Oaks, PA, Tire Plant	1969–1972
Passenger Tire Compound Development	1972–1974

International Technical Support (Resident)	
Colombia	1998-1974
Brazil	1975-1978
Radial HD and LT Compound Development	1978-1981
Tire Group Materials and Systems Development (which became Uniroyal Goodrich Tire Co. in 1986)	1981-1985
Radial Passenger Compound Development/Laboratory and System Information	1985-1991
International Technical Support	
Korea, Philippines, Indonesia, India	1991-1994
DSM Copolymer	1994-Present
Senior Development Technologist	1994-2000
Technical Service Manager	2000-Present

Professional Organizations:

Akron Rubber Group Chairman	1985-1986
American Chemical Society	1966-Present
Councilor	1998-Present
Divisional Activities Committee	1998-2003
Macromolecular Secretariat Secretary General	1998-2004
Rubber Division ACS Chairman	1998-1995
Rubber Division ACS Past Chairman	1998-1996
Rubber Division ACS Councilor	1998-Present
United States Delegate to the International Rubber Conference Committee	1996-Present
Intersociety Polymer Education Council Board	1998-Present
President	2003-Present
ASTM	1981-Present
D-11 Rubber	
D-24 Carbon Black	
F-9 Tires	

Recognitions:

Who's Who in Science in America
Who's Who in the Midwest
Who's Who in America
Men of Achievement

john.long@dsm.com

Oscar Noel

Oscar Noel has been involved with R&D in the areas of fiberglass-reinforced plastics, PVC, thermoplastics, TPEs, and rubber for 30 years. He is presently a Senior Technical Manager at Luzenac America in Centennial, CO. He is responsible for product development pertaining to the use of talc in thermoplastics, TPEs, and thermoset rubbers. Mr. Noel is member of the ACS Rubber Division and Southern Rubber Group. He has lectured on the function of talc in rubber as part of the Rubber Division's educational symposium and made several technical presentations to various local rubber groups. Mr. Noel has published numerous technical publications and holds one patent. He has an M.S. from the University of Colorado in chemical engineering, where his graduate work in the area of polymer blends was supported by an NFS Scholarship and a Goodyear Tire & Rubber Fellowship.

oscar.noel@america.luzenac.com

Leonard Outzs

B.S. in Chemistry from Purdue University
39 years in the rubber business
Seven years in tires as a compounder
Eighteen years in hose and belts as a Research Manager
Fifteen years in Technical Service with DuPont Dow Elastomers
Memberships and active in NORG, ARG, SRG, ACS Rubber Division, ACS, and ASTM
Married with two grown children and three grandkids
Interests and hobbies: golf, woodworking, and grandkids

Leonard.l.outzs@dupont-dow.com

Thomas Powell

Educational Background:
B.S., Chemistry, University of Akron, 1973
M.S., Polymer Science, University of Akron, 1991

Thesis: Dynamic Mechanical Properties of Polypropylene Oxide Reinforced with Amorphous Silica

Professional Experience:
1973–1989: B. F. Goodrich Co. (Tire Division) and Uniroyal Goodrich Tire Company. Final position was as Tire Scientist. Experienced in raw materials analysis, tire manufacturing, and new tire product development for radial passenger tires and HD radial truck tires.

1989–Present: Degussa Corporation, currently as Technical Market Manager for pigments, with technical market and application responsibilities for pigment grade carbon blacks and silica for the printing ink, nonimpact printing, and adhesives markets. Experienced in carbon black, organosilane, and silica applications for rubber compounds in the tire and industrial rubber industry. Also experienced in carbon black and silica applications for the plastic industry.

Professional Organizations:
Secretary, ASTM D24 on Carbon Black
Member, Society of Plastic Engineers
Delegate to ISO TC 45/SC3/WG3 on Carbon Black

Thomas.powell@degussa.com

Charles P. Rader, Ph.D.

Charles P. Rader is a retired Marketing Technical Service Principal from Advanced Elastomer Systems, L.P. (AES) in Akron, Ohio. A rubber compounder and chemist by background, Dr. Rader served AES and Monsanto (a co-owner of AES) for 40 years in a broad variety of research, marketing, and management positions. A graduate of the University of Tennessee (B.S. in chemistry, M.S., Ph.D.), Dr. Rader has more than 200 publications and patents in the areas of physical chemistry, organic chemistry, food technology, rubber technology, and thermoplastic elastomers. He is a past chair of the Rubber Division, American Chemical Society, and for the past 27 years he has been an Associate Editor of *Rubber Chemistry & Technology*.

Telephone: 330-836-6940
Charles_Rader@msn.com

Ronald J. Schaefer

President
Dynamic Rubber Technology

Medina, Ohio, Ron Schaefer is President and founder of Dynamic Rubber Technology, a consulting firm specializing in the development and testing of rubber compounds for dynamic applications. He has over thirty years' experience in the rubber industry. Ron was formerly employed by Zeon Chemicals, where he performed compound development on vibration isolation and damping products. Prior to this, he was employed by B. F. Goodrich in the Corporate Research, Tire, and Chemical divisions.

Ron has authored papers on dynamic properties, vibration isolation and damping, automobile engine mounts, and tire performance. He has several patents and is a member of ACS, SAE, and the Akron NEO and Detroit Rubber

Groups. Ron holds a B. S. degree in chemistry from West Liberty College and has graduate credit in polymer science and chemistry from the University of Akron.

rschaeferdrt@msn.com

Kelvin K. Shen

Dr. Kelvin K. Shen is a Technical Manager of Fire Retardant Industry at Luzenac/Borax. He holds a Ph.D. degree in organic chemistry from the University of Massachusetts at Amherst, and he did his thesis work on organometallic chemistry and X-ray crystallography at Brookhaven National Laboratory. He was a Research Fellow at Yale and the California Institute of Technology, and he taught at California State University at Los Angeles before joining Borax in 1972. He has worked in fire retardant research and marketing since 1979. He was the Technical Manager of Borax Business Development Group before joining the Luzenac/Borax Polymer Group in 2003.

kelvin.shen@borax.com

John G. Sommer

John Sommer, a registered professional engineer (Ohio) with an M.S. in polymer chemistry, founded Elastech after a long career with GenCorp. His professional activities resulted in over 50 publications and 16 US patents, one of which became the technology base for a new company. Other commercialized patents involved aerospace and coatings. He has made over 200 presentations on elastomer-related topics at universities, technical societies, and at manufacturers of elastomers and elastomer products. Major topics include molding, compounding, testing, physical properties, and design.

He was appointed an Adjunct Professor of Polymer Engineering to teach mold design at the University of Akron, and he received a distinguished service award from the University of Wisconsin (Milwaukee) for teaching molding and other courses there for more than ten years. Other awards include the University of Dayton Engineering Award of Excellence, and the Melvin Mooney Award from the Rubber Division, ACS, for repeated and significant contributions to rubber technology. In 2003, he published a 472-page book, *Elastomer Molding Technology*.

James F. Stevenson

Jim Stevenson has over 35 years of experience in processing rubber and plastic materials. During 19 years at GenCorp (formerly General Tire) in Akron, Ohio, Stevenson held several technical and management positions and headed the Rubber Extrusion Laboratory. Over the past seven years at Honeywell International, where he holds the position of Fellow, Stevenson has worked on advanced polymer, metal powder, and ceramic processing for aerospace applications.

During 1966, Stevenson was Director of Research at Trexel, a start-up company, where he initiated work on injection molding of microcellular foam (MuCell). Before joining GenCorp, he was an Associate Professor of Chemical Engineering at Cornell University. He was a founding member of the Cornell Injection Molding Project and conducted research on polymer flow.

Since 1991, Stevenson and coworkers have presented the short course "Rubber Extrusion Technology" more than 20 times to an audience exceeding 1000 participants. He has written book chapters on extrusion of rubber and plastic and of composite materials. His edited book, *Innovation in Polymer Processing: Molding*, was published in 1996. He has published 60 papers on polymer processing and flow and holds more than 12 patents. Stevenson earned his B.S. in chemical engineering from Rensselaer Polytechnic Institute and M.S. and Ph.D. degrees from the University of Wisconsin–Madison. His graduate work was on stretching flow of polymers.

His work on rubber extrusion includes extruder operation, die design, feeding and takeaway operations, instrumentation, and process control. He has experience with equipment ranging from 2-inch extruders for vinyl gaskets, to 3.5-inch extruders for industrial rubber products, to dual hot feed and triplex cold feed pin barrel extruders for tire components. His work on directed flow technology resulted in a head capable of forming curved weatherstrip and of quick die changes. Other areas of expertise include process simulation, injection-, transfer-, and compression molding, modeling reaction kinetics, and measurement of flow properties.

james.stevenson@honeywell.com

Byron To

Education: Bachelor of Science in Chemical Engineering from the University of Maryland in 1966

Current Position: Marketing Technical Service Manager, Flexsys America L.P.

Current R&D Interests:

Productivity increase through compound modifications

The effects of compound ingredients migration and dispersion on properties and performance.

Highlights of Professional Experience:

Byron's 29 years with Monsanto started as a process engineer at their plant in Nitro, West Virginia. He was responsible for developing the process to manufacture the sulfenamide accelerator TBBS.

In 1974 Byron took an overseas assignment with Monsanto located in Singapore as Technical Service Specialist for the rubber industry in South East Asia, and he was the pioneer of the Monsanto Technical Service Function in the Asia Pacific Region.

After five years of service in Asia, Byron returned to the United States in 1979 and continued his work in the Technical Service group with the Monsanto Rubber Chemicals Division at the Akron, Ohio, Technical Center until the formation of the new company, Flexsys, in May, 1995, where he had responsibility for providing technical service support to the Asia Pacific and Latin America regions.

With Flexsys, Byron continues his work in the rubber industry as Marketing Techncial Service Manager in their Global Business Group.

byron.h.to@flexsys.com

Walter H. Waddell

Walter is currently the Product Support Manager in Butyl Polymers Technology at ExxonMobil Chemical Company. He has also served as Associate Professor of Chemistry at Carnegie-Mellon University, Section Head at The Goodyear Tire & Rubber Co., and Senior Scientist at PPG Industries. He received a B.S. in chemistry from the University of Illinois, Chicago and a Ph.D. in chemistry from the University of Houston, and he served as a Research Associate at Columbia University. Walter has received several nationally recognized awards, including a Research Fellow Award from the National Institutes of Health, the Sparks-Thomas Award from the American Chemical Society, a Distinguished Corporate Inventor Award, and the Melvin Mooney Distinguished Technology Award from the American Chemical Society. He has 20 patents and 120 publications and has presented 75 seminars at universities and scientific meetings.

walter.h.waddell@exxonmobil.com

Meng-Jiao Wang

Meng-Jiao Wang graduated from Shandong University of China and received a Dr. Sc. degree in physical chemistry from the University of Haute Alsace, France. He was a chief engineer at Beijing Research and Design Institute of Rubber Industry, China, and visiting scientist at the Institute of Polymer Science, the University of Akron, Ohio, USA, and the German Institute of Rubber Technology, Hanover, Germany. From 1989 to 1993, he was a senior scientist at the applied Research and Technical Service Department for Fillers and Rubber Chemicals of Degussa AG, Germany, and in 1993 he joined the Business and Technology Center of Cabot Corporation, Billerica, MA, USA, as a scientific research fellow.

He has over 120 publications in several international journals on rubber and is a coauthor and coeditor of seven books, including *Carbon Black, Science and Technology* (in English). He is also a co-owner of 27 patents about rubber fillers and technology.

meng-jiao_wang@cabot-corp.com

Author's Biography

John S. Dick

John Dick has over 39 years of experience in the rubber industry. He was with B. F. Goodrich and later Uniroyal Goodrich Tire Co. as a Section Manager and Development Scientist in R&D, until 1991 when he joined Monsanto's Rubber Instruments Group (now Alpha Technologies) as a Senior Marketing Technical Service specialist. Mr. Dick has authored over 80 journal and magazine publications, including four books on rubber technology. He has consulted, presented technical papers, or given seminars in over 38 countries. Some of Mr. Dick's publications have been translated into eight languages. He received the Monsanto Master Technical Service Award, the ACS Rubber Division "Best Paper Award," and the University of Akron and University of Wisconsin Appreciation Awards for teaching rubber compounding and testing courses in their continuing education programs. He is a Fellow in the American Society for Testing and Materials (ASTM), receiving the Award of Merit in 1990 and a Distinguished Service Award in 2005. Also, he has represented the United States as a delegate to the International Standards Organization (ISO) for the last 30 years. He was appointed in 1992 to be leader of the US delegation to ISO TC-45 on Rubber. He teaches ten rubber technology courses in the University of Akron and University of Wisconsin continuing education departments. He is a member of the American Chemical Society, the Society of Rheology, and ASQ with a CQE and CQA. He is also a representative to the RMA and has received recognition in Who's Who in America. Mr. Dick received his B.S. degree from Virginia Polytechnic Institute and an M.A. from the University of Akron. He is married with a son and a daughter and his hobbies include photography and amateur radio.

John.Dick@dynisco.com
www.rubberchemist.com

Subject Index